ENCYCLOPEDIA OF
WEATHER AND CLIMATE
VOLUME II

ENCYCLOPEDIA OF
WEATHER AND CLIMATE

VOLUME II

M–Z

MICHAEL ALLABY

☑®
Facts On File, Inc.

For Ailsa
—M.A.

To my late wife, Jen, who gave me
inspiration and support for almost 30 years.
—R.G.

Encyclopedia of Weather and Climate

Copyright © 2002 by Michael Allaby

Facts On File, Inc.
132 West 31st Street
New York NY 10001

Library of Congress Cataloging-in-Publication Data

Allaby, Michael
Encyclopedia of weather and climate / Michael Allaby
p. cm.
Includes bibliographical references and index.
ISBN 0-8160-4801-0 (Volume I)
ISBN 0-8160-4802-9 (Volume II)
ISBN 0-8160-4071-0 (set) (alk. paper)
1. Meteorology—Encyclopedias. 2. Climatology—Encyclopedias. I. Title.
QC854 .A45 2001
551.6'03—dc21
2001023103

Facts On File books are available at special discounts when purchased in bulk quantities for businesses, associations, institutions or sales promotions. Please call our Special Sales Department in New York at 212/967-8800 or 800/322-8755.

You can find Facts On File on the World Wide Web at http://www.factsonfile.com

Text design by Joan M. Toro
Cover design by Cathy Rincon
Illustrations by Richard Garratt

Printed in the United States of America.

VB FOF 10 9 8 7 6 5 4 3 2 1

This book is printed on acid-free paper.

CONTENTS

ENTRIES M-Z

M

m   *See* METER.

macchia   *See* CHAPARRAL.

mackerel sky   CIRROCUMULUS cloud in which the individual units of cloud are swept by the high-level wind into long, parallel rows, so they form an orderly pattern reminiscent of the pattern of scales on the back of a mackerel fish. It frequently forms ahead of a WARM FRONT along which the rising warm air is unstable. This indicates that the front is likely to produce showers, possibly heavy ones, with bright intervals between them. Traditionally, a mackerel sky has also been taken to indicate a strengthening of the wind. *See also* MARES' TAILS.

macroburst   A strong downdraft of air that emerges from the base of a CUMULONIMBUS cloud. It spreads to the sides when it strikes the ground surface, generating winds of up to 130 mph (209 km h^{-1}) that can damage property within a range of more than 2.5 miles (4 km). A macroburst may continue for up to 30 minutes. It is similar to a MICROBURST, but on a larger scale.

macroclimate   The CLIMATE that is typical of a very large area, such as a continent or the entire Earth. This is the largest of the scales used by climatologists; the others are MICROCLIMATE, LOCAL CLIMATE, and MESOCLIMATE. In the widely used climatological classification introduced by the Japanese climatologist M. M. Yoshino, a macroclimate is designated *Mc*. The study of macroclimates is macroclimatology.

macroclimatology   *See* MACROCLIMATE.

macrometeorology   The scientific study of the atmosphere at the largest scale, including the GENERAL CIRCULATION and the development and behavior of AIR MASSES and large weather systems.

macrotidal   *See* TIDAL RANGE.

Madden–Julian Oscillation (MJO)   An atmospheric disturbance that was first described in 1972 by R. A. Madden and P. R. Julian. It begins over the Indian Ocean and travels eastward as a wave with a 30– to 60-day period. An MJO causes a warming in the lower atmosphere, and several MJO cycles can amplify the effect.

(There is more technical information about the MJO at http://www.ncdc.noaa.gov/coare/meetings/scigrp/lsa/lsa5_rep.html.)

maestrale   A cold northerly or northwesterly wind that blows over northern Italy, most commonly near Genoa, and land bordering the Adriatic Sea. It is similar to the MISTRAL and its name also means "masterful."

maestro   A northwesterly wind that blows across Italy and over the Adriatic Sea and onto the western shores of Sardinia and Corsica. It creates fine weather and is most common in summer.

magnetic declination   The difference in the direction of the north and south magnetic poles and the north and south true or geographic poles. This difference occurs because the magnetic poles are not located at the geographic poles. The magnetic North Pole is located in the islands of northern Canada, at approximately 77.3° N 101.8° W and the magnetic South Pole is at approximately 65.8° S 139.0° E. The declination varies from place to place and from time to time, because the magnetic poles slowly change their location. In Seattle, for example, the declination is about 21° east, but in central Wisconsin it is only 2° east and in Door County, Wisconsin, it is about 3.5° west.

(You can learn more about magnetic declination and how to correct for it when using a compass from Chris M. Goulet, "Magnetic Declination: Frequently Asked Questions," at www.cam.org/~gouletc/decl_faq.html and from "Magnetic Declination Maps" at feature.geography.wisc.edu/sco/maps/m_magnet.html.)

magnetic wind direction   The WIND DIRECTION as measured by a magnetic compass. The compass indicates the direction of the magnetic North Pole, which is located in the islands of northern Canada, at approximately 77.3° N 101.8° W (the magnetic South Pole is at approximately 65.8° S 139.0° E). Consequently, the magnetic wind direction is not the same as the wind direction measured in relation to true or geographic north. The difference between the direction of magnetic north and south and true north and south, called the MAGNETIC DECLINATION, varies at different positions on the Earth's surface and at different times (because the magnetic poles move). WIND VANES are usually oriented to geographic north and south, so they indicate the true wind direction. A person who is told the wind direction as this is indicated by a wind vane and who needs to apply this information by using a magnetic compass (for example, a sailor at sea or the pilot of an airplane) must remember to make the necessary correction.

magnetopause   The boundary that lies between the MAGNETOSPHERE and the bow shock wave where the SOLAR WIND interacts with the PLASMA of the magnetosphere. The magnetopause is a boundary current carrying charged particles at the outer edge of the mag-

netosphere. When a plasma, such as the solar wind, flows across a magnetic field, its particles are deflected in a direction perpendicular to their original velocity and to the local magnetic field. It is this reaction that generates the boundary current.

magnetosheath   A region of the MAGNETOSPHERE that lies between the bow shock wave, where the SOLAR WIND reacts with the PLASMA of the magnetosphere, and the MAGNETOPAUSE. It is the region where some solar-wind particles penetrate the shock wave. A small proportion of magnetosheath particles penetrate the magnetopause, producing a boundary layer beneath it.

magnetosphere   The region of space that lies above the EXOSPHERE. It consists of a PLASMA, constantly maintained by bombardment by the SOLAR WIND. The charged particles of the magnetospheric plasma are concentrated in the two VAN ALLEN RADIATION BELTS. Gravity has a negligible effect on particles within the magnetosphere, but they are strongly affected by electromagnetic forces and their movement is often oriented in the direction of the magnetic field. There is a bow shock wave at a distance of 10–15 Earth radii where the solar wind and magnetosphere interact and the solar-wind particles are deflected. On the side facing the Sun the magnetosphere extends to a distance of 8–13 Earth radii. On the opposite side of the Earth the magnetosphere is stretched into a "magnetotail" that extends for a distance of 200 to more than 500 Earth radii and that is 50–60 Earth radii in diameter. Although the density of particles in the magnetosphere is extremely low, the total volume of the magnetosphere is between 100,000 and 1,000,000 times greater than that of Earth itself. (One Earth radius is 3,959 miles [6,371 km].)

magnetotail   *See* MAGNETOSPHERE.

Magnus, Olaus (1490–1557) *Swedish Priest and naturalist* Olaus Magnus is the latinized name of Olaf Mansson, the author of a work on natural history that contained the earliest European drawings of ice crystals and snowflakes.

Olaus was born in October 1490 at Linköping, Sweden, where his father, Magnus Peterson, was a prominent citizen. In those days, Swedish people did not have family names. *Olaus Magnus* meant "Olaus the son of Magnus." He had an elder brother, Johannes Magnus (Johannes, son of Magnus, 1488–1544). Olaus attended a school in Linköping and then he and Johannes spent nearly seven years traveling around Europe together to complete their education.

From 1518 to 1520 Magnus served as the deputy to the papal vendor of indulgences. Indulgences were printed forms that pardoned people for sins they had committed (or in some cases for sins they had not yet committed but might). The official vendor filled in the name of the sinner. The sale of indulgences was an important source of church revenue. Magnus was ordained a priest in 1519. He was a vicar in Stockholm in 1520 and dean of Strengnäs Cathedral in 1522.

In 1523 the king, Gustav Vasa, sent Magnus to Rome on a diplomatic mission and from there on other missions to several Dutch and German cities and finally, in 1528, to Poland, where he was to meet Sigismund, the grandson of Gustav Vasa and king of both Poland and Sweden. The Swedish Reformation took place while Magnus was in Poland. He remained firmly Catholic, however, and was expelled from Poland for this reason. In 1530 he severed his links with the king of Sweden and his property there was confiscated.

Magnus took refuge in Danzig (Gdansk) in 1534 and later the same year moved to Italy, where he and Johannes lived together, for the first three years in Venice and then in Rome. Johannes was then archbishop of Sweden. When he died in 1544 the pope appointed Olaus Catholic archbishop of Sweden in his place. Magnus always hoped to return to Sweden in this capacity and wrote to the king repeatedly, asking for permission, but to no avail.

He had a distinguished ecclesiastical career, associating with several of the leading figures of his day. As well as being archbishop of Sweden, he was the director of the religious house of St. Brigitta, in Rome.

He had other interests as well, all of them related to his homeland. Olaus was a skilled cartographer. In 1539 he published, in Venice, his *Carta marina*. This was the first detailed and reasonably accurate map of Scandinavia.

His drawings of ice crystals and snowflakes—things unfamiliar to his friends in Rome—were in his *Historia de gentibus septentrionalibus*. This was published in 1555 and became very popular throughout

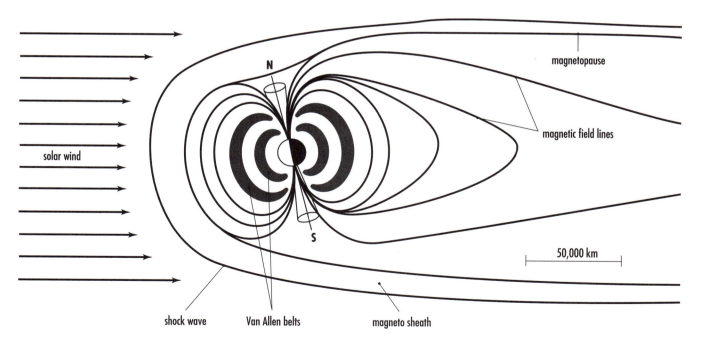

Magnetosphere. The impact of the solar wind causes a bow shock wave on the side of the magnetosphere facing the Sun. On the opposite side the magnetosphere is stretched into a long magnetotail. The charged particles are concentrated in the two Van Allen radiation belts.

Europe. There were many editions and translations. The first English translation of it, called *History of the Goths, Swedes and Vandals,* appeared in 1658.

The *Historia* was his very personal account of the history and daily life of the people of the north, told with great affection and pride, together with many details of the natural history of northern Europe. It remains one of the most important sources of information about life in Scandinavia in the early 16th century. Magnus based his book on experiences he had gathered between 1518 and 1520, the two years he spent traveling in his capacity of deputy to the vendor of indulgences.

Magnus died in Rome on August 1, 1557.

mai-u (plum rains)   The very heavy rains that fall during the first half of June along the valley of the Yangtze River, China. They are associated with very hot, humid air and thick, low cloud and mark the BURST OF MONSOON.

maize rains   The heavy and prolonged rains that fall between February and May in East Africa.

mallee scrub   *See* CHAPARRAL.

mammatus   A supplementary cloud feature that sometimes forms on the underside of a large ANVIL. It appears as many smooth, udder-shaped protrusions from the cloud base. These form when ice crystals at the top of the anvil sublime (*see* SUBLIMATION) into the dry air above them. Sublimation absorbs latent heat, chilling local areas in the cloud. These become denser and sink through to the bottom of the anvil. Mammatus forms only in very large storm clouds, so it is an indication of severe storms, possibly generating TORNADOES.

mango showers (blossom showers)   Rain showers, produced by occasional thunderstorms, that fall in April and May in southern India.

maquis   *See* CHAPARRAL.

march   A variation over a specified period, such as the changes in weather associated with the seasons of the year, which are sometimes described as the *march of the seasons.* The daily march of temperature is the rhythmic cycle of temperature change in the course of 24 hours.

March many weathers   An English expression that encapsulates the variability of the weather as winter is giving way to spring. This is one of several folk sayings about March weather. The saying "If March comes in like a lion it goes out like a lamb; if it comes in like a lamb it goes out like a lion" refers to windiness. Mists cannot form in strong winds. This is reflected in "As many mists in March as there are frosts in May." "March windy, April rainy, clear and fair May will be" is an English saying that has comparable French and German counterparts.

mares' tails   CIRRUS FIBRATUS cloud that appears as long, wispy strands that curl at the ends. Cirrus clouds extend vertically for only a short distance. Their appearance is wispy partly because they are too thin to obscure the sky above them fully and partly because the ICE CRYSTALS of which they are composed are carried by the wind. This is what sweeps them into long streamers. Where the wind weakens, crystals start to fall. Then they enter drier air and vaporize. This action produces the curls at the ends of the tails.

The longer the filaments of the tails, the stronger is the wind producing them. Although this high-level wind is stronger than the surface wind beneath the clouds and often blows from a different direction, cirrus frequently forms near the top of an approaching warm front and the high-level wind that shapes it is blowing behind the front. If the high-level wind is strong, the surface wind behind the front is also strong and so the appearance of mares' tails is often a sign that the wind will strengthen within the next few hours. Sailors knew this long ago, and the link is recorded in the saying

Mackerel sky and mares' tails
Make lofty ships carry low sails.

See also MACKEREL SKY.

Maria   A TYPHOON that struck the southern Chinese provinces of Guangdong and Hunan on September 1, 2000. A cargo ship sank in Shanwei Harbor, more than 7,000 homes were damaged, and at least 45 people were killed. The cost of the damage was estimated at $223 million.

Marilyn A HURRICANE that struck the U.S. Virgin Islands and Puerto Rico on September 15 and 16, 1995, not long after LUIS. It generated winds of more than 100 mph (160 km h⁻¹) and destroyed 80 percent of the houses on Saint Thomas. It killed nine people.

marin A southeasterly wind of the SIROCCO type that is common in the land bordering the Gulf of Lions, in southern France. It blows ahead of DEPRESSIONS, sometimes reaching gale force. Usually it produces warm, cloudy weather and heavy rain. Because it is an onshore wind it can cause danger to ships. There is a gap in the high ground near the town of Carcassonne. After the marin has passed through this gap it becomes the AUTAN.

marine forecast A weather forecast that is prepared for the crews of ships at sea. Updated marine forecasts are broadcast at regular intervals. Forecasts issued by the British Meteorological Office, which cover the sea areas around the British Isles, including the Republic of Ireland, begin with a summary of conditions at 13 coastal weather stations, the positions of which are known to mariners. This is followed by a general summary of the way weather systems are expected to develop over the 24-hour forecast period. The summary states whether an area of high or low pressure will be centered in the area and, if so, the location of the center and the pressure at the center. More detailed forecasts, of pressure, change in pressure, wind direction and speed, precipitation, and visibility, are then broadcast for each coastal sea area in turn. Forecasts for more distant sea areas, as far as the central Atlantic and north into the Arctic, are relayed to ships from weather satellites. In addition, warnings of severe weather such as storms and gales are issued.

marine meteorology The branch of METEOROLOGY that specializes in the study of weather conditions over the open ocean, coastal seas, coastal land areas, and islands. Reports and forecasts are prepared by marine meteorologists primarily for the use of ships and aircraft.

marine subarctic climate In the STRAHLER CLIMATE CLASSIFICATION a climate in his group 3, comprising climates controlled by polar and arctic AIR MASSES. Marine subarctic climates occur in latitudes 50°–60° N

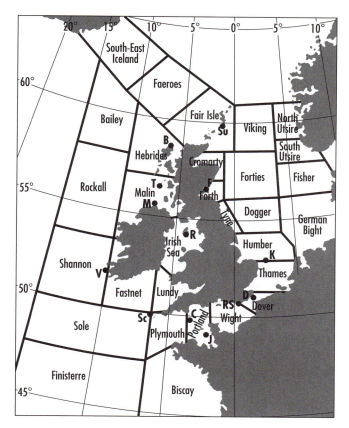

Marine forecast. Depicted are the sea areas around the British Isles. Weather forecasts for coastal shipping cover each of these areas individually. The letters show the location of coastal weather stations. Each forecast begins with reports from these stations. They are Tiree (T), Butt of Lewis (B), Sumburgh (Su), Fife Ness (F), Smith's Knoll Automatic (K), Dover (D), Royal Sovereign (RS), Jersey (J), Channel Light-Vessel Automatic (C), Scilly (Sc), Valentia (V), Ronaldsway (R), and Malin Head (M).

and 45°–60° S, along coasts and on islands exposed to FRONTAL ZONES involving maritime polar and continental polar air masses. Precipitation is fairly heavy and there is a fairly small temperature range through the year. This is a polar or tundra climate designated *ET* in the KÖPPEN CLASSIFICATION.

marine west coast climate In the STRAHLER CLIMATE CLASSIFICATION a climate of group 2, comprising climates controlled by both tropical and polar AIR MASSES. Marine west coast climates occur in latitudes 40°–60° in both hemispheres along western coasts that are exposed to frequent storms produced by FRONTAL SYSTEMS associated with cool, moist maritime polar air masses. Precipitation is distributed fairly evenly through the year, but with a maximum in winter. The climate is cloudy and the temperature range through the year is small. This type of climate includes temperate rainy or humid mesothermal climates, designated *Cfb* in the KÖPPEN CLASSIFICATION if the summer is warm and *Cfc* if it is short and cool.

Mariotte, Edmé (c. 1620–1684) French *Physicist* Edmé Mariotte was born at Dijon in about 1620 and spent most of his life there. He was ordained as a priest and was also a scientist. It was as a reward for his scientific work that he was appointed prior of the abbey of Saint-Martin-sous-Beaune, near Dijon.

When the French Academy of Sciences was founded in 1666, Mariotte was one of its first members. His interests were wide and he wrote on many scientific topics, including vision ("Nouvelle découvertes touchant la vue" [New discoveries touching vision], 1668), color ("Traité des couleurs" [Treatise on colors], 1681), and plants ("De la végétation des plantes" [On the growth of plants], 1679 and 1696), but his most important work concerned the behavior of fluids. He wrote about freezing ("Expériences sur la congélation de l'eau" [Findings on the freezing of water], 1682) and in 1679 he wrote "De la nature de l'air" (On the nature of air).

In this article Mariotte reported his observation that "the diminution of the volume of the air proceeds in proportion to the weights with which it is loaded." This states the relationship between the volume a gas occupies and the pressure under which it is held. It is the law discovered much earlier by ROBERT BOYLE, which Mariotte had discovered quite independently, but Mariotte had noticed something else. He found that air expands when it is heated and contracts when it is cooled, so that the relationship between pressure and volume remains true only so long as the temperature remains constant. Boyle had overlooked this. The relationship is known in English-speaking countries as *Boyle's law* and in French-speaking countries as *Mariotte's law*. The importance of Mariotte's contribution justifies supporting the French view.

Many of Mariotte's papers were included in the first volume of the *Histoire et mémoires de l'Académie*, published in 1733. His collected papers were published in the Netherlands in 1717 and again in 1740.

He died in Paris on May 12, 1684.

maritime air Air that forms AIR MASSES over all of the oceans (*see* SOURCE REGION). Maritime air is moist and its temperature is less extreme than that of CONTINENTAL AIR forming in the same latitude.

maritime arctic air *See* ARCTIC AIR.

maritime climate (oceanic climate) A climate that is produced by MARITIME AIR. It occurs in areas close to the sea. Because of the influence of the sea, the seasonal TEMPERATURE RANGE is smaller than that of an area with a CONTINENTAL CLIMATE and PRECIPITATION is higher. In Fiji, for example, the annual temperature range is 4.9° F (2.7° C) and the average annual precipitation is 119 inches (3,026 mm). Although its latitude makes the Falkland Islands (Malvinas) colder and drier than Fiji, the corresponding figures are 3.6° F (2.0° C) and 25.6 inches (651 mm).

Bulawayo, Zimbabwe, is in about the same latitude as Fiji, and Saskatoon, Saskatchewan, is in about the same latitude as the Falklands. Both these cities are in the interior of continents. In Bulawayo, the annual temperature range is 40° F (22° C) and the average annual precipitation is 23.5 inches (597 mm). In Saskatoon the annual temperature range is 88° F (49° C) and the average annual precipitation is 15 inches (381 mm).

In middle latitudes, the PREVAILING WINDS are westerlies. These carry AIR MASSES from west to east. Consequently maritime climates occur on the western coasts of continents (*see* MARINE WEST COAST CLIMATE). Despite their proximity to the ocean, the eastern coasts of middle-latitude continents are affected by air that has crossed the continent to reach them and that produces climates closer to the continental type. Islands, such as Fiji and the Falklands, have climates of the most extreme maritime type and places like Saskatoon, which is located near the center of a large continent, have an extreme continental climate. In other places the maritime or continental influence is less extreme and its extent can be calculated as the CONTINENTALITY or OCEANICITY of the climate.

maritime polar air *See* POLAR AIR.

maritime tropical air *See* TROPICAL AIR.

Mars atmosphere Compared with that of Earth, the atmosphere of the planet Mars is cold and thin. The surface atmospheric pressure is about 6 mb (on Earth it is 1,013.2 mb). The composition of the atmosphere by volume is: CARBON DIOXIDE, 95.3 percent; nitrogen, 2.7 percent; argon, 1.6 percent; and oxygen, 0.13 percent. There are traces of water vapor, CARBON MONOXIDE, NEON, KRYPTON, and XENON.

Mars is about half the size of Earth. Its average diameter is 4,212 miles (6,779 km). The average diameter of Earth is 7,918 miles (12,740 km). Mars is also farther than Earth from the Sun. Mars is 1.52 astronomical units (AU) from the Sun—1 AU is the average distance between Earth and the Sun—so Mars is 1.52 times more distant from the Sun than Earth. The average distance between Mars and the Sun is 141,642,000 miles (227,940,000 km) and between Earth and the Sun it is 92,961,000 miles (149,600,000 km).

The global mean surface temperature is 218 K (−67.27° F [−55.15° C]), but this ranges from a minimum of 140 K (−208° F [−133° C]) to a maximum of 300 K (80° F [27° C]) in dark regions of the Tropics in summer. The atmospheric pressure is too low for water to exist on the surface as a liquid. It changes directly between the solid and gaseous phases (*see* SUBLIMATION). There may be a large reservoir of water in the form of ice at the poles and below ground as PERMAFROST. There may also be a reservoir of carbon dioxide in the form of limestone and other carbonate rocks. There is a permanent ice cap made from solid carbon dioxide at the Martian South Pole and an ice cap of water ice at the North Pole. Both poles expand and retreat with the seasons, as carbon dioxide and water vapor are added to and removed from each in turn.

Martian air is very dusty. The smallest particles, 0.1 μm to 1 μm in size, are especially abundant, and they scatter light (*see* SCATTERING) just as dust particles do on Earth, but on Mars the effect is to make the sky pink at all times of day. These particles are thrown into the air by the impact of bigger sand grains, about 100 μm across, that bounce along the surface (the process is called *saltation*) driven by the constant wind. DUST STORMS are common and have a significant climatic effect by altering the ALBEDO at the polar ice caps and shading the surface from sunlight.

In addition to many local dust storms there are major dust storms that occur every year. Mars has an orbit of high ECCENTRICITY (0.093), and the planet is at PERIHELION during the Southern Hemisphere summer. The dust storms commence in the Southern Hemisphere spring, as the surface warms and the ice cap retreats. The sublimation of carbon dioxide and water vapor absorbs LATENT HEAT from the ground beneath the ice, chilling it just as sunshine is warming the adjacent surface. This produces large temperature differences locally, causing strong CONVECTION that raises dust. As the southern summer progresses the dust storms spread, eventually to envelop the entire planet.

(You can learn more about Mars from humbabe.arc.nasa.gov/mgcm/faq/faq.html and humbabe.arc.nasa.gov/mgcm/faq/marsfacts.html.)

Martin A CYCLONE that struck the Cook Islands in November 1997. It killed nine people.

mass flux The rate at which the volume of a given mass of air changes. Air is very elastic; therefore, a PARCEL OF AIR expands, is compressed, and its shape is deformed when it moves against bodies that are more or less dense than it is. Nevertheless, the total amount of gas (its mass) remains constant. Consequently, when the parcel of air expands, contracts, or is deformed some of its mass flows out of or into the original volume. The rate at which this happens is the mass flux.

mass mixing ratio *See* MIXING RATIO.

mass wasting *See* WEATHERING.

mathematical climate A very simple system for classifying climates according to the height of the Sun above the horizon. The world was divided into three latitudinal belts. The belt closest to the equator was originally known as the winterless zone and later as the Torrid Zone, the high-latitude belts in each hemisphere were known as the summerless zones and later as the Frigid Zones, and separating the winterless and summerless zones were the intermediate zones, later called the Temperate Zones. The boundaries separating the zones were the ARCTIC and ANTARCTIC CIRCLES and the TROPICS of Cancer and Capricorn.

matorral *See* CHAPARRAL.

Maunder, Edward Walter (1851–1928) English *Astronomer* The English astronomer who first identified the period from 1645 to 1715, now known as the *MAUNDER MINIMUM,* during which the recorded number of SUNSPOTS and AURORAS was extremely low.

Maunder worked at the Royal Observatory, in Greenwich, London, but he had no formal qualification as an astronomer. The youngest son of a Methodist minister, he graduated from Kings College London (since 1900 a part of the University of London, but then an independent institution) and then went to work at a bank in London. In 1873, a vacancy for a photographic and spectroscopic assistant occurred at the Royal Observatory. This was a position within the British Civil Service, for which there was an entry examination. Maunder passed the examination and was appointed to the position.

In 1891, another new member joined the staff. Annie Scott Dill Russell (1868–1947), a brilliant graduate of Girton College, Cambridge, arrived as a "lady computer." She and Maunder married and then collaborated in writing many articles about the Sun and popular articles on astronomy.

Maunder was given the job of photographing sunspots and measuring their areas and positions. As he did so, he discovered that the solar latitudes in which sunspots appear vary in a regular fashion during the course of the 11-year sunspot cycle.

While he was engaged in photographing and measuring sunspots his attention was drawn to the work of the German astronomer Gustav Spörer, who had identified a period from 1400 to 1510 when very few sunspots were seen. This period is now known as the *SPÖRER MINIMUM.* Maunder began searching through old records at the observatory to see whether Spörer was correct and whether there were any other such periods. It was this search that led to his discovery of the Maunder minimum.

He was a keen and accurate observer who experimented to discover the smallest object that he could see without the help of a lens. This demonstrated that objects seen on the surface of the Sun or any other distant object in fact must be very large and contain much fine detail that is invisible from Earth. He used this line of reasoning to challenge the existence of channels, or "canals," on Mars, although at that time these were widely accepted as genuine and his was a minority opinion that was brushed aside.

Similarly, his discovery of the sunspot minimum was overlooked. Other scientists thought he placed too much reliance on old records that, in their view, were likely to be incomplete or inaccurate.

Maunder minimum   The period from 1645 to 1715 when very few SUNSPOTS were observed. The period was identified by the English astronomer EDWARD WALTER MAUNDER by searching through old records. He found a period of 32 years during which not a single sunspot was recorded in the Sun's Northern Hemisphere and several periods of 10 years during which no sunspots were reported anywhere on the Sun. Fewer sunspots were seen over the entire 70 years than are seen today in an average single year. This was not the only period of sunspot absence. The German astronomer Gustav Spörer had already alerted Maunder to a similar period, now known as the *SPÖRER MINIMUM.* Maunder published several papers on his discovery, but these attracted little attention. Scientists knew that the number of sunspots increased and decreased over an 11-year cycle and interpreted this as meaning the Sun behaved in a very regular fashion. Maunder contradicted this idea by showing that in the recent past solar behavior had departed dramatically from the regular cycle, and he was ignored.

There is no doubt that the Maunder minimum really occurred. Sunspots had always interested astronomers, who recorded them meticulously, and by the middle of the 17th century they used excellent telescopes. Several of the most eminent scientists commented on the absence of sunspots. The number of sunspots also affects the frequency of AURORAS, or northern lights, which can be seen without instruments. These were also much rarer during the Maunder minimum than they are now. In Scandinavia, where they are seen almost every night, they were so uncommon that people regarded them as omens. When one was seen in England in 1716, at the end of the minimum, EDMUND HALLEY, the Astronomer Royal, wrote a paper explaining it, in which he said it was the first he had ever seen.

Measurements of the proportion of carbon-14 present in wood dated to the period confirmed the phenomenon. Carbon-14 is formed in the atmosphere by collisions between cosmic rays and nitrogen atoms, and the intensity of cosmic radiation is affected by the strength of the Sun's magnetic field. When there are few sunspots the solar field is weak and more cosmic radiation enters the atmosphere, increasing the formation of carbon-14. Carbon-14 formation increased during the Maunder minimum.

Both the Spörer and Maunder minima correspond precisely with periods of unusually low average temperatures. These occurred during the LITTLE ICE AGE and mark its coldest episodes. During the minima average temperatures in Europe were 1.8° F (1° C) lower than they are today and Alpine glaciers advanced. Conditions were much warmer from about 1510 to 1645, between the two minima.

Since the Spörer and Maunder minima were recognized other, similar episodes have been identified, as well as sunspot maxima, when there were more sunspots than usual. In each case, a major change in the number of sunspots coincides with a period of warmer or cooler climate. There was a warm period during the Middle Ages, for example, that began around 600 A.D. and reached its peak between 1100 and 1300. Such changes are now known to have occurred at intervals over the last 5,000 years.

(For more information, read H. H. Lamb, *Climate, History and the Modern World* [New York: Routledge, 1995] and John A. Eddy "The Case of the Missing Sunspots," *Scientific American,* May 1977, pp. 80–89. Information from the Mount Wilson Observatory is also available at www.mtwilson.edu/.MWO/Science/HK_Project/Maunder/.)

Maury   A TYPHOON that struck northern Taiwan on July 19, 1981. It caused floods and landslides in which 26 people died.

Maury, Matthew Fontaine (1806–1873) American *Naval officer, oceanographer, and meteorologist* Matthew Fontaine Maury was born on January 14, 1806, in Spottsylvania County, Virginia, and spent his youth in Tennessee. His elder brother was a naval officer and Matthew dreamed of following in his footsteps. In 1825, when he was 18 years old, Maury realized his ambition and joined the U.S. Navy as a midshipman. He was assigned to the U.S.S. *Vincennes* and embarked on a four-year cruise that took him around the world. This was followed by other extended voyages to Europe and to the western coast of South America.

Maury returned to the United States in 1834. In the same year he married Ann Hull Herndon and the couple settled in Fredericksburg. Matthew spent the years from 1834 until 1841 writing descriptions of ocean voyages and works on navigation. He also wrote essays urging naval reforms.

In 1839 Maury was injured in a stagecoach accident and rendered permanently lame. No longer fit for active service, he was appointed superintendent of the Depot of Charts and Instruments in 1841. He remained in this post until 1861. It should have given him a quiet, easy life with ample leisure, but Maury threw himself into it. By the time he left, he had transformed this obscure department into the United States Naval Observatory and the Hydrographic Office.

He was especially interested in ocean currents and the winds that drive them. He issued captains specially prepared logbooks in which they were asked to record their observations of winds and currents. He charted the course of the GULF STREAM and described it as "a river in the ocean." Knowledge of the locations and course of currents allowed captains to sail with them rather than against them, thus shortening the duration of voyages. In 1850 he charted the depths of the North Atlantic Ocean, in order to facilitate the laying of the transatlantic cable.

Maury demonstrated very clearly that the comprehensive study of meteorology at sea called for international cooperation. He had already achieved international recognition when, in 1853, he was able to play a leading role in organizing a conference on oceanography and meteorology in Brussels, which he attended as the United States representative. Two years later, in 1855, he published *Physical Geography of the Sea,* which was the first textbook in oceanography. Also in 1855 he was promoted to the rank of commander.

When war broke out in 1861, Maury threw in his lot with the South. Maury resigned from the navy on April 20, three days after his native Virginia had seceded from the Union. A few days later he was made a commander in the Confederate States Navy. He became head of coastal, harbor, and river defenses, and he invented an electric torpedo for harbor defense and experimented with electric mines. Because of his international reputation, he was sent to England in 1862 to purchase naval supplies.

At the end of the war Maury went into voluntary exile. He settled for a time in Mexico, where he became commissioner of immigration to the emperor Maximilian and attempted to found a Virginian colony. The emperor abandoned the colonization project in 1866 and Maury moved to England.

By 1868 tempers had cooled and Maury returned home to take up the post of professor of meteorology at the Virginia Military Institute. He settled in Lexington, where he died on February 1, 1873. His body was buried temporarily in Lexington and then moved to Hollywood Cemetery, Richmond, where it remains.

Maury was completely forgiven for having supported the losing side in the Civil War. There is a Maury Hall at the Naval Academy in Annapolis, Maryland, and in 1930 he was elected to the Hall of Fame for Great Americans.

maximum thermometer   A thermometer that records the highest temperature reached during the time since it was last reset. The thermometer uses mercury in a tube that has a constriction. As the temperature rises, the force with which mercury expands is sufficient to push it past the constriction. As the temperature falls, however, the mercury is unable to pass the constriction and return to the bulb, so it continues to indicate the highest temperature attained. The thermometer is reset by shaking it to jerk the mercury through the constriction and back into the bulb.

maximum wind level   The altitude at which the greatest WIND SPEED occurs. This information is of importance to aircraft navigators.

maximum zonal westerlies   The strongest westerly component of the winds that occur in middle and high latitudes and the region where they are found. Winds in these latitudes blow predominantly from west to east because of the CORIOLIS EFFECT. The HADLEY CELL circulation transports air from the equator into higher latitudes. As it moves away from the equator, the air retains its original motion derived from the rotation of the Earth, so by the time it reaches middle latitudes it is moving eastward. This movement is slowed by friction with the surface and vertical air movements. Maximum speeds are found in the upper TROPO-SPHERE, between the levels at 200-mb and 300-mb pressure. In winter the maximum is close to latitude 30° N and 30° S and the average wind speed is above 80 knots (92 mph [148 km h⁻¹]). In summer the maximum is at about latitude 40° N and 40° S and wind speeds average about 30 knots (34.5 mph [55.5 km h⁻¹]).

may   Either the month or a country name for the hawthorn (*Crataegus monogyna*), which flowers in the month of May. In temperate regions May is a transitional month between spring and summer and consequently it can have wide variations in temperature. In Britain, May frosts are fairly common, but so are pleasantly warm days. This variability gives rise to the old country saying "Ne'er cast a clout till may is out." A *clout* (cloth) is any warm winter garment, but opinions differ as to whether *may* refers to the month or to the hawthorn blossom.

mb   *See* MILLIBAR.

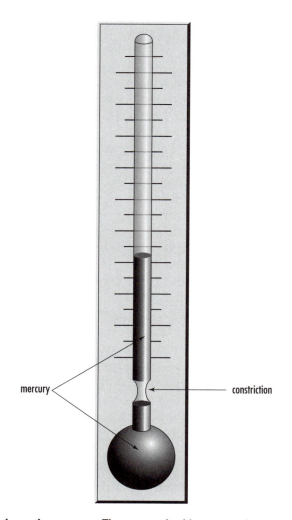

Maximum thermometer. The mercury is able to expand past the constriction, but not to contract past it. Consequently it registers the highest temperature reached.

MCR   *See* BEETLE ANALYSIS.

MCSST   *See* MULTICHANNEL SEA SURFACE TEMPERATURE.

mean   The average value of a set of values. The arithmetic mean, which is the mean that is most often referred to by the word *mean*, is calculated by adding all the values together and dividing their sum by the number of values.

$$X_a = \Sigma x/n$$

where X_a is the arithmetic mean, Σx is the sum of all the variables, and n is the number of variables. The geometric mean (X_g) is given by the nth root of the sum of the variables:

$$X_g = {}^n\sqrt{\Sigma x}$$

mean chart   A meteorological chart on which the average values for particular features are drawn as ISOPLETHS.

mean free path   The average distance that a molecule travels before colliding with another molecule. It is proportional to the density of the medium through which the molecule travels and is also related to its VISCOSITY. The free path, which is the actual distance a particular molecule travels between two collisions, cannot be known, because calculating it would require knowledge of the position and motion of all other molecules in a vicinity. Consequently the mean free path is calculated as a probability. In air at sea-level pressure and 0° C (32° F) the mean free path of an oxygen molecule is 6 μm (2×10^{-6} inch).

mean temperature   The air temperature measured at a particular place over a specified period, such as a day, month, or year, and then converted to a MEAN. Mean temperatures can also be shown, and plotted as ISOTHERMS, for large regions, continents, and the whole world.

mechanical turbulence   TURBULENT FLOW that is produced in moving air that encounters physical obstacles, such as buildings or trees. Mechanical turbulence causes EDDIES, with the result that the wind may blow from almost any direction close to the ground, especially in cities. It is why the wind direction must be measured in the open.

med   *See* MEDIOCRIS.

medieval warm period   A time when the global climate was warmer than it was in the centuries preceding and following it. Temperatures began to rise in about A.D. 800, and as early as A.D. 600 in Greenland, and the peak occurred between A.D. 1100 and A.D. 1300, after which temperatures fell as the world entered the LITTLE ICE AGE. The TREE LINE was up to 650 ft (198 m) higher in central Europe than it was in the 17th century.

In the early part of the warm period Viking ships sailed the northern seas and Norsemen established colonies in Iceland and Greenland (in the 980s). In about A.D 1000 Viking sailors whose ships had been blown off course accidentally discovered the North American coast. They called it *Vinland*, or "Wineland," but difficulties with local people prevented them from colonizing it. By that time surface temperatures in Norwegian fjords were probably about 7° F (4° C) warmer than they are now.

During the peak of the period, settled farms were established in northern Norway and expanded onto higher ground elsewhere in Europe. The period also coincided with the building of cathedrals in Europe and with the Crusades, both of which may have been facilitated by the warmer conditions and resulting abundance of food. Summer temperatures were probably about 1.3°–1.8° F (0.7°–1.0° C) warmer than the 20th-century average in England and 1.8°–2.5° F (1.0°–1.4° C) warmer in central Europe. The warm period passed its peak in Greenland during the 12th century and conditions began to deteriorate in Europe in the early 14th century.

mediocris (med)   A species of CUMULUS clouds (*see* CLOUD CLASSIFICATION) that have fairly small protuberances and are of moderate vertical extent. The name of the species is a Latin word that means "of middle height."

Mediterranean climate   In the STRAHLER CLIMATE CLASSIFICATION a climate of group 2, comprising climates controlled by both tropical and polar AIR MASSES. Mediterranean climates occur in latitudes 30°–45°

in both hemispheres and are characterized by seasonal alternations between conditions typical of WEST COAST DESERT CLIMATES and those of MARINE WEST COAST CLIMATES. Winters are wet and summers dry. In winter, maritime polar air masses predominate, producing FRONTAL SYSTEMS with storms and abundant rain. In summer, maritime tropical air masses predominate, producing dry weather. DROUGHTS are common. There is a moderate range of temperature through the year. This type of climate includes temperate rainy or humid mesothermal climates and is designated *CsA* in the KÖPPEN CLASSIFICATION if the summer is dry and hot, and *Csb* if the summer is dry and warm.

Mediterranean regime   *See* RAINFALL REGIME.

medium-range forecast   A weather forecast that covers a period of five to seven days. It is compiled in the same way as a SHORT-RANGE FORECAST.

megathermal climate   *See* PERHUMID CLIMATE.

Meli   A CYCLONE that struck Fiji on March 27, 1979. It killed at least 50 people and destroyed about 1,000 homes.

meltemi   *See* ETESIAN.

melting   *See* FREEZING.

melting band   A region in certain clouds, especially NIMBOSTRATUS, where melting snowflakes become coated in a layer of water. The snow then reflects RADAR waves more strongly than the ICE CRYSTALS and SNOWFLAKES above or the RAINDROPS below, and so the region appears on radar screens as a bright, horizontal band. The existence of melting bands confirms that in middle and high latitudes most of the rain that reaches the ground is melted snow.

melting level   The height at which MELTING BANDS occur. It is the level at which the temperature is slightly above freezing, so that falling ICE CRYSTALS and SNOWFLAKES begin to melt.

melting point   The temperature at which the solid and liquid phases of a substance are in equilibrium at a given pressure (*see* STANDARD ATMOSPHERE). The melting point of pure water at the standard sea-level pressure of 1013.25 mb (760 mm or 29.92 inches of mercury) is 32° F (0° C, 273.15 K). When water melts, 80 cal g^{-1} (334 joules per gram) of LATENT HEAT is absorbed.

meniscus   *See* SURFACE TENSION.

Mercator projection   The method by which some of the most familiar maps are produced. It bears the name of the Flemish geographer Gerardus Mercator (the latinized name of Gerhard Kremer, 1512–94). He did not invent the projection, but in 1569 he used it to produce a map of the world and he was the first person to apply it to a nautical chart.

Cartographers face the difficulty that the Earth is a sphere, but for convenience of use maps and charts need to be drawn on a flat plane. This means a way must be found to represent the three-dimensional surface of the Earth accurately on a two-dimensional sheet of paper. The form the cartographer chooses to translate three dimensions into two is called the *projection*.

Mercator used a cylindrical projection, as though a cylinder were wrapped around the globe, surface features projected onto the inside of the cylinder, and then the cylinder was unrolled as a flat sheet. In this projection, lines of latitude and longitude appear as straight lines intersecting at right angles, just as they do on the spherical globe. Its great advantage is that a navigator can represent the desired track between two points as a straight line that intersects all the lines of latitude and longitude at the same angle. This is called *a rhumb line* and its use simplifies navigation, even though it is not the shortest distance. The shortest distance between two points on the surface of a sphere is a great circle, which is the shorter arc of a circle passing through the two points, the center of which is at the center of the Earth. A great circle appears as an arc on a Mercator projection and is difficult to draw accurately. The disadvantage of the Mercator projection is that it does not represent all areas to the same scale and the shapes of some areas are distorted.

mercury barometer   A BAROMETER that measures atmospheric pressure as the weight of a column of air that presses down on the exposed surface of mercury.

1. *reservoir type*

air pressure   air pressure

mercury

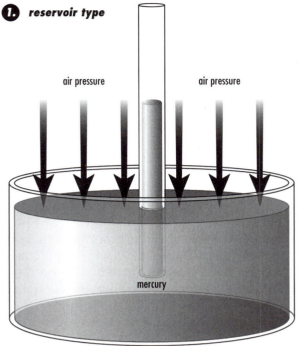

2. *j-shaped type*

air pressure

mercury

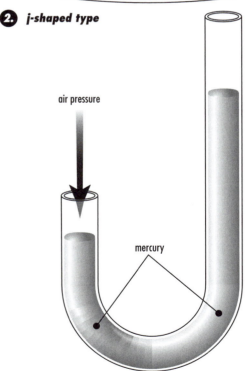

Mercury barometer. In the reservoir type the end of an evacuated glass tube is inserted into a reservoir of mercury. Air pressure forces mercury up the tube. In the J-shaped type there is no reservoir. Pressure is measured by the distance mercury is pushed up the longer arm of the tube.

A glass tube that is sealed at one end is filled with mercury. The tube is then inverted and supported vertically so its open end is below the surface of the mercury held in a reservoir. The weight of the atmosphere above the exposed surface pushes the mercury some distance into the tube. At standard sea-level pressure (*see* STANDARD ATMOSPHERE) the height of the column of mercury in the tube is 29.92 inches (760 mm). In an alternative design, the mercury is contained in a J-shaped tube and there is no reservoir. The taller arm of the tube is sealed and the end of the shorter arm is open to the air. Pressure is read from the difference in the height of the two columns of mercury.

meridional circulation   A large-scale movement of air that flows in a north–south or south–north direction, following the meridians (lines of longitude). The meridional circulation transfers warm air away from the equator and cool air toward the equator. Its existence was first postulated by EDMUND HALLEY and refined by GEORGE HADLEY in the 18th century. The description of the meridional circulation was finally completed in the 19th century by WILLIAM FERREL, as the THREE-CELL MODEL.

meridional flow   Movement of air that flows in a north–south or south–north direction, parallel to the meridians (lines of longitude), on a smaller scale than that of the MERIDIONAL CIRCULATION. Meridional flow occurs when waves develop in the MIDLATITUDE WESTERLIES (*see* INDEX CYCLE). Its effect is to carry warm air into latitudes higher than it reaches at other times and to carry cool air closer to the equator.

meridional index   The component of air movement that is parallel to the lines of longitude (meridians). It is averaged along the whole of a line of latitude and takes no account of the direction of the meridional flow (north or south).

meridional wind   A wind, or component of a wind, that blows parallel to the lines of longitude (meridians). The TRADE WINDS have a strong meridional component. Those near the surface blow toward the equator at almost 7 mph (11 km h⁻¹). In the tropical upper TROPOSPHERE, meridional winds blow away from the equator. These meridional winds are the horizontal components of the HADLEY CELL circulation.

mesoclimate The CLIMATE of a large area that can be defined by a particular physical characteristic. The typical climate of a grass-covered plain, such as the Great Plains of North America and the steppes of central Asia, or of a large mountain range constitutes a mesoclimate. A mesoclimate extends vertically to about 6,000 m (20,000 feet). Within the area experiencing a mesoclimate there are many MICROCLIMATES and LOCAL CLIMATES. In the widely used climatological classification introduced by the Japanese climatologist M. M. Yoshino, mesoclimates are designated Ms_1 to Ms_3.

Ms_1 plain
Ms_2 mountains
Ms_3 basin area

The study of mesoclimates is known as *mesoclimatology*.

mesoclimatology *See* MESOCLIMATE.

mesocyclone A mass of air that spirals upward at the center of a very large CUMULONIMBUS storm cloud that has developed a SUPERCELL. Wind shear deflects air that is rising convectively through the cloud. This causes the rising air to start rotating cyclonically about its own vertical axis, starting high in the cloud, where the upcurrent starts to level off as it enters the ANVIL. Rotation then spreads downward until the entire system of upcurrents is turning, with the diameter of the VORTEX decreasing progressively as it descends lower. Reducing the diameter accelerates the ANGULAR VELOCITY to conserve the ANGULAR MOMENTUM. If the vortex extends through the base of the cloud it becomes a FUNNEL CLOUD, and if the funnel cloud reaches the ground it becomes a TORNADO.

mesometeorology The study of weather systems that extend horizontally for about 1–100 km (0.6–60 miles). Satellite images make it possible for METEOROLOGY to be conducted at this scale. These show the clouds associated with such phenomena as FRONTAL SYSTEMS, SQUALL LINES, and TROPICAL CYCLONES as well as showing large individual CUMULONIMBUS clouds associated with THUNDERSTORMS and the GUST FRONTS they produce. Through its widespread use of

satellite images, modern weather forecasting relies heavily on mesometeorology.

mesopause The boundary that separates the MESOSPHERE from the THERMOSPHERE, two layers of the upper atmosphere. Its base is about 50 miles (80 km) above sea level, and it extends to a height of about 56 miles (90 km). The temperature within the mesopause remains constant with height and is usually about –130° F (–90° C) in winter, although it can be lower. In summer the temperature can rise to –22° F (–30° C).

mesopeak The name that is sometimes given to the temperature maximum that occurs at the STRATOPAUSE, at a height of about 80 miles (128 km). Temperatures rise with increasing height through the upper STRATOSPHERE to a peak of about 80° F (27° C [300 K]) in the stratopause. They then fall again through the MESOSPHERE.

mesophyll *See* TRANSPIRATION.

mesophyte A plant that is adapted to conditions that are neither extremely wet nor extremely dry.

mesoscale In CLIMATOLOGY and METEOROLOGY, a scale that extends horizontally from 1 to 100 km (0.6 to 60 miles). Meteorology at this scale (*see* MESOMETEOROLOGY) became feasible once satellite images showed the clouds associated with entire weather systems.

mesosphere The layer of the atmosphere that lies above the STRATOPAUSE. Its lower boundary is about 30 miles (50 km) above the surface and it extends to a height of about 50 miles (80 km). Temperature remains constant with height above the stratopause, then decreases above about 34 miles (55 km) to reach about –130° F (–90° C) at the base of the MESOPAUSE. The temperature at the mesopause can fall as low as –148° F (–100° C) in winter, and in summer the continuous action of sunlight in breaking apart and ionizing air molecules can raise the temperature to –22° F (–30° C). Pressure in the mesosphere falls from about 1 mb at the top of the stratopause to about 0.01 mb at the top of the mesopause.

mesotherm A plant that grows in a MESOTHERMAL climate.

mesothermal An adjective that describes a midlatitude climate in which the mean temperature in the coldest month is higher than –3° C (26.6° F). The term was introduced in connection with the KÖPPEN CLIMATE CLASSIFICATION.

mesothermal climate *See* HUMID CLIMATE.

mesotidal *See* TIDAL RANGE.

Meteor 3 A Russian satellite, launched in 1991, carrying a TOTAL OZONE MAPPING SPECTROMETER (TOMS) that is used to measure the emission and subsequent spread of sulfur dioxide from volcanoes.

meteoric water Water that falls from the sky as precipitation and then moves downward through the soil until it joins the GROUNDWATER. *See* JUVENILE WATER.

meteorogram A diagram that shows the way weather conditions have changed. Variable meteorological phenomena, such as temperature, humidity, and pressure, are plotted against time.

meteorological drought A DROUGHT that is defined as a decrease in PRECIPITATION. This is not the same as an AGRICULTURAL DROUGHT. Meteorological droughts are classified by the PALMER INDEX.

meteorological equator The mean latitude of the EQUATORIAL TROUGH, which is 5° N. This coincides with the THERMAL EQUATOR.

meteorological minima The lowest values for VISIBILITY, CLOUD BASE, and other relevant features of the weather that are prescribed for specified types of flying operations.

Meteorological Office The British government agency that gathers meteorological data and compiles and issues weather forecasts. It was formed in 1854 as a small department within the Board of Trade. Commanded by Admiral ROBERT FITZROY, it was to provide information on the weather and sea currents for the use of mariners. It began issuing storm warnings to seaports and weather forecasts to the press in 1861. These services were suspended in 1866, on the advice of a committee of the Royal Society, but were reinstated in

1879. During the First World War meteorological services were established in all branches of the military, and in 1920 they were joined together under the control of the Air Ministry. Naval meteorological services were taken over by the Admiralty in 1937. In April 1990 the Meteorological Office became an Executive Agency of the Ministry of Defense, and on April 1, 1996, it was made a Trading Fund, allowing it to enter into commercial contracts with its customers. The Hadley Centre for Climate Prediction and Research was opened in May 1990 as part of the Meteorological Office.

(There is more information about the Meteorological Office at www.met-office.gov.uk/sec1/sec1pg2.html.)

meteorology The scientific study of the atmospheric phenomena that produce weather, and especially the application of this study to the forecasting of weather. The word is derived from the Greek word *meteorologia,* which is derived in turn from *meteoron,* which means "of the atmosphere," and *meteoros,* which means "lofty."

People have always been fascinated by the weather, and predicting it has always been important. Farmers planning to sow a crop needed to know whether their seedlings would survive, fishermen needed to know whether a storm was developing, and ordinary people needed to know whether they should take a coat when they left home. It is not surprising, therefore, that meteorology originated a very long time ago, in ancient Greece. The first book on the subject was written by ARISTOTLE in about 340 B.C. Called *Meteorologica,* it gave us the name of the science.

Archimedes discovered that when one body is immersed in another it displaces its own volume. This is known as ARCHIMEDES' PRINCIPLE and meteorologists apply it when they calculate the BUOYANCY of a rising PARCEL OF AIR.

Since then, many of the greatest scientists have contributed to the study. GALILEO GALILEI invented a THERMOMETER and his associate EVANGELISTA TORRICELLI invented the BAROMETER. ROBERT BOYLE, JACQUES CHARLES, and JOSEPH GAY-LUSSAC discovered the relationship between the temperature and pressure of a gas and from that discovery the GAS LAWS were derived. BLAISE PASCAL proved that atmospheric pressure decreases with height. The astronomer EDMUND

HALLEY was the first person to propose an explanation for the TRADE WINDS, which led GEORGE HADLEY and, much later, WILLIAM FERREL to complete the explanation of the GENERAL CIRCULATION of the atmosphere. Thomas Jefferson was one of many enthusiasts who kept detailed weather records. BENJAMIN FRANKLIN demonstrated the electrical nature of THUNDERSTORMS. The German-born English astronomer Sir William Herschel (1738–1822) discovered invisible INFRARED RADIATION. This made it possible to study the heat balance of the atmosphere and the GREENHOUSE EFFECT.

The invention of reliable instruments made it possible to measure temperature, pressure, humidity, and wind speed and direction, but it was still possible to study the weather only on a local scale and weather forecasting had to rely on local observations. It was not possible to examine conditions over a large area in time to compile a forecast, because no information could travel faster than the speed of a horse. That changed in the middle of the 19th century with the introduction of TELEGRAPHY, invented by JOSEPH HENRY and SAMUEL MORSE and the Morse code to translate the letters of the alphabet and numerals into sets of on and off signals.

LEWIS RICHARDSON devised a way to forecast the weather mathematically. The mathematics he used were those that apply to quantities that are constantly changing. This branch of mathematics, called the *calculus,* was invented independently by the English physicist and mathematician Sir Isaac Newton (1642–1727) and the German mathematician Gottfried Wilhelm Leibniz (1646–1716). Richardson's method was too complex to be used in his own lifetime, but his ideas strongly influence modern forecasting techniques. These rely on satellite images and measurements as well as observations from surface stations, and they are able to use large, powerful supercomputers to perform the hundreds of thousands of mathematical calculations that are required to produce a forecast.

Meteosat   A series of European meteorological satellites in geostationary orbits, the first of which was launched by the United States in 1977. Meteosat satellites are now launched by the European Space Agency. At present there are eight Meteosat satellites in orbit. The second Meteosat generation will comprise three satellites in GEOSTATIONARY ORBITS. The first, *MSG-1,* is planned to be launched in October 2000 by an Ari-

ane launcher. MSG will start to replace the first-generation Meteosat program in 2003 and the process will be completed in 2012.

(There is more information about Meteosat, including the wavelengths on which they transmit data, images and data that can be downloaded, and future plans, at http://www.eumetsat.de.)

meter (m)   The SYSTÈME INTERNATIONAL D'UNITÉS (SI) UNIT of length, which is defined as a length equal to 1,650,763.73 wavelengths in vacuum corresponding to the transition of an atom of krypton-86 between levels $2p_{10}$ and $5d_5$; 1 m = 3.28084 feet.

methane (CH_4)   A GREENHOUSE GAS with a GLOBAL WARMING POTENTIAL (GWP) of about 21 that is present in the atmosphere at a concentration of about 2 parts per million by volume. It is the second most important greenhouse gas after CARBON DIOXIDE (not counting water vapor) because although its GWP is fairly low it is much more abundant than other greenhouse gases. It enters the air as a by-product of digestive processes in ruminant mammals (cattle, sheep, goats, etc.) and termites, as natural emissions from coal formations, from landfills and decomposing organic wastes, and, in highly variable amounts, from wetland rice cultivation.

methyl chloroform (CH_3CCl_2)   A chemical compound that was once used as a solvent. It is very toxic to humans and contributes to an enhanced GREENHOUSE EFFECT, having a GLOBAL WARMING POTENTIAL of about 700. It is also a source of free chlorine atoms that contribute to the depletion of stratospheric OZONE. At the fourth meeting of the signatories to the MONTREAL PROTOCOL ON SUBSTANCES THAT DEPLETE THE OZONE LAYER, held in Copenhagen in November 1992, it was agreed that the use of methyl chloroform should cease by January 1996.

Meuse Valley incident   An air pollution disaster that occurred between December 1 and December 5, 1930, between the towns of Seraing and Huy near Liège in the valley of the river Meuse in southern Belgium. The affected area was about 24 km (15 miles) long and 2.4 km (1.5 miles) wide and surrounded by hills 100 m (330 feet) high. There were many factories in the area, including steel mills, power plants, lime kilns, and glassworks, as well as plants refining zinc and manufacturing sulfuric acid and fertilizer. All of them used coal-burning furnaces as a source of power, and most of the local inhabitants burned coal in their homes. Cold weather and a KATABATIC WIND flowing down from the hills combined to produce FOG that was trapped, together with the polluting chimney emissions, beneath an INVERSION. This produced dense SMOG. People suffered from chest pains, shortness of breath, and coughing produced by more than 30 contaminants; those of sulfur compounds were the most serious. The worst affected were those with a previous history of respiratory complaints and the elderly. Several hundred people became ill and more than 60 died. Many cattle also had to be slaughtered. This was the first recorded major air pollution incident.

microbarograph   A BAROGRAPH that is designed to record very small changes in atmospheric pressure. Microbarographs are used when it is necessary to record pressure changes very accurately.

microburst   A strong downdraft that occurs below a fairly weak CONVECTION CELL some distance from the center of a CUMULONIMBUS storm cloud. In 1975 Prof. Theodore Fujita of the University of Chicago (*see* FUJITA TORNADO INTENSITY SCALE) was the first person to recognize and name the phenomenon. The downdraft blows downward commonly at 25–55 mph (40–88 km h^{-1}). When it strikes the ground the air spreads to all sides as a strong horizontal wind. The effect is very local, extending for no more than 2.5 miles (4 km).

It is probable that the airflow in a microburst forms a VORTEX, with a downward spiral at the center, intensifying the downdraft, and an upward spiral around the edge. This is called a *ring vortex*. The upward curl in the airflow produces a ring of low air pressure. The ring vortex has been confirmed experimentally and has been observed in the dust or spray raised by microbursts.

The sequence of events in the development of a microburst begins with a shaft of rain or VIRGA that is about 0.6 mile (1 km) in diameter and has very sharply defined edges. This generates a strong downdraft of air. The difference in BUOYANCY inside and outside the downdraft produces forces that start the air rotating. As the downdraft hits the BOUNDARY LAYER close to ground level, friction causes the ring vortex to form.

The downdraft then forces the ring vortex to expand and as it does so it begins to spiral upward. Several ring vortices can form concentrically in a very strong microburst.

Microbursts represent a serious hazard to aircraft flying below about 1,000 feet (300 m), and since 1990 the Federal Aviation Authority has required most commercial pilots in the United States to undergo a program of education and training in the recognition of microbursts and techniques for avoiding or negotiating them. The area that is affected by a microburst is similar to the length of a runway at a major commercial airport. Aircraft are at greatest risk immediately after take-off or during the approach to landing.

As it flies through a microburst, the aircraft first experiences a strong headwind from the horizontal wind. About 10 seconds later the aircraft reaches the central part of the downdraft, where it experiences a strong downward force. A few seconds after that it experiences a tailwind as it reaches the horizontal wind on the other side of the center. The headwind increases the airspeed (speed in relation to the air, not the ground) of the aircraft, which also increases lift. The aircraft tends to climb and accelerate simultaneously. When it reaches the downdraft its airspeed decreases and there is a loss of lift. The tailwind causes a further loss in airspeed and reduction in lift. Overall, the effect is first to lift and accelerate the aircraft and then to reduce its airspeed and throw it hard toward the ground. The difference between its airspeed in the headwind and that in the tailwind is usually about 58 mph (92.5 km h^{-1}), but in about 5 percent of the microbursts that have been documented the airspeed difference reaches about 105 mph (169 km h^{-1}).

Microbursts can also capsize small boats. If one occurs over a forest fire it can produce a fire storm.

(You can learn more about microbursts at www.nssl.noaa.gov/~doswell/microburst/Handbook.html.)

microclimate   The CLIMATE of a very small area when this can be clearly distinguished from the climate of the surrounding area. A forest clearing has a climate that differs from that of the surrounding forest, so this constitutes a microclimate. A city street lined on both sides by tall buildings has a microclimate, as does the side of a hill. On a still smaller scale, on land covered by low-growing vegetation, such as grass or a field crop, there is a microclimate between the ground and the top of the vegetation. In the widely used climatological classification introduced by the Japanese climatologist M. M. Yoshino, microclimates are designated M_1 to M_6.

M_1 cornfield
M_2 forest clearing
M_3 city canyon
M_4 hill slope
M_5 ice fields
M_6 grass cover

Microclimate is the smallest of the scales used by climatologists; the others are MACROCLIMATE, LOCAL CLIMATE, and MESOCLIMATE. The study of microclimates is known as *microclimatology*.

microclimatology   *See* MICROCLIMATE.

micrometeorology   The scientific study of the atmospheric conditions that prevail inside a MICROCLIMATE. It includes the study of phenomena that are very important locally, such as the behavior of air as it crosses a particular area of high ground, convective movements and cloud formation that result from uneven heating of the ground, and the TURBULENT FLOW that is produced by particular surfaces.

microtherm   A plant that grows in a MICROTHERMAL climate.

microthermal   An adjective that describes a midlatitude climate in which the mean temperature in the coldest month is lower than –3° C (26.6° F). The term was introduced in connection with the KÖPPEN CLIMATE CLASSIFICATION.

microthermal climate   *See* MOIST SUBHUMID CLIMATE.

microtidal   *See* TIDAL RANGE.

microwave limb sounder (MLS)   An instrument that is carried on the UPPER ATMOSPHERE RESEARCH SATELLITE. It is a spectrometer that measures in the microwave wave band and detects concentrations of ozone (O_3), water vapor (H_2O), and chlorine monoxide (ClO) in the STRATOSPHERE, and O_3 and H_2O in the MESOSPHERE.

microwaves   Electromagnetic radiation that has a WAVELENGTH of 1 mm to 10 cm (0.04–4 in). Many satellite instruments use microwave radiation to measure sea ice and various features of the atmosphere. *See* MICROWAVE LIMB SOUNDER.

microwave sounding unit (MSU)   An instrument that is carried on the TIROS-N (*see* TELEVISION AND INFRARED OBSERVATION SATELLITE) series of NATIONAL OCEANIC AND ATMOSPHERIC ADMINISTRATION (NOAA) satellites. An MSU measures the emissions of microwave radiation from molecular oxygen in the TROPOSPHERE. The resultant readings are used to calculate atmospheric temperature with an estimated accuracy of ±0.01° C (±0.02° F). The satellites are in POLAR ORBITS that cross every part of the surface of the Earth several times each day. The continuous record of atmospheric temperature measured by the MSUs began in January 1979. When the signal from a strong warming in 1998, caused by the 1997–98 EL NIÑO, is removed, the MSU record shows no statistically significant temperature trend over its period of operation.

middle cloud   Cloud that forms with a base between 2,000-m and 6,000-m (6,500 and 20,000-ft) altitude. ALTOCUMULUS and ALTOSTRATUS are classified as middle clouds.

middle-latitude desert and steppe climates   In the STRAHLER CLIMATE CLASSIFICATION a climate in group 2, comprising climates controlled by both tropical and polar AIR MASSES. Climates of these types occur in latitudes 35°–50° in both hemispheres in continental interiors where mountains prevent maritime tropical or polar air masses from penetrating. The climates are dominated by continental tropical air masses in summer and continental polar air masses in winter. These produce a very large range of temperature through the year, with hot summers and cold winters. Desert climates are designated *BWk* in the KÖPPEN CLASSIFICATION if they are cool and *BWk'* if they are cold. Steppe climates are designated *BSk* if they are cool and *BSk'* if they are cold.

midlatitude westerlies The PREVAILING WINDS of the middle latitudes, between about 30° and 60° in both hemispheres. They affect a region that is bounded on one side by subtropical air and on the other by polar air and throughout the TROPOSPHERE. The upper winds blow in the same direction as the surface winds: from the southwest in the Northern Hemisphere and from the northwest in the Southern Hemisphere. Their velocity and frequency are both at a maximum at about latitude 35°. The upper winds are centered on a mean latitude of 45°. There is very little MERIDIONAL FLOW, except when the westerly flow is interrupted during the INDEX CYCLE.

Mie scattering The change in direction of incoming solar radiation that occurs when the radiation interacts with particles of a size similar to the wavelength of the radiation. DIFFRACTION, REFLECTION, and REFRACTION combine to cause the change in direction. Mie scattering is predominantly in a forward direction and all wavelengths are affected. This form of scattering was first described in detail in 1908, by the German physicist Gustav Mie (1868–1957).

migratory An adjective that is used to describe a pressure system that is embedded in the general westerly airflow of middle latitudes and that therefore travels with it, from west to east.

Mike A TYPHOON that struck the Philippines on November 14, 1990. It killed 190 people and rendered 120,000 homeless.

Milankovich cycles Variations in the amount of solar radiation the Earth receives that are due to cyclical changes in the orbit of the Earth about the Sun and in the rotation of the Earth on its axis. Most climatologists accept that these changes affect the climate, in particular by triggering the onset and ending of ICE AGES.

There are three such cycles, with different periods. Each has only a minor effect in itself, but their influence is significant when all three coincide. They were first described in 1920, by MILUTIN MILANKOVICH.

The first cycle involves changes in the Earth's orbit. This describes an ellipse with the Sun at one focus, so the distance between the Earth and Sun varies through

the year that it takes to complete one orbit (*see* APHELION and PERIHELION). Because the Sun is not at the center of the path described by the Earth's orbit, the orbit is said to be *eccentric*. The shape of the ellipse is not constant, however. Gradually its eccentricity increases and then decreases, changing over a cycle with a period of about 100,000 years, the time it takes

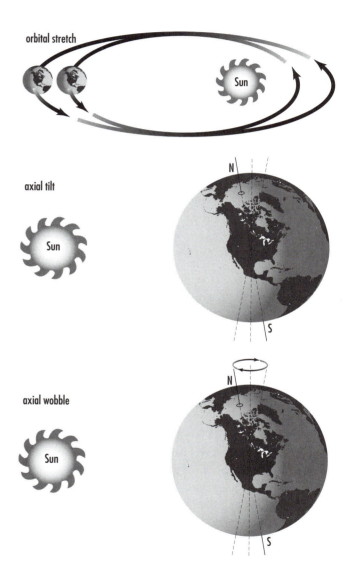

Milankovich cycles. Three cyclical variations in the orbit and rotation of the Earth that trigger the onset and ending of glaciations. Orbital stretch is a change in the eccentricity of the Earth's orbit. Axial tilt is the variation in the angle at which the north–south axis of the Earth is tilted with respect to the vertical. Axial wobble is the way the Earth's axis slowly turns around the vertical.

to return to any point in the cycle. At its most eccentric the Earth is farther from the Sun at both aphelion and perihelion than it is when the orbit is less elliptical. At maximum eccentricity there is a difference of 30 percent between the distance between the Earth and Sun at perihelion an aphelion.

The second and third cycles affect the rotation of the Earth about its own axis. This axis is not at right angles to the PLANE OF THE ECLIPTIC—it is oblique. If it were at right angles (normal to the ecliptic), the Sun would be directly overhead at the equator at noon on every day in the year and there would be no SEASONS. At present the angle between the rotational axis and the plane of the ecliptic is about 23.45°, but this is not constant. Over a cycle of about 42,000 years its obliquity moves from 22.1° to 24.5° and back again. This does not alter the amount of radiation Earth receives from the Sun, but it does affect the way it is distributed. The greater the obliquity the more warmth high latitudes receive in summer, and the less they receive in winter.

Like a toy gyroscope or spinning top, the Earth also wobbles as it spins, taking about 25,800 years to complete each cycle, or full wobble. This alters the dates of the SOLSTICES and EQUINOXES. At present, perihelion occurs around January 3 and aphelion around July 4. In about 10,000 years from now these dates will be reversed because of the wobble in the axis. Then, the Northern Hemisphere will receive more solar radiation in midsummer and less in midwinter than it does now, and the Southern Hemisphere will receive less in midsummer and more in midwinter. These changes are also known as the *PRECESSION OF THE EQUINOXES*.

Milankovich calculated the history of these astronomical cycles over the last several hundred thousand years. He found that the dates when the cycles were in phase, so all of them were exerting their maximum and minimum effects at the same time, coincided with the start and end of ice ages. This observation was confirmed in 1976, when studies of sediment cores taken from the ocean floor showed that climate changes had occurred at the times predicted by the Milankovich mechanism.

(For more information about the Milankovich cycles see geography.miningco.com/library/weekly/aa121498. htm and deschutes.gso.uri.edu/~rutherfo/milankovitch.html.)

Milankovich, Milutin (1879–1958) Serbian *Mathematician and climatologist*   The first person to develop fully an astronomical theory to account for major climatic changes, Milankovich studied in Vienna but in 1904 moved to take up a post at the University of Belgrade, where he remained for the rest of his career.

Sir John Herschel (1792–1871) was the first of several scientists to suggest a link between changes in climate and in the amount of solar radiation received by the Earth. In 1864, James Croll (1821–90) suggested it is changes in solar radiation that trigger ICE AGES. The idea was fairly vague, however, and Milankovich determined to test it. Astronomical changes occur with great regularity—think of the accuracy with which solar eclipses can be predicted. This meant that if Milankovich could identify those factors that alter the amount of radiation Earth receives he would be able to calculate mathematically how they had exerted their influence at various times in the past. He found three cyclical changes that seemed relevant and calculated their effects over hundreds of thousands of years. In 1920, he published his results, and he elaborated on them in subsequent years. Although his theory aroused considerable interest, for a long time there was no firm evidence to support it. Confirmation was found in 1976, however. The astronomical changes he identified are now known as the *MILANKOVICH CYCLES* and their influence on the initiation and ending of ice ages is widely accepted.

millet rains   The heavy rains that fall in East Africa between October and December, during the season when millet is sown.

milli atm cm   *See* MILLI ATMOSPHERES CENTIMETER.

milli atmospheres centimeter (milli atm cm)   A unit that is used to measure volcanic emissions. It is identical in concept to the DOBSON UNIT, in that it measures the quantity of a substance present in the air as the thickness of the layer it would form if it were the only constituent of the air and subjected to sea-level pressure. The atmospheric concentration of sulfur dioxide, for example, is usually about 15 milli atm cm. This means that if all the sulfur dioxide contained in the column of air from the measuring station to the top of the atmosphere were compressed into a layer of gas at sea-level pressure, that layer would usually be about 0.015 cm (0.006 in) thick.

millibar (mb) The unit in which atmospheric pressure is reported in weather forecasts. It is equal to one-thousandth of a BAR: 1 mb = 0.0145 pound per square inch = 0.75 mm of mercury (0.03 inch of mercury).

Mindel glacial A GLACIAL PERIOD of the European Alps that is the equivalent of the ANGLIAN GLACIAL of Britain and the Elsterian glacial of northern Europe. It partly coincides with the KANSAN GLACIAL of North America. It is named after an alpine river. The Mindel began about 350,000 years ago and ended about 250,000 years ago. It was followed by the MINDEL–RISS INTERGLACIAL; the interglacial preceding it has no name.

Mindel–Riss interglacial An INTERGLACIAL period in the European Alps that is the equivalent of the HOX-NIAN INTERGLACIAL of Britain and the Holsteinian interglacial of northern Europe and that partly coincides with the YARMOUTHIAN INTERGLACIAL of North America. It lasted from about 250,000 years ago to about 200,000 years ago, between the MINDEL GLACIAL and the RISS GLACIAL. The Mindel–Riss interglacial was formerly known as the *Great interglacial,* but this name is no longer used.

minimum thermometer A THERMOMETER that records the lowest temperature reached since it was last reset. The thermometer contains a fluid with a low density, most commonly colored alcohol. Inside the thermometer tube there is a small strip of metal, often in the shape of a dumbbell, called the *index.* When the temperature falls, the liquid contracts toward the bulb. As the upper surface of the liquid reaches the top of the index, the index is drawn along the tube by SURFACE TENSION. When the temperature rises, the liquid flows past the index, leaving it in the position it reached when it was drawn toward the bulb. The tip of the index farthest from the thermometer bulb therefore registers the lowest temperature attained. To reset the thermometer it is held vertically with the bulb upper-most. The index then sinks to the top of the liquid. Because the index can move along the tube by gravity, the thermometer must be mounted horizontally.

mirage An optical phenomenon that is caused by the REFRACTION of light as it passes from cool to warm or warm to cool air. The most familiar example is the shimmer resembling a wet surface that appears on a hot day on a road ahead of the observer, but the most famous example is the palm-fringed lake that appears to thirsty travelers in a desert. The image is not an optical illusion and can be photographed. Air in contact with a hot road or desert surface is at a higher temperature than air above it. As light from a higher level approaches the surface it is refracted in the opposite direction to the curvature of the Earth—away from the

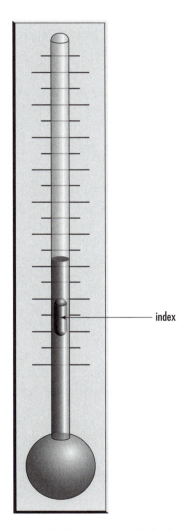

Minimum thermometer. As the temperature falls the index is drawn toward the bulb. As the temperature rises the liquid flows past the index, leaving it in position.

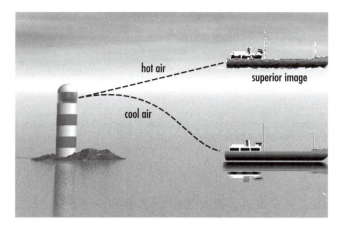

Mirage. Light bends when it crosses a boundary between two bodies of air with different densities. This effect causes us to see a distant object apparently shifted to an incorrect position. This is a mirage. The object is real, but displaced downward to form an inferior image (left) or upward to form a superior image (right).

surface. A human observer assumes light always travels in a straight line, and so a low-level image of something at a much higher level appears. In the case of the road shimmer, the image is of the sky. In the case of the desert image, the palm trees are real, but the traveler also sees a second image of them. This is often inverted and looks like a reflection, implying the existence of water. The "water" and the "reflected" trees constitute the mirage. Because both this mirage and road shimmer produce images below the position of the object that is its source, these are called *inferior images.*

Superior images can also occur; in these cases, light passes from warmer to cooler air and is refracted in the same direction as the curvature of the Earth. Again, the observer assumes light travels in a straight line and sees the image at a position higher than its source. This can make it possible to see a ship that is below the horizon, and sometimes it can make an object appear to be above the surface, in which case it is said to be *looming.*

Air may lie in layers at different temperatures, causing light to be refracted separately at each boundary. This can distort images, even to the extent of producing the impression of a mountain above the sea, and can also enlarge them. *See* FATA MORGANA.

Mireille A TYPHOON, rated as category 4 on the SAFFIR/SIMPSON SCALE, that struck Kyushu and Hokkaido, Japan, on September 27, 1991. It generated winds of up to 133 mph (214 km h⁻¹) and killed 45 people.

Miriam *See* JOAN.

missing carbon Carbon that cannot be accounted for in the global CARBON CYCLE. Every year between about 5 and 6 billion tonnes (5.5 and 6.6 billion tons) of carbon is released into the atmosphere in the form of CARBON DIOXIDE as a result of burning FOSSIL FUELS and about 0.6 to 2.6 billion tonnes (0.7 to 2.9 billion tons) is released through changes in land use in the Tropics. Between 3.0 and 3.4 billion tonnes (3.3 and 3.7 billion tons) accumulates in the atmosphere. Of the remaining 3.2–6.1 billion tonnes (3.5–6.7 billion tons), between 1.2 and 2.8 billion tonnes (1.3 and 3.1 billion tons) is absorbed by the oceans. This leaves between 1.4 and 2.4 billion tonnes (1.5 and 2.6 billion tons) unaccounted for: This is the missing carbon. It amounts to 25–40 percent of the amount released through human activities. Increasing the atmospheric content of carbon has a fertilizing effect on plants, and some of the missing carbon is absorbed by vegetation, but not all of it can be utilized in this way, because it would produce a bigger increase in plant growth than has been observed. Locating the SINK for this carbon is important, because unless scientists are able to balance the global carbon budget they cannot predict how the atmospheric concentration of carbon dioxide is likely to change in years to come, and it is on that estimate that calculations of global warming are based. So far the missing carbon remains undiscovered.

(You can learn more about the missing carbon and ideas about where it might be from "Bomb Tests Help Duke Researcher Find Possible Repository of 'Missing' Carbon" at www.dukenews.duke.edu/Environ/RAD-CARB.html; David Herring and Robert Kannenberg, "The Mystery of the Missing Carbon" at earthobservatory.nasa.gov/Study/BOREAS/; and "The CO_2 Fertilization Factor and the 'Missing' Carbon Sink" at www.bren.ucsb.edu/~keller.papers/climch33.html.)

mist Liquid PRECIPITATION in which the droplets are 0.0002–0.002 inch (0.005–0.05 mm) in diameter. These are large enough to be felt on the face of a person walking slowly through the mist. VISIBILITY in mist is more than 1,094 yards (1 km). Mist is usually associated with STRATUS cloud. If the droplets increase in size, mist becomes DRIZZLE.

mistral A cold, northerly wind that blows over southern Europe bordering the Mediterranean, but especially along the lower part of the Rhône River valley to the Gulf of Lions, France. It can occur at any time of year but is most frequent in winter and spring. Its name means "masterful" and it has been known to blow trains over. Many houses in the Rhône Valley have no windows or doors in their north-facing walls. It can reach more than 80 mph (130 km h⁻¹) and produce freezing temperatures. It is caused by a flow of POLAR AIR driven by a large anticyclone to the north, sometimes far to the north, or the AZORES HIGH, and a depression centered south of Toulon.

Mitch A HURRICANE, which weakened to a TROPICAL STORM, that struck Central America in late October 1998. It formed on October 21 in the southwest Caribbean, then moved toward Honduras and intensified to category 5 hurricane status on the SAFFIR/SIMPSON SCALE. On October 29 it was downgraded to a tropical storm and moved inland and southward. The storm was declared over on November 1. The high rainfall associated with it caused appalling damage from flooding, in which an estimated 11,000 people died, the highest death toll from a hurricane for 200 years. During a period of 41 hours, from 1500 on October 29 to 0700 on October 31, a total of 27.5 inches (698 mm) of rain fell on Honduras and between 1800 on October 27 and 2100 on October 31 the rainfall was 35.3 in (896 mm).

mixed cloud A cloud that contains both water droplets and ICE CRYSTALS. The cloud base is below the FREEZING LEVEL, but the cloud also extends above it.

mixed layer *See* MIXING DEPTH.

mixed nucleus A CLOUD CONDENSATION NUCLEUS that consists of material of two different types. One substance is hygroscopic and the other possesses the property of WETTABILITY. The efficacy of a mixed nucleus is inferior to that of a HYGROSCOPIC NUCLEUS and superior to that of a wettable particle.

mixing condensation level The lowest height at which water vapor condenses in a layer of air that is thoroughly mixed. Vertical mixing of the air averages the temperature and MIXING RATIO throughout the mixed layer, and the mixing condensation level occurs where the POTENTIAL TEMPERATURE is equal to the DEW POINT TEMPERATURE of the mixed air.

mixing depth The distance between the surface of the Earth and the height beyond which the vertical motion of air by CONVECTION ceases. This varies greatly, according to the intensity of convectional motion. Convection mixes the atmospheric constituents, thereby diluting pollutants in a large but variable volume of clean air. Consequently, the greater the mixing depth the better the air quality is likely to be and when the mixing depth is shallow pollutants concentrate in it. The air below the mixing depth is known as the *mixed layer*.

mixing law When a volume of fluid containing several ingredients is mixed, each becomes evenly distributed throughout the total volume. Once this condition is reached, further mixing has no effect. This can be demonstrated by adding a small amount of pigment (such as food coloring) to water and then stirring the mixture. At first the colorant occupies discrete areas in the water, but very soon it becomes evenly distributed, the color of the mixture is uniform throughout, and further stirring does not alter the color.

mixing length The average distance that a particle in an EDDY travels perpendicular to the mean path of a TURBULENT FLOW of air. Eddies transport heat, water vapor, and momentum, so the mixing length is a measure of the distance to which these properties affect the air to each side of the eddy.

mixing ratio (mass mixing ratio) The ratio of the mass of any particular gas present in the air to a unit mass of dry air. This is usually measured as grams of the gas in one kilogram of air without the gas. It is most often used to refer to the amount of water vapor present, as grams of water vapor in one kilogram of dry air: grams H_2O/kg air. Because it is measured in units of mass it is not affected by changes of temperature or pressure. The mixing ratio of water vapor is difficult to measure directly, but it can be determined from the RELATIVE HUMIDITY. Mixing ratio should not be confused with SPECIFIC HUMIDITY. *See also* SATURATION MIXING RATIO.

MJO *See* MADDEN–JULIAN OSCILLATION.

MLS *See* MICROWAVE LIMB SOUNDER.

mock moon *See* SUN DOG.

mock sun *See* SUN DOG.

model A simple description of the way a complex process is believed to function. An example is the THREE-CELL MODEL of the circulation of the atmosphere, which explains how warm air moves away from the equator and is replaced by cool air from higher latitudes. This model is easy to visualize and understand, but it is very approximate and lacks detail, so its value is limited. More detailed models comprise sets of mathematical relationships that aim to simulate the processes that take place in a system, such as the atmosphere, and computers perform the necessary calculations. The simulation simplifies the system and accelerates or slows the rate at which the processes occur. This facilitates the study of the system. CLIMATE MODELS are of this type.

moderate breeze In the BEAUFORT WIND SCALE, force 4, which is a wind that blows at 13–18 mph (21–29 km h⁻¹). In the original scale, devised for use at sea, a force 4 wind was defined as "or that in which a man-of-war with all sail set, and clean full would go in smooth water from." On land, a moderate breeze causes loose leaves and dry scraps of paper to blow about.

moderate gale In the BEAUFORT WIND SCALE, force 7, which is a wind that blows at 32–38 mph (51–61 km h⁻¹). In the original scale, devised for use at sea, a force 7 wind was defined as "or that to which a well-conditioned man-of-war could just carry double-reefed topsails, jib, etc. in chase, full and by." On land, people walking into the wind feel it exerting a strong pressure in a moderate gale.

moist adiabat *See* SATURATED ADIABAT.

moist climate A climate in which the amount of annual precipitation exceeds the annual POTENTIAL EVAPOTRANSPIRATION. In the THORNTHWAITE CLIMATE CLASSIFICATION this is any climate with a MOISTURE INDEX greater than 0. In this classification moist climates belong to categories A, B, and C_2.

moist subhumid climate In the THORNTHWAITE CLIMATE CLASSIFICATION, a climate in which the MOISTURE INDEX is between 0 and 20 and the POTENTIAL EVAPOTRANSPIRATION is 11.2–22.4 inches (28.5–57 cm). The climate is designated C_2. In terms of THERMAL EFFICIENCY, this is a microthermal climate (C'_1 to C'_2).

moisture Water that is present in the atmosphere. In CLIMATOLOGY, moisture is the amount of PRECIPITATION associated with a particular type of climate, or the effectiveness of that precipitation, which is the amount of precipitation minus the amount of water that evaporates from the ground surface. This can be measured as the MOISTURE FACTOR. In the THORNTHWAITE CLIMATE CLASSIFICATION the effectiveness of precipitation is used to calculate a PRECIPITATION-EFFICIENCY INDEX. In METEOROLOGY, moisture is either the amount of WATER VAPOR present in the atmosphere or the total amount of water present as ice, liquid, or vapor in a given volume of air. The amount of water vapor in the air is known as the *HUMIDITY* of the air.

moisture balance *See* WATER BALANCE.

moisture factor A measure of the effectiveness of PRECIPITATION, which is made by dividing the amount of precipitation in centimeters by the temperature in degrees Celsius for the period under consideration.

moisture index (*Im*) In the THORNTHWAITE CLIMATE CLASSIFICATION, a value that is calculated to show the monthly surplus or deficit of water in the soil. It is given by

$$Im = 100(S - D)/PE$$

where *S* is the monthly water surplus, *D* the monthly water deficit, and *PE* is the POTENTIAL EVAPOTRANSPIRATION. It can also be calculated from

$$Im = 100(r/PE - 1)$$

where *r* is the annual precipitation.

moisture inversion A layer of air through which the HUMIDITY increases with height.

mol *See* MOLE.

molar heat capacity *See* HEAT CAPACITY.

mole (mol) The SYSTÈME INTERNATIONAL D'UNITÉS (SI) UNIT of amount of a substance, which is equal to the amount of any substance that contains as many elementary units as there are atoms in 0.012 KILOGRAM of carbon-12. The elementary units may be atoms, molecules, ions, electrons, or other particles or groups of particles, but they must be specified.

Molina, Mario José (born 1943) Mexican *Atmospheric chemist* Professor Molina was the joint winner of the 1995 Nobel Prize for Chemistry, shared with PAUL CRUTZEN and F. SHERWOOD ROWLAND. The prize was awarded for their work in identifying the threat to the OZONE LAYER from chlorofluorocarbon compounds (CFCs).

Mario Molina was born in Mexico City on March 19, 1943, the son of a lawyer. He was educated at a boarding school in Switzerland and then in Mexico City, where he graduated in 1965 with a degree in chemical engineering from the Universidad Nacional Autonoma de Mexico. In 1967 he obtained a postgraduate degree from the University of Freiburg, Germany, and his Ph.D. from the University of California at Berkeley in 1972, where Sherwood Rowland was one of his supervisors. While at Berkeley Molina met and married Luisa Tan, a fellow graduate student.

He held teaching and research posts at the Universidad Nacional Autonoma de Mexico, the University of

California at Irvine, and from 1982 to 1989 the Jet Propulsion Laboratory of the California Institute of Technology, in Pasadena. In 1989 he moved to the Department of Earth, Atmospheric and Planetary Sciences and Department of Chemistry at Massachusetts Institute of Technology (MIT). He was named MIT Institute Professor in 1997.

The first warnings about what might happen to the ozone layer appeared in 1970. This attracted scientific attention and in 1974 Molina and Rowland published a paper in the scientific journal *Nature* describing the results of their studies of the chemical characteristics of CFC compounds. At the time, CFCs were used widely as propellants in spray cans; in refrigerators, freezers, and air conditioners; and in foam plastics. Molina and Rowland calculated that these very stable and therefore long-lived compounds could cross the TROPOPAUSE and that once they were in the STRATOSPHERE CFC molecules would be broken apart by their exposure to ULTRAVIOLET RADIATION. This chemical degradation would release chlorine, which would engage in reactions that deplete OZONE. Other atmospheric scientists received their paper with some skepticism, but in 1985 stratospheric ozone depletion was detected. It was eventually found that CFCs were involved, just as Molina and Rowland had indicated.

More recently, Mario Molina has studied the pollution of the TROPOSPHERE. In particular, he is keen to find ways to reduce pollution levels in large cities that suffer from traffic congestion, such as his native Mexico City. He serves on many committees, including the President's Committee of Advisors in Science and Technology, the Secretary of Energy Advisory Board, the National Research Council Board on Environmental Studies and Toxicology, and the boards of the U.S.–Mexico Foundation of Science and other nonprofit environmental organizations.

(You can learn more about Mario Molina at www.sciam.com/1197issue/1197profile.ht.)

moment of a force *See* TORQUE.

moment of inertia The equivalent of mass in calculations that involve a body rotating about an axis. It is calculated by multiplying the magnitude of each element of the body's mass by the square of its distance from the rotational axis.

momentum   The product (*mv*) of the mass (*m*) and speed (*v*) of a body that is moving in a straight line. For a moving body that is following a curved path, the equivalent term is ANGULAR MOMENTUM.

monsoon   A reversal in the direction of the PREVAILING WIND that occurs twice a year over much of the TROPICS, producing two seasons with distinctly different weather. The word *monsoon* is derived from the Arabic word *mausim,* which means "season." This became *monção* in Portuguese and *monssoen* in Dutch, from which language the word entered English.

A monsoon is one of two very clearly defined seasons in the Tropics. One monsoon is very wet and the other is very dry.

Winds are the movement of air from areas of high surface pressure to areas of low surface pressure. If the wind changes direction, therefore, that indicates that there has been a change in the distribution of pressure. This is the general explanation for the monsoons.

During winter, as the land cools and chills the air above it, surface air pressure builds over large continents to produce very extensive anticyclones. Air is subsiding, and as it sinks and warms by compression (*see* ADIABATIC) its RELATIVE HUMIDITY falls. It is very dry when it reaches the surface and flows outward from the high-pressure region. This produces the dry winter monsoon.

At high altitude, near the TROPOPAUSE, air is moving into the region and sinking at the center. This produces a region of CONVERGENCE and low pressure

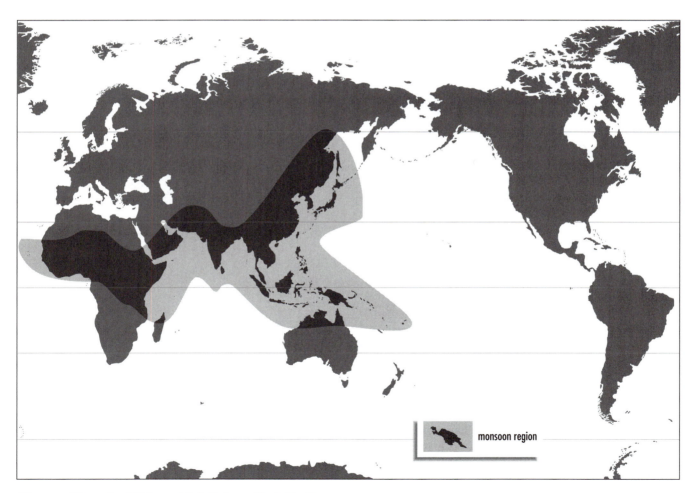

monsoon region

Monsoon. **The parts of Africa and Asia that are affected by the monsoons.**

above the region of high pressure and DIVERGENCE at the surface. Outside this region there are areas of low surface pressure and high pressure near the tropopause. These are over the sea, where the air temperature remains warmer than it is over land, because of the slower rate at which water cools, and warm air is rising.

In summer the situation reverses. The land warms more quickly than the sea, producing a large area of low surface pressure. When the pressure over the continental interior is lower than the pressure over the sea, air moves from sea to land. As it crosses the ocean, water evaporates into it, so it is close to SATURATION by the time it reaches land. It rises as it moves over the continent, and its water vapor condenses, releasing LATENT HEAT to sustain the instability of the air. This produces the giant CUMULONIMBUS clouds and torrential storms of the wet summer monsoon.

The mechanism is very similar to that which produces LAND AND SEA BREEZES, but it operates on a much larger scale and this type of circulation can develop only where conditions are fairly BAROTROPIC and a large landmass lies adjacent to a warm sea. These necessary conditions are met in Asia, but only partly met in other continents. Tropical Africa and the eastern part of the Amazon basin in South America experience seasonal changes of a monsoon type, and there are distinct dry and wet seasons in North America, but the reversals of wind direction are not strongly marked.

It is the Asian monsoons that produce the most extreme weather. This is because of the combined effects of the extremely large Asian landmass and the Himalayas. At Bombay, India, an average of 4 inches (104 mm) of rain falls during the dry, winter monsoon that lasts from October until May. During the wet, summer monsoon, lasting from June until September, the city receives 67 inches (1,707 mm).

The Himalayas form a barrier that confines the summer monsoon to the region to the south of the mountains, but air over the Tibetan plateau is being strongly heated, producing high pressure and divergence at high altitude. At the same time, the INTERTROPICAL CONVERGENCE ZONE (ITCZ) moves northward, to between 25° N and 30° N. Together these establish a distribution of pressure and temperature that generates an easterly JET STREAM, which

increases the rainfall over southeastern Asia, the Arabian Sea, and the Horn of Africa. In autumn, the temperature difference between the land and sea weakens and the ITCZ moves south. Behind it the winds are predominantly westerly. These disrupt the circulation. The jet stream disappears and is replaced by the POLAR JET STREAM, centered over the mountains and carrying westerly high-level winds. The surface PRESSURE GRADIENT is from north to south, and the surface winds are from the north or northeast. The rain ceases and the winter monsoon commences, carrying very dry air down from the mountains and out across the lands to the south.

People in Southeast Asia and especially in India watch eagerly for the start of the summer monsoon. The arrival of the rain produces relief from the very hot, humid conditions that precede it, but for farmers the rain is essential. If the monsoon fails, or even if its start is delayed, harvests are poor. The rains arrive suddenly, as the BURST OF MONSOON. Generally, the later they arrive the less rain the monsoon as a whole produces.

There is much that scientists still do not know about the Asian monsoon. A major project to study the monsoon, based at Nagoya University, Japan, began in 1996. It is called the *GEWEX Asian Monsoon Experiment* (GAME) (GEWEX is the Global Energy and Water cycle EXperiment).

(You can learn more about GAME at www.ihas. nagoya-u.ac.jp/GAME/.)

monsoon air   *See* AIR MASS.

Monsoon Drift   An ocean current that flows through the Arabian Sea. It breaks away from the NORTH EQUATORIAL CURRENT off the southernmost tip of India, flows parallel to the western coast of India, then turns in about latitude 15°–20° N to flow parallel to the southern coast of the Arabian Peninsula, finally joining the Somalia Current. This current flows past the eastern coast of Africa, through the Mozambique Channel separating Africa and Madagascar, and joins the AGULHAS CURRENT.

monsoon fog   A FOG that occurs along some coasts during the summer (wet) MONSOON. It is caused by ADVECTION when monsoon winds carry warm, very moist air across a cold land surface.

monsoon low   An area of low surface atmospheric pressure that develops over a continent during the summer and over the adjacent ocean during the winter. This is the pattern of pressure distribution associated with MONSOON conditions.

monsoon regime   *See* RAINFALL REGIME.

monsoon trough   A TROUGH of low pressure that forms during the summer (wet) MONSOON at about latitude 25° N and extends just to the south of the Himalayas, through Pakistan, northern India, and Bangladesh. Warm, moist air circulating in a CYCLONIC direction around the trough encounters the Himalayas and other mountain ranges. It is forced to rise, producing very heavy OROGRAPHIC RAIN.

month degrees   A measure of the conditions for plant growth that are associated with a particular type of climate. It is calculated by subtracting 43° F (6° C) from the MEAN TEMPERATURE for each month and adding together the remainders (the number of degrees by which the temperature is above or below 43° F). The resulting total represents the number of degrees by which the mean temperatures are above or below 43° F in the course of the year; 43° F is the minimum temperature for growth of most plant species. Month degrees are used in some CLIMATE CLASSIFICATIONS.

Montreal Protocol on Substances That Deplete the Ozone Layer   An international agreement that was reached in 1987 under the auspices of the United Nations Environment Program. The protocol aims to reduce and eventually eliminate the release into the atmosphere of all synthetic substances that deplete stratospheric ozone. The provisions made in the protocol for achieving this have been strengthened through amendments adopted in London in 1990, Copenhagen in 1992, Vienna in 1995, and Montreal in 1997.

(The text of the protocol is available in English and French from www.unep.ch/ozone/treaties.htm and detailed information about the protocol is at www.unep.ch/ozone/home.htm.)

moon dog   *See* SUN DOG.

moon pillar   A column of light that occasionally appears above or below the Moon, when the Moon is low in the sky. It is the nighttime equivalent of a SUN PILLAR and forms under similar conditions, by the reflection of moonlight by the undersides of slowly falling ICE CRYSTALS of the types known as *plates* and *capped columns.*

Morse code   A code that was devised by SAMUEL MORSE for the transmission of telegraph signals. He had developed it by 1838 and it was adopted throughout the world. When the telegraph began to be used to transmit meteorological data from outlying weather stations to central offices where they were assembled into weather reports and used to compile forecasts, Morse code was used to send them. It was also the code used to transmit radio communications to and from ships at sea and to and from aircraft. Some ships and aircraft were still using it in the 1940s. Although it was designed to be sent and received as sound, the code could also be sent as flashes of light from a powerful focused and directed lamp. In 1995 the U.S. Coast Guard abandoned the code for communication with ships at sea. That marked the end of its general use for telegraphic communication, but it is sometimes used when conditions are poor and its simplicity and reliability make it attractive to amateur radio enthusiasts.

The code is made up of what are known conventionally as dots (•) and dashes (—). These represent units. One unit is equal in duration to one dot; one dash is equal to three units. One unit pause marks the space between the components of a character, a pause of three units separates the letters in a word, and a pause of six units separates words.

There are two versions of Morse code, the International and the American. Additional pauses are inserted between components of some characters in the American code and some components are equal to four or six units. There is no code for an exclamation mark (!) or ampersand (&) in the international code.

Morse, Samuel Finley Breese (1791–1872) American *Artist and inventor*   The man responsible for building the first telegraph line in the world and devising a code by which telegraph messages could be sent simply, efficiently, and reliably was born on April 27, 1791, at Charlestown, Massachusetts, which is now part of Boston. His father, Jedidiah, was a clergyman and geographer. Samuel studied art at Yale University, where he developed a keen interest in painting miniature portraits. He graduated in 1810 and then persuaded his parents to allow him to travel to England to

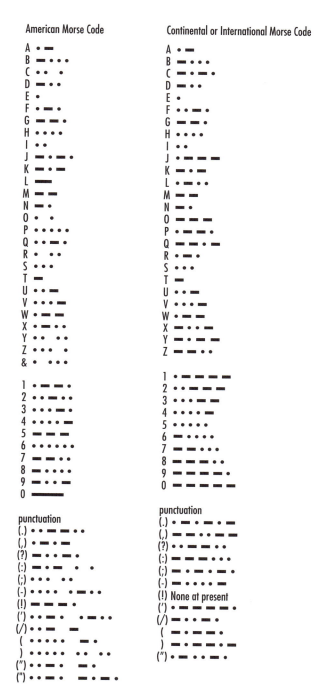

Morse code. The telegraphic code devised by Samuel Morse consists of patterns of long and short signals. It is easy to learn and efficient at conveying messages against background noise.

Samuel Morse. The American painter and inventor who established the world's first telegraph line and devised the code used to transmit messages by telegraph. *(John Frederick Lewis Collection, Picture and Print Collection, The Free Library of Philadelphia)*

study historical painting. He stayed in England from 1811 until 1815.

After his return, his talent and gift for making friends generated enough portrait commissions to allow him to earn a living, but he never became wealthy and Americans were not impressed by his historical paintings. He settled in New York, where his reputation grew, and he was one of the founders of the National Academy of Design and its first president, from 1826 to 1845. Morse taught art at the University of the City of New York (now New York University) and twice ran unsuccessfully for election as mayor of New York, on an anti-Catholic, antiimmigrant ticket.

Electricity and electromagnetism were just becoming known and aroused great interest. In 1832, while returning from one of his visits to Europe to study art, Morse fell to discussing electrical experiments with Charles Thomas Jackson (1805–80), a fellow passenger on the ship. Various people had suggested the possibility of transmitting messages electrically, and his conversations with Jackson gave Morse the idea to make a device that would do so. Unfortunately, his knowledge of electricity was inadequate to the task, but a university colleague drew his attention to the work of JOSEPH HENRY, whom he met later. Henry had already given the idea of the telegraph considerable thought, and he was unstinting in his help and advice, answering all Morse's questions.

By about 1835 Morse had made a telegraph that worked. Henry had designed one earlier, but Morse believed his was the first. Then began his real task. Having failed to build a "Morse line" in Europe, he set out to persuade a reluctant Congress to appropriate the $30,000 it would cost to build a telegraph line between Baltimore and Washington, D.C., a distance of 40 miles (64 km). In 1843 he finally succeeded and the line was built in 1844. Morse used his own code, the MORSE CODE, which he had developed by 1838, to transmit the first message: "What hath God wrought?" Within a few years TELEGRAPHY was being used to transmit meteorological data, a development that led directly to the first weather reports and forecasts.

After the feasibility of a telegraph line had been demonstrated, Morse was caught up in a succession of legal actions brought by his partners and other inventors, including Jackson, who claimed priority for the invention. Eventually Morse established his patent rights and was rewarded by many European governments. In old age he became a philanthropist. He died in New York on April 2, 1872, and in 1900, when it opened, Samuel Morse was made a charter member of the Hall of Fame for Great Americans on the campus of New York University.

mother cloud A cloud from which other clouds have been produced and that is seen at the same time as the clouds to which it gave rise.

mother-of-pearl cloud *See* NACREOUS CLOUD.

Mount Agung (Gunung Agung) A volcano on the island of Bali, Indonesia, that erupted violently in March 1963, ejecting large amounts of particulate material, together with SULFUR DIOXIDE and SULFATE AEROSOL. Some of the sulfate entered the lower STRATOSPHERE, and the wind patterns at the time carried it into the Southern Hemisphere. Within a very

short time aerosol droplets were detected between Bali and Australia. The absorption of solar energy by the stratospheric aerosols warmed the lower stratosphere by 11°–12° F (6°–7° C) and surface temperatures fell by about 1° F (0.5° C). After about six months the aerosols had spread around the world, and they remained in the air for several years, producing spectacular sunsets, but no measurable climatic effect.

mountain breeze A KATABATIC WIND that blows at night in some mountain regions. It most commonly occurs when conditions are calm and the sky is clear and is more frequent in winter. Once the Sun has set, heat absorbed during the day by the surface of the mountainsides starts to be radiated away. The ground cools and chills the layer of air in contact with it. This air is then denser than the warmer air farther down the mountainside, and it flows downhill by gravity, producing a cool breeze at lower levels and especially in the valley. Mountainsides where mountain and VALLEY BREEZES occur regularly are often preferred for growing fruit, because the constant air movement prevents the static conditions in which frost can form.

mountain climate A climate that differs from the climate typical of the latitude in which it occurs by reason of the high elevation of the land. The climatic effect of increasing elevation is broadly similar to that of increasing latitude, but there is wide variation in the types of mountain climates. Generally, mean temperatures are lower and conditions are windier. PRECIPITA-

Mountain breeze. **The circulation of the air produces a cool wind that flows down the side of a mountain at night.**

TION is greater at lower levels on a mountainside, but above the permanent SNOW LINE it decreases, because the air has lost most of its moisture. Air DENSITY decreases with elevation, thereby reducing the capacity of the air to retain heat. Consequently, the contrast in temperature between day and night and between places exposed to full sunshine and those in shade is greater on mountains than it is at sea level, regardless of latitude. Mountains are often shrouded in cloud. At the surface this is identical to FOG, but it is caused by ORO-GRAPHIC LIFTING and makes most mountain climates much foggier than low-level climates. Mountainsides also experience KATABATIC WINDS, which can warm the air below the summit and remove snow rapidly. Beyond such generalizations, however, the climate of a particular mountainside is affected by such factors as the direction it faces, the amount of shelter and shading it receives from surrounding mountains, and FUNNEL-ING effects on the PREVAILING WIND.

mountain-gap wind (canyon wind, gorge wind, jet-effect wind) A local wind that occurs where the prevailing wind is funneled through the space between two mountains and accelerated, so it is markedly faster than most winds in the region where it occurs.

mountain meteorology The study of the effects mountains have on the atmosphere and the weather conditions these produce.

Mount Aso-san A volcano on the Japanese island of Kyushu. The volcano is 5,223 ft (1,593 m) high and has one of the biggest calderas (craters) in the world, measuring 17 miles (27 km) from north to south and 10 miles (16 km) from east to west. Mount Aso-san erupted violently in 1783, injecting a large amount of dust and AEROSOL into the STRATOSPHERE. The eruption was followed by unusually cold weather from 1784 to 1786.

Mount Katmai A volcano in Alaska that erupted violently in 1912, ejecting an estimated 5 cubic miles (21 km³) of dust high into the atmosphere. It was the most violent volcanic eruption of the 20th century. During the months that followed observatories in California and Algeria recorded a 20 percent drop in solar radiation and the weather was unusually cool, although temperatures had been somewhat lower than normal before the eruption. A BISHOP'S RING was seen after the eruption.

Mount Pinatubo A volcano on the island of Luzon, Philippines, which erupted in 1991. It was the second most violent volcanic eruption of the 20th century (after that of MOUNT KATMAI). The volcano had not erupted for 600 years. Activity began in April and on June 14–16 the mountain split into pieces in a series of explosions. The eruption ejected dust and SULFATE AEROSOL to a height of 25 miles (40 km). This spread over most of the world. The amount of material in the STRATOSPHERE reached a peak in September 1991, after which the amount began to decrease over the Tropics. It did not peak until the spring of 1992 over latitudes 40°–60° N and remained fairly constant until the end of 1992 over latitudes 40°–60° S. It could still be detected over Hawaii and Cuba in January 1994. Absorption of solar radiation by material in the stratosphere caused a cooling of about 0.7°–1.25° F (0.4°–0.7° C) in the TROPOSPHERE that lasted through 1992 and 1993.

Mount Spurr A volcano in Alaska that erupted on June 27 and August 18, 1992, injecting dust and SULFUR DIOXIDE into the atmosphere. There is no record of any climatic effect.

Mount Tambora A volcano in the Dutch East Indies (now Indonesia) that erupted in April 1815. The eruption was one of the most violent in the last few thousand years, and scientists calculated that it released at least 3.6 cubic miles (15 km³) of dust and sulfuric acid AEROSOLS that spread to form a veil over much of the Northern Hemisphere. The particles may have added to some that were still present from earlier volcanic eruptions at Saint Vincent in the West Indies and Awu in Sulawesi. Temperatures fell by up to 2° F (1° C) in many areas and wind patterns were distorted. The following year, 1816, came to be known as "the year with no summer." Snow, driven by a northeasterly wind, fell over a wide area of eastern North America in June 1816, and from June 6 to June 11 the ground inland was covered in snow as far south as Pittsburgh. Connecticut had frosts at some time in every month of 1816, and there were some June days when the temperature did not rise above freezing in Québec City. Harvests failed in many parts of North America and

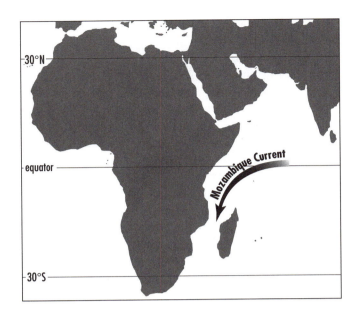

Mozambique Current. **The current carries warm water into the Mozambique Channel, between Mozambique and Madagascar.**

Europe, but the altered wind patterns produced a fine summer in the north of Scotland and there was a heat wave in Ukraine. The bad weather of 1816 may have been partly responsible for the worst outbreak of typhus Europe had ever experienced, lasting from 1816 until 1819.

Mozambique Current A branch of the SOUTH EQUATORIAL CURRENT that flows around the northern end of Madagascar and continues as a warm ocean current in a southwesterly direction through the Mozambique Channel off the eastern coast of Africa. To the south of Madagascar it becomes the AGULHAS CURRENT.

MSU *See* MICROWAVE SOUNDING UNIT.

mud rain Rain that contains a quantity of fine soil particles large enough to discolor the rain water and leave a mudlike deposit on surfaces.

muggy *See* CLOSE.

multichannel sea surface temperature (MCSST) A procedure in which sea-surface temperatures are cal-

culated from data received from an ADVANCED VERY HIGH RESOLUTION RADIOMETER. First the data are checked to identify points referring to clouds or AEROSOLS. These are removed. Using the remaining data, sea-surface temperature (SST) is calculated by an ALGORITHM such as $SST = a_0 + a_1 T_1 + a_2 T_2$, where T_1 is the AVHHR brightness temperature at the wave band 3.55–3.93 μm; T_2 is the brightness temperature at the wave band 10.3–11.3 μm or 10.5–11.5 μm, depending on the channel used; and a_0, a_1, and a_2 are coefficients that convert the T values into sea-surface temperatures.

mutispectral scanner An instrument carried on LANDSAT satellites that obtains images of Earth's surface with a spatial resolution of 24 feet (80 m). It is used to monitor surface changes, for example, in vegetation, coastlines, ice sheets, glaciers, and volcanoes.

multivariate analysis In statistics, the analysis of a number of measurements of different variable characteristics (such as temperature, pressure, air density, and humidity) all of which refer to the same subject.

Muroto II A TYPHOON that struck Japan in September 1961. It caused a STORM SURGE of 13 ft (3.9 m) that produced floods in Osaka in which 32 people died.

softer rock worn away by wind and chemical erosion

Mushroom rock. **Chemical weathering wears away the base of the rock, leaving a large rock standing on a slim stem. The shape resembles a mushroom.**

mushroom rock (pedestal rock, rock mushroom) An unstable rock formation that is often seen in deserts. It consists of a large rock that is mounted on a much narrower stem, creating a mushroom shape. The most famous example is Pedestal Rock, Utah. The shape is the result of erosion. The wind blows away surrounding dust and sand and wears away softer rock, leaving a column of harder rock exposed. When it rains, water is retained rather longer near the base of the rock than it is higher up on the surface of the rock.

Chemicals dissolve from the rock, form crystals as they dry, and are then blown away by the wind. This process of chemical WEATHERING erodes the base of the rock faster than it erodes the rock above it.

mutatus A cloud development in which the shape of the cloud is changing fairly rapidly because of new cloud masses that are growing from it.

mutual climatic range *See* BEETLE ANALYSIS.

N   *See* NEWTON.

N   *See* NITROGEN.

N_A *See* AVOGADRO CONSTANT.

nacreous cloud (mother-of-pearl cloud) Bright cloud, white usually tinged with pink, occasionally seen in high latitudes when the Sun is just below the horizon (shortly before dawn or after sunset). It resembles the mother-of-pearl that lines the inside of some seashells. Nacreous clouds form in winter in the lower to middle STRATOSPHERE, at heights of about 12–19 miles (20–30 km) above sea level, in air at a temperature of about –112° F (–80° C). They are composed of ice crystals and develop on the crests of waves generated by the low-level movement of air across mountains. POLAR STRATOSPHERIC CLOUDS are a variety of nacreous cloud.

nadir The point on the surface that lies directly beneath an observational satellite.

NADW *See* NORTH ATLANTIC DEEP WATER.

Namu A TYPHOON that struck the Solomon Islands on May 19, 1986. It killed more than 100 people and rendered more than 90,000 homeless.

NAO *See* NORTH ATLANTIC OSCILLATION.

n'aschi A northeasterly wind that blows in winter along the Iranian coast, bordering the Persian Gulf, and on the coast of Pakistan, bordering the Arabian Sea.

National Hurricane Center (NHC) The center in Miami, Florida, that is part of the NATIONAL WEATHER SERVICE and maintains a continuous watch on TROPICAL CYCLONES over the Atlantic, Caribbean, Gulf of Mexico, and eastern Pacific Ocean throughout the hurricane season, which lasts from May 15 through November 30. The NHC prepares and issues HURRICANE WATCHES and HURRICANE WARNINGS for the public and specialist information for other users. Outside the hurricane season the NHC runs training courses for managers of emergency services from those parts of the United States and other countries that are affected by tropical cyclones. The NHC also conducts research aimed at improving hurricane forecasting techniques and runs programs to raise public awareness of hurricanes, their dangers, and methods to remain safe.

(You can learn more about the NHC at www.nhc.noaa.gov/.)

National Meteorological Center The government center, in Washington, D.C., that specializes in numerical modeling of the weather. Its models are used in the preparation of forecasts and in studies of the behavior of the atmosphere.

National Oceanic and Atmospheric Administration (NOAA) An agency of the United States government, within the Department of Commerce, that is responsible for issuing short-term and long-term weather forecasts, operating U.S. meteorological satellites, conducting research, and performing such other tasks as are necessary in pursuit of these functions. Its tasks are defined as describing and predicting changes in the Earth's environment and conserving and managing wisely the coastal and marine resources of the United States in order to ensure sustainable economic opportunities.

National Rainfall Index (RI) An index that is calculated country by country by weighting the national average precipitation according to the long-term averages of all the individual stations in the country. The resulting scale can be used together with other national indices, such as that for agricultural production, and it allows comparisons to be made between countries and years. Its disadvantage is that because it is calculated from precipitation, wetter areas have an undue influence on it, making it less useful when measuring the severity of drought.

National Severe Storms Forecast Center The establishment based in Kansas City, Missouri, that issues outlooks for the intensity of CONVECTION several times every day, covering all of the United States. Its convection outlooks provide warning of severe THUNDERSTORMS up to 24 hours in advance, and where storms are likely to generate TORNADOES, appropriate warnings are issued for roughly six- or eight-hour periods.

National Severe Storms Laboratory The laboratory at which scientists study all aspects of severe weather. The laboratory is located at Norman, Oklahoma, but staff also work in Colorado, Nevada, Washington, Utah, and Wisconsin. The laboratory is part of the NATIONAL OCEANIC AND ATMOSPHERIC ADMINISTRATION.

(You can learn more about the National Severe Storms Laboratory at www.nssl.noaa.gov/.)

National Snow and Ice Data Center (NSIDC) The organization that collects data and supports research relating to polar conditions and all aspects of

ice, including PERMAFROST, PALEOGLACIOLOGY, and ICE CORES. It keeps records of the extent of snow cover, AVALANCHES, GLACIERS, and ICE SHEETS. The center distributes data and information and publishes reports. It is part of the University of Colorado Cooperative Institute for Research in Environmental Sciences and is affiliated with the NATIONAL OCEANIC AND ATMOSPHERIC ADMINISTRATION and the National Geophysical Data Center.

(You can learn more about the NSIDC at nsidc.org/index.html.)

National Weather Service The federal agency of the United States government, forming part of the NATIONAL OCEANIC AND ATMOSPHERIC ADMINISTRATION, that is responsible for preparing and issuing weather forecasts and information on a range of hydrological matters, including river levels, tides, STORM SURGES, and TSUNAMIS. Its head office is at Silver Spring, Maryland, and it maintains five major operating centers in various parts of the country. In addition to the published and broadcast forecasts for which it provides the meteorological data, the service makes its information available on-line.

(For more information, the National Weather Service home page is at www.nws.noaa.gov/index.html. Additional information about its services is at www.nws.noaa.gov/oso/fospage.html, and technical information on its communications networks is at www.nws.noaa.gov/oso/histgate.shtml.)

natural period *See* PERIOD.

natural season Five periods, each lasting for at least 25 days, that are characterized by weather of a distinct type. These periods were identified by HUBERT LAMB from his studies of records for the British Isles from 1898 through 1947.

Spring–early summer lasts from early April until the middle of June. The weather is variable, but long settled spells are likely, with ANTICYCLONES in late May and early June.

High summer lasts from the middle of June until early September. It is characterized by long spells of weather of different types, varying from year to year, but persistent CYCLONIC conditions are more common than persistent ANTICYCLONIC conditions.

Autumn lasts from the middle of September until the middle of November. Long spells of settled weather occur. During the first half of the period these are mainly anticyclonic, and in the second half they are cyclonic and often stormy.

Early winter lasts from the third week in November until the middle of January. The weather is variable, but when long spells occur they usually are of mild, stormy weather.

Late winter and early spring last from the third week in January until the end of March. There are long spells of weather, but types that vary from one year to another. In some years this season resembles the middle of winter and in others there is springlike weather from late in February.

nautical mile   A unit of length that was introduced early in the 17th century; it was based on a suggestion by the English mathematician and inventor Edmund Gunter (1581–1626). Gunter thought navigation at sea might be simplified if distances were related directly to degrees of latitude. To achieve this he proposed establishing a unit of length that is equal to the distance subtended by one minute of arc. Consequently, one nautical mile is the average meridian (north–south) length of one minute of latitude. It is necessary to take an average length, because the distance subtended by one minute of arc of latitude varies slightly with latitude because the Earth is not perfectly spherical. The International Hydrographic Conference of 1929 recommended that this distance be measured at latitude 45°, to produce a nautical mile of 6,076 feet (1,852 m). This is the length that is most widely used. In Britain, however, the nautical mile is measured at latitude 48°, which gives a length of 6,080 feet (1,854 m), so that 1 English nautical mile = 1.00064 international nautical miles. A speed of 1 nautical mile per hour is known as *1 KNOT*.

naval meteorology   The branch of METEOROLOGY that is concerned with weather conditions over the oceans and the interactions between the oceans and atmosphere.

navigable semicircle   The side of a TROPICAL CYCLONE where the winds are lightest and where they tend to push ships out of the path of the approaching storm. The difference in wind speeds is due to the fact that the storm itself is moving. Consequently, on one side of the storm its own speed of motion must be added to the wind speed of the storm, and on the other side it must be deducted from it. Because the circulation around the storm is CYCLONIC and the storms move in a generally easterly direction, driven by the TRADE WINDS, then turn away from the equator, the navigable semicircle is on the side of the storm closer to the equator. This is the southern side in the Northern Hemisphere and the northern side in the Southern Hemisphere.

NDVI   *See* NORMALIZED DIFFERENCE VEGETATION INDEX.

Ne   *See* NEON.

neap tides   *See* TIDES.

near-infrared mapping spectrometer   An instrument carried on some satellites that takes readings in the near-infrared part of the spectrum (*see* NEAR-

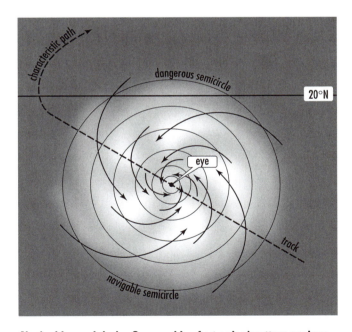

Navigable semicircle. On one side of a tropical pattern cyclone the wind blows in the opposite direction to the direction in which the storm is moving. This pattern reduces the effective wind speed and the winds tend to blow ships behind the center of the storm, rather than into its path. This is the navigable semicircle. The opposite side is the dangerous semicircle.

INFRARED RADIATION and WAVE BAND). The chemical composition, structure, and temperature of the atmospheres of planets and satellites can be calculated from the data that are produced, as well as details of the mineral and geochemical composition of the surface.

near-infrared radiation INFRARED RADIATION that has shortest WAVELENGTH in the infrared WAVE BAND, of about 1–3 μm. WATER VAPOR is a very efficient absorber of near-infrared radiation.

near UV *See* ULTRAVIOLET RADIATION.

neb *See* NEBULOSUS.

Nebraskan glacial A GLACIAL PERIOD in North America that began about 800,000 years ago and ended about 600,000 years ago. It was preceded by predominantly warm conditions throughout pre-Nebraskan time from the start of the PLEISTOCENE EPOCH 12.6 million years ago. The Nebraskan was followed by the AFTONIAN INTERGLACIAL.

nebulosus (neb) A species of clouds (*see* CLOUD CLASSIFICATION) that form a layer or veil with no clearly distinguishable features. The species occurs with the cloud genera CIRROSTRATUS and STRATUS. The name of the species is the Latin word that means "mist."

Nell A TYPHOON that struck the Philippines on December 25 and 26, 1993. It killed at least 47 people.

Neogene The period of geological time (*see* GEOLOGICAL TIME SCALE) that began about 23.3 million years ago. It comprises the latter part of the TERTIARY subera and includes the MIOCENE and PLIOCENE epochs.

neon (Ne) A colorless gas that is one of the NOBLE GASES and forms almost no compounds. It is present in the atmosphere, accounting for 0.00182 percent by volume. The atmospheres of the other planets and satellites of the solar system have much less neon. Neon vaporizes at –410.89° F (–246.05° C) and freezes at –416.61° F (–248.67° C). Neon is extracted from the air and used in neon lamps. The gas is sealed in a tube and an electric current is passed through it at 60–90 volts. This ionizes (*see* IONIZATION) the neon atoms around the cathode (negative electrode), causing the emission of a red light.

nephanalysis The study of cloud types and formations in satellite images to obtain information about the weather systems that produce them. This allows scientists to calculate the type and intensity of those systems and to track their movement. *Nephos* is the Greek for "cloud." In Greek mythology, Nephele was a phantom created by Zeus, who became a cloud goddess. She married Athamas. He tired of her, preferring Ino, who hated the children of Nephele, Phrixus and Helle. To help her children escape from Ino, who wished to destroy them, Nephele gave them a ram with a golden fleece that carried them to safety.

nephcurve In NEPHANALYSIS, a line that is drawn to mark the boundary between clouds and clear sky, areas of PRECIPITATION, clouds of different types, or clouds at different heights.

nephelometer An instrument that is used in remote sensing. It uses a laser beam to measure the SCATTERING of light by atmospheric particles.

nephology The scientific study of clouds.

nephoscope An instrument that is used to measure the direction of movement of a cloud and its angular velocity around a point on the surface directly beneath it. If the height of the cloud is known, its linear direction can also be calculated. One widely used type, the Besson comb nephoscope, comprises a number of pointed rods, like the teeth of a comb, attached to the top of a long, vertical rod. The device is rotated until the cloud appears to be moving between the points, allowing the direction of movement to be measured.

von Neumann, John (1903–1957) Hungarian–American *Mathematician and physicist* John (originally Janos) von Neumann was born in Budapest, Hungary, on December 28, 1903. He was educated privately until he was 14, when he entered the gymnasium (high school) but continued to receive extra tuition in mathematics. Von Neumann left Hungary in 1919 to escape the chaotic conditions that followed the defeat of the Austro–Hungarian Empire in World War

I. He studied at the universities of Berlin (1921–23) and Zürich (1923–25) and graduated from Zürich in chemical engineering. He received a doctorate in mathematics from the University of Budapest in 1926 and then continued his studies at the University of Göttingen, where he worked with Robert Oppenheimer (1904–67). From 1927 until 1929 von Neumann worked as an unpaid lecturer at the University of Berlin. He then moved to the University of Hamburg and in 1930 migrated to the United States. He was appointed a visiting professor at the University of Princeton and in 1931 became a full professor of mathematics at the newly formed Institute of Advanced Studies at Princeton. Von Neumann later held a number of advisory posts for the U.S. Government, but he remained at the institute for the rest of his life.

Von Neumann made many contributions to mathematics and theoretical physics. He invented the study of game theory.

In the 1940s von Neumann became interested in meteorology and began to study the methods used in NUMERICAL FORECASTING. He devised a technique for analyzing the stability of those methods and in 1946 he established the Meteorology Project at Princeton. In April 1950 the group led by von Neumann made the first accurate numerical forecast, using the ENIAC computer.

In 1952 he designed and supervised the construction of MANIAC-1, which was the first computer that was able to use a flexible stored program. His work at Princeton on MANIAC-1 influenced the design of all the programmable computers that followed, including those used in weather forecasting.

Von Neumann received many honors and awards, including the Medal of Freedom, the Albert Einstein Award, and the Enrico Fermi Award, all of which were presented to him in 1956. He was already in poor health and he died in Washington, D.C., of cancer on February 8, 1957.

neutral atmosphere An atmosphere in which the ENVIRONMENTAL LAPSE RATE is equal to the DRY ADIABATIC LAPSE RATE. This condition results from convectional INSTABILITY. CONVECTION tends to redistribute heat until the two lapse rates equalize.

neutral stability The condition in which the atmosphere is stratified and the POTENTIAL TEMPERATURE neither increases nor decreases with height. If a PARCEL OF AIR moves vertically in either direction it enters a region where its DENSITY is similar to that of the surrounding air. Consequently, gravity does not restore it to its original level and it remains where it is.

neutron probe *See* SOIL MOISTURE.

neutrosphere The shell around the Earth that is formed by the atmosphere from the surface to the base of the IONOSPHERE, at a height of about 43–56 miles (70–90 km). Within this shell most of the constituents of the atmosphere are electrically neutral.

nevados A cold, KATABATIC WIND that blows down the mountainsides and into the high valleys of Ecuador. It is caused by nighttime radiative cooling, which chills the surface air. This starts to flow gravitationally and is chilled further by contact with the snow- and ice-covered surface across which it moves.

newton (N) The derived SYSTÈME INTERNATIONAL D'UNITÉS (SI) UNIT of force, which is equal to the force needed to accelerate a mass of 1 KILOGRAM at 1 METER per SECOND per second (1 N = 1 kg m s^{-2}). The name of the unit was adopted in 1938 in honor of the English physicist and mathematician Sir Isaac Newton (1642–1727).

NEXRAD *See* NEXT GENERATION WEATHER RADAR.

Next Generation Weather Radar (NEXRAD) A network of 175 DOPPLER RADAR installations located at weather stations, airports, and military airfields throughout the United States and operated by the National Weather Service. Each unit operates at frequencies of 2.7–3.0 GHz, emits a beam 1° wide, and has an antenna 25 feet (7.6 m) in diameter enclosed within a casing resembling a golf ball and mounted on top of a tall tower. It takes the radar five minutes to scan 360° in AZIMUTH and from 0° to 20° in elevation. It can detect and measure rainfall at a distance of up to 286 miles (460 km) and can measure the rate of rotation of a storm at up to 143 miles (230 km).

(More information can be found at http://home.earthlink.net/~djbwx/projects/nexrad.html.)

NHC *See* National Hurricane Center.

Nigel *See* ERIC.

nightglow Radiation that is emitted at night in the MESOSPHERE, mainly at heights of 50–60 miles (80–100 km). It is due to chemical reactions among the constituents of the air at that height, and it becomes stronger as the night progresses. Atomic oxygen emits yellow–green light, at a WAVELENGTH of 555.7 nm; sodium emits yellow light at 589 nm; and HYDROXYL and oxygen molecules emit radiation at wavelengths outside the visible spectrum. Even the visible light is too faint to be visible from the ground, although instruments on satellites can detect it.

nimbostratus (Ns) A genus of low, gray, fairly uniform clouds (*see* CLOUD CLASSIFICATION) that often deliver steady, continuous rain or snow. Although its base is low, it usually extends vertically to above 6,500 feet (2 km), and it is thick enough to obscure the Sun and Moon completely, making daylight dull and nights exceedingly dark. The cloud is shapeless and dark and sometimes appears to be faintly illuminated from the inside. This is due to breaks in the nimbostratus that reveal paler STRATUS above it. Nimbostratus often forms SCUD.

Like stratus, which it closely resembles, nimbostratus forms when moist, stable air is forced to rise and its water vapor condenses. Consequently, it occurs on WARM FRONTS and also on mountains as a result of OROGRAPHIC LIFTING. There are no species or varieties of nimbostratus.

nimbus The Latin word for "cloud," which originally meant a bright cloud or the halo or aureole that surrounds the head of a holy person. In his 1803 classification of clouds, LUKE HOWARD used the word to imply "rain" (the Latin for "rain" is *pluvia*), and *nimbus* is now used in the form *nimbo-* as a prefix attached to the cloud genus *stratus,* to give *nimbostratus,* and as a suffix to the genus *cumulus* to give *cumulonimbus.* In both cases it associates the generic names with the idea of precipitation.

Nimbus satellites A series of U.S. weather satellites, the first of which, *Nimbus 1,* was launched on August 24, 1964, to be followed by six more; *Nimbus 7* was launched in 1978. Nimbus satellites carry equipment for AUTOMATIC PICTURE TRANSMISSION, a television camera system for mapping clouds, and an infrared radiometer that allows pictures to be taken at night. The Nimbus series are second-generation satellites that followed the TIROS series.

Nina A TYPHOON that struck the Philippines on November 26, 1987, and caused a STORM SURGE. It killed 500 people in Sorsogon Province, Luzon.

nitrogen (N) A colorless gas that is the major constituent of the atmosphere, accounting for 78.08 percent by volume. Nitrogen freezes at –345.75° F (–209.86° C) and boils at –320.44° F (–195.80° C). There are two natural ISOTOPES, ^{14}N and ^{15}N; the latter constitutes about 3 percent of the total amount of nitrogen. Chemically, nitrogen is fairly inert at the pressures and temperatures prevailing in the atmosphere, but it is oxidized by LIGHTNING and converted into ammonia and nitrates by bacteria (*see* NITROGEN FIXATION). Nitrogen is an essential component of all proteins and nucleic acids (deoxyribonucleic acid [DNA] and ribonucleic acid [RNA]), and it is therefore essential for life.

nitrogen cycle The sequence of reactions by which NITROGEN moves between the atmosphere, soil, living organisms, and returns to the air. It is one of the major cycles of elements.

The energy of LIGHTNING causes gaseous nitrogen to react with oxygen in the vicinity of the spark. This reaction produces NITROGEN OXIDES in a series of steps. The oxides then dissolve in raindrops to form weak nitric acid (HNO_3). The nitric acid is washed to the ground, where it forms nitrates (NO_3).

Certain BACTERIA and cyanobacteria possess the ability to utilize gaseous nitrogen (*see* NITROGEN FIXATION). This also leads to the formation of nitrates by the process of nitrification.

Nitrates are soluble and plants are able to absorb the soil solution through their roots. This is how they obtain the nitrogen they need to synthesize proteins. Animals cannot manufacture proteins directly from nitrates. They obtain the proteins that their bodies need by feeding on plants (herbivores), on other animals (carnivores), or on both (omnivores).

Both plants and animals release waste products. These, together with their own bodies when they die, provide food for a hierarchy of soil animals, fungi, and bacteria. Through the process of decomposition, these organisms convert plant and animal proteins into ammonia (NH_3). Ammonia dissolves in water and plants are able to use it directly in the form of ammonium ($NH4$). However, it can also combine with carbon dioxide (CO_2) to form ammonium carbonate (($NH_4)CO_3$). Further nitrification by bacteria converts the ammonium carbonate into nitrites, in an oxidation reaction that releases energy:

$$(NH_4)CO_3 + 3O_2 \rightarrow 2NHO_2 + CO_2 + 3H_2O + energy$$

Nitrous acid (HNO_2) is unstable. It reacts with magnesium (Mg) or calcium (Ca) to form nitrites ($MG(NO_2)_2$ or $Ca(NO_2)_2$). Other bacteria then nitrify the magnesium or calcium nitrite to form nitrates once more. This is also an oxidation reaction that releases energy.

$$2Ca(NO_2)_2 + 2O_2 \rightarrow 2Ca(NO_3)_2 + energy$$

The bacteria that perform nitrification reactions use the energy that these release to move and digest food.

A final set of reactions returns nitrogen to the air as a gas. This process is called *denitrification,* and it is performed by another bacterial species, *Thiobacillus denitrificans.* Some species release gaseous nitrogen (N_2). Others release ammonia (NH_3). Ammonia is very soluble, but it boils at –29° F (–34° C), and consequently it vaporizes rapidly if it goes out of solution. Once in the air it is oxidized to nitrate, but most of the ammonia remains in the soil, where it is captured by nitrifying bacteria, continuing in the cycle until it reaches *T. denitrificans* and is released as nitrogen gas.

The amount of nitrogen that leaves the air through nitrogen fixation is precisely balanced by the amount that is returned by denitrification. Denitrification completes the cycle.

This is the natural cycle, however, and it does not tell the whole story. The manufacture of nitrogen fertilizer involves fixing gaseous nitrogen industrially. The fertilizer is then released into the environment either as nitrates or as compounds that are converted to nitrates in the soil. Burning fuel at high temperatures and pressures also fixes nitrogen by supplying the energy needed to oxidize it to nitrogen oxides. Internal combustion engines are now a major source of nitrogen oxides, which react to form OZONE and PHOTOCHEMICAL SMOG as well as contributing to ACID RAIN. The

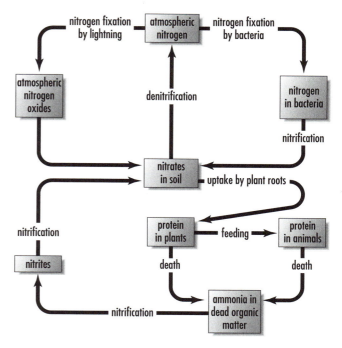

Nitrogen cycle. **Atmospheric nitrogen is fixed by lightning and by bacteria. It is then used by plants and animals and finally returns to the atmosphere.**

amount of nitrogen that is fixed and released industrially is now comparable to the amount moving through the natural cycle. This "industrial nitrogen" then joins the natural cycle, increasing the volume of nitrogen that is engaged in it.

nitrogen fixation The process by which gaseous nitrogen is converted into a soluble compound. This happens in three ways: by LIGHTNING, by BACTERIA, and by industrial processes.

Lightning supplies the energy that is needed to make nitrogen react with oxygen. The resulting oxides are soluble and are washed to the surface by PRECIPITATION in the form of weak nitric acid (HNO_3).

The amount of nitrogen fixed in this way is minute compared with the amount that is fixed by bacteria. Several groups of bacteria fix nitrogen. All of them possess the enzyme nitrogenase, which catalyzes a reaction in which nitrogen (N_2) reacts with hydrogen (H_2) to produce ammonia at ordinary temperature and pressure:

$$N_2 + 3H_2 + energy \rightarrow 2NH_3$$

Colonies of *Rhizobium* species form nodules on the roots of leguminous and some nonleguminous plants. Other species, including *Azotobacter* and *Clostridium* spp., are free-living in the soil, as are the nitrogen-fixing sulfur bacteria belonging to *Chromatium, Rhodospirillum, Chlorobium,* and other genera. There are also species of *Nostoc* and *Anabaena* and other species of cyanobacteria (formerly known as *blue– green algae*) that fix nitrogen in lakes and ponds.

Nitrogen is fixed industrially to make nitrogen-based fertilizer. There are several techniques for doing this, all of which use a large amount of energy. In the Haber process, devised in 1908 by the German chemist Fritz Haber (1886–1934), air is heated to between 750° F and 930° F (400° and 500° C), at a pressure of about 677–845 inches of mercury (20–25 MPa, 200–250 atmospheres, or about 1 1/4–1 1/2 tons per square inch) in the presence of a catalyst. Under these conditions nitrogen reacts with hydrogen to form ammonia by the reversible reaction $N_2 + 3H_2 \rightarrow 2NH_3$.

nitrogen oxides (NO$_x$) Nitrogen forms seven oxides: NITROUS OXIDE or dinitrogen oxide (N_2O), nitric oxide or nitrogen oxide (NO), nitrogen trioxide (N_2O_3), nitrogen dioxide (NO_2), dinitrogen tetroxide (N_2O_4), dinitrogen pentoxide (N_2O_5), and nitrogen trioxide (NO_3). Not all of these compounds are stable and only N_2O, NO, and NO_2 are climatologically important.

Unlike the other oxides, N_2O does not react with the hydroxyl radical (OH). Reactions with OH remove most pollutants from the lower atmosphere by converting them to harmless compounds that are washed to the ground by PRECIPITATION. Because it remains unaffected, N_2O is left to absorb INFRARED RADIATION, making it a GREENHOUSE GAS, with a GLOBAL WARMING POTENTIAL of 310. Its long lifetime also allows it to drift into the STRATOSPHERE, where potentially it could become indirectly involved in OZONE depletion. N_2O is produced mainly by bacteria.

The term *nitrogen oxides* (NO$_x$) refers specifically to NO and NO_2. The burning of fossil fuels (*see* CARBON CYCLE) and plant material is the principal source for atmospheric NO$_x$, especially vehicle exhausts and the burning of waste after forest clearance in the Tropics.

These gases are catalysts for the formation of ozone in the TROPOSPHERE but can cause the destruction of ozone in the stratosphere. NO and NO_2 are reversibly interchangeable through the gain or loss of an oxygen atom. This interchange takes place during the PHOTOLYTIC CYCLE that both makes and destroys ozone. NO$_x$ also play a critical role in the reactions that lead to the formation of PHOTOCHEMICAL SMOG.

Below about 15 miles (25 km), NO$_x$ catalyze ozone formation. Above that height they catalyze its destruction. Stratospheric NO$_x$ are produced by the oxidation of N_2O (nitrous oxide) present in the stratosphere:

$$N_2O + O \rightarrow 2NO$$

$$NO + O \rightarrow NO_2$$

The oxygen atoms result from the photolytic destruction of oxygen. After that, the reactions are

$$NO + O_3 \rightarrow NO_2 + O_2$$

$$O + NO_2 \rightarrow NO + O_2$$

These can be summarized as

$$2O_3 \rightarrow 3O_2$$

These reactions are not important in stratospheric ozone chemical processes because only a small amount of N_2O reaches the stratosphere. In the 1970s there were fears that large fleets of commercial supersonic airliners would be developed. These aircraft would fly in the lower stratosphere and release NO in their exhausts, potentially causing ozone depletion. It did not happen because the supersonic fleets were not built. Today NO$_x$ are important only because of their contribution to the formation of tropospheric ozone and photochemical smog.

nitrous oxide (N$_2$O) A GREENHOUSE GAS with a GLOBAL WARMING POTENTIAL of 310 that is present in the atmosphere at a concentration of 314 parts per billion by volume (measured in 1998). This concentration is increasing at a rate of 0.25 percent a year. It is released into the air naturally from soils in wet tropical forests and dry savannah grasslands and from the oceans. It is also released from certain industrial processes, including the manufacture of fertilizers, nitric acid, and nylon (adipic acid), and from automobiles fitted with three-way catalytic converters. Once in the air a molecule of N_2O remains there on average for 120 years and disappears by being broken apart by the energy of sunlight once it reaches the STRATOSPHERE. ICE CORES from the GREENLAND ICE CORE and EURO-CORE PROJECTS show the atmospheric concentration of N_2O changed more or less in step with climate changes since the end of the last ice age.

NOAA *See* NATIONAL OCEANIC AND ATMOSPHERIC ADMINISTRATION.

NOAA-class satellites *See* TELEVISION AND INFRARED OBSERVATION SATELLITE.

noble gases (inert gases) The chemical elements HELIUM (He), NEON (Ne), ARGON (Ar), KRYPTON (Kr), XENON (Xe), and RADON (Rn), which together compose group 0 (or group VIII) of the periodic table. All of them are chemically unreactive, although xenon and krypton do form a few compounds. Argon is a major atmospheric constituent (*see* ATMOSPHERIC COMPOSITION). The others, apart from radon, are present in trace amounts. Radon is present in the air locally but has a very short HALF-LIFE.

noctilucent cloud Cloud that is occasionally seen on summer nights in high latitudes. It shines with light reflected from the Sun, which is well below the horizon. The cloud forms in the upper MESOSPHERE, at a height of about 50 miles (80 km). How it forms is uncertain, but it may be by the DEPOSITION of traces of water vapor onto particles swept in from space.

noise A signal that conveys no useful information and may obscure valuable data. The noise is said to be *white* if the signal producing it is random. Satellite observations of the surface are subject to noise caused, for example, by winds that produce ocean waves and alter the emission of radiation from the ocean surface. LAND CONTAMINATION is a form of noise.

nomogram A diagram that shows the relationship between two or more values or scales of measurement. It is produced in order to facilitate calculation by simplifying conversions between, for example, imperial and metric units or the FAHRENHEIT and CELSIUS TEMPERATURE SCALES. A nomogram can also show two

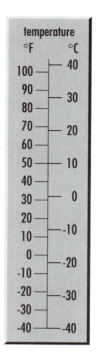
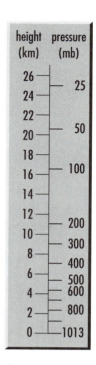

Nomogram. **A nomogram is a diagram showing the relationship between two or more values. These examples compare Fahrenheit and Celsius temperatures and the change of atmospheric pressure with height.**

types of measurement side by side, such as the change of atmospheric pressure with altitude.

nonfrontal depression An area of low atmospheric pressure that is not associated with a FRONTAL SYSTEM. Most tropical DEPRESSIONS are nonfrontal and are caused either by differential heating of the surface or by disturbances such as EASTERLY WAVES.

nonselective scattering The SCATTERING of all WAVELENGTHS of radiation equally as the radiation passes through the atmosphere. This is caused by particles that are bigger than the wavelength of the radiation. Because all wavelengths are scattered equally, nonselective scattering produces a white sky.

nonsupercell tornado *See* LANDSPOUT.

Nordenskjöld line A line that marks the boundary between boreal forest and tundra and therefore indicates the highest latitude at which full-size trees are able to grow (small, stunted trees form part of the tundra vegetation). The line is drawn to link places where the mean temperature for the warmest month is high enough to produce tree growth and the mean temperature for the coldest month is not so low as to kill trees. The line was first drawn by the Swedish geologist and polar explorer Otto Nordenskjöld (1869–1928), who made detailed records of his expeditions to Greenland and the Yukon, as well as to the southern tip of South America and Antarctica. Nowadays the line separating forest and tundra can be seen on satellite images, and so the Nordenskjöld line is little used.

(You can read about some of the adventures of Dr. Nordenskjöld at www.south-pole.com/p0000091.htm.)

nor'easter (**northeast storm**)   A storm that produces northeasterly winds of up to hurricane force (75 mph [121 km h⁻¹]), in eastern North America from Virginia to the Canadian Maritime Provinces. A nor'easter develops when a deep DEPRESSION forms over the North Atlantic, off Cape Hatteras, North Carolina, and then moves northward. As it travels along the coast, the CYCLONIC circulation around the center of low pressure draws MARITIME POLAR AIR over the area to the east of the Appalachians. Nor'easters are most common between September and April, and as well as the fierce winds they produce freezing or near-freezing temperatures and heavy falls of snow.

normal   The mean value of any meteorological value, such as TEMPERATURE or PRECIPITATION, calculated from measurements made at a particular place over a long period. This value is then taken to be the standard for that place. For example, according to measurements made over 51 years, the normal summer daytime temperature in Miami, Florida, is 86° F (30° C) and the average winter rainfall is 10 inches (250 mm). In mathematics, the word *normal* means "at right angles."

normal chart (**normal map**)   A chart on which the distribution of NORMAL features of the weather is plotted.

normal distribution   *See* GAUSSIAN DISTRIBUTION.

normalize   To produce a dimensionless ratio between two quantities by dividing one quantity by a more fundamental quantity in the same dimensions. Any quanti-

ty divided by itself is equal to 1. Consequently dividing a dimension by itself also yields 1 and the dimension disappears.

Normalized Difference Vegetation Index (**NDVI**)   An index that measures the amount of actively photosynthesizing plant biomass in a landscape. It is shown in maps compiled from AVHRR data. It is the ratio $(I - R) \div (I + R)$, where I is the amount of radiation received by the satellite in the near-infrared wavelength of 0.725–1.10 μm and R is the amount of radiation at the red wavelength of 0.58–0.68 μm. Healthy vegetation reflects most of the near-infrared radiation falling on it but absorbs most of the red radiation (which is used in PHOTOSYNTHESIS). In the name *difference* refers to the first part of the calculation, $I - R$. *Normalized* refers to the fact that this is divided by $I + R$.

(For more information see Claire L. Parkinson, *Earth From Above: Using Color-Coded Satellite Images to Examine the Global Environment* [Sausalito, Calif.: University Science Books, 1997].)

normal map   *See* NORMAL CHART.

Normand's theorem   A rule that was proposed in 1924 by the meteorologist C. W. B. Normand. The rule states that the height of the LIFTING CONDENSATION LEVEL can be shown by the intersection of two lines on an AEROLOGICAL DIAGRAM. If the DRY ADIABAT is drawn from a point on the surface at the ambient tem-

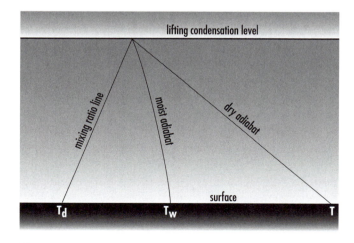

Normand's theorem. **Lines drawn on an aerological diagram can be used to indicate the height of the lifting condensation level.**

perature T and the saturation MIXING RATIO is drawn as a line from the surface DEW POINT TEMPERATURE T_d, the two lines intersect at the lifting condensation level. If a SATURATION ADIABAT (or moist adiabat) is drawn from the point of intersection to the surface, the point where it meets the surface marks the WET-BULB TEMPERATURE T_w. The theorem is derived from the fact that the amount of energy consumption that accounts for the difference between the saturation and dry adiabats is the same as that which accounts for the difference between the dry- and wet-bulb temperatures.

norte (papagayo)   A cold, northerly wind that blows in winter down the eastern coast of Mexico. It can reduce temperatures to below freezing and cause frosts at elevations above 4,000 feet (1,220 m). It is a southern continuation of a NORTHER.

North American high   A weak area of high surface pressure that covers most of North America in winter.

North Atlantic Current   *See* NORTH ATLANTIC DRIFT.

North Atlantic Deep Water (NADW)   The cold, saline dense water that drives the ATLANTIC CONVEYOR. It forms mainly in two places, in the Norwegian Sea to the northeast of Iceland and in the Labrador Sea, between Canada and Greenland.

The NADW begins close to the edge of the SEA ICE. When seawater freezes, the crystallization process expels the salt that was held by the water molecules while they were in the liquid phase. Seawater ice is fresh and the expelled salt increases the salinity of the seawater adjacent to the ice. The average salinity of the Atlantic Ocean is 34.9 parts per thousand (‰), but that of the water adjacent to the newly formed sea ice is up to 35.5‰. Its increased load of salts makes the water denser. The water is slightly warmer than 35° F (2° C). Seawater with a salinity of 35‰ freezes at about 28.6° F (−1.91° C) and its density is greatest at just over 32° F (0° C). It is denser than the water farther from the ice and so it sinks beneath it and its place is taken by warmer, less saline surface water that flows northward.

A ridge extends along the ocean floor between Greenland and Scotland. The NADW fills the basin on

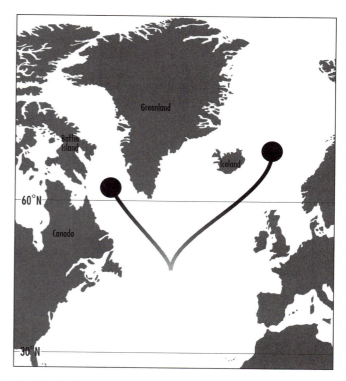

North Atlantic Deep Water. At two places, in the Labrador Sea and the Norwegian Sea, dense water sinks to the ocean floor. This is the North Atlantic Deep Water.

the northern side of the ridge, then spills over the ridge between Iceland and Scotland. It is then known as *Iceland Scotland Overflow Water* (ISOW). It is denser than the water of the deep ocean and it sinks all the way to the floor at a depth of about 10,000 feet (3,000 m). As more water sinks behind it, the deep water forms a current that flows southward at an average speed of about 0.5 inch per second (1.4 cm s⁻¹), closely following the edge of the North American continent (which is several hundred miles from the coast). The water takes more than 20 years to reach the equator.

Farther south, more water joins the NADW. This flows from the Mediterranean Sea. Because of its warm climate, the rate of EVAPORATION is very high over the surface of the Mediterranean and the sea loses more water by evaporation than it receives from rainfall and from the rivers that flow into it. Evaporation makes the Mediterranean water very saline and therefore dense. The evaporative loss draws in water from the Atlantic through the Strait of Gibraltar. The inflow enters at the surface and floats above the saltier Mediterranean water, which sinks to the floor. The inflow is balanced by the outflow of salty water through the Gibraltar Strait at a depth of 3,300–6,600 feet (1,000–2,000 m). Two currents flow through the Gibraltar Strait. A surface current flows into the Mediterranean and a deep current flows beneath it, out of the Mediterranean, moving at about 6.5 feet per second (2 m s⁻¹). The water of the deep current joins the NADW.

The total flow rate of the NADW is about 530–700 million cubic feet per second (15–20 million m³ s⁻¹). Because it formed at the surface and sank fairly quickly the water is rich in dissolved oxygen (but poor in nutrients for marine organisms). It ventilates the entire ocean and remains near the ocean floor for less than 200 years before returning to the surface. Eventually the NADW reaches the WEST WIND DRIFT current and becomes part of it.

The NADW plays an important role in regulating the climate, especially the climate of northwestern Europe. From time to time in the past its flow has changed and there is evidence linking these changes to climatic events, including the LITTLE ICE AGE, MEDIEVAL WARM PERIOD, a cold period that occurred during the Dark Ages, and a warm period that occurred in Roman times. These changes seem to follow a cycle with a period of about 1,500 years (see STOCHASTIC RESONANCE). When the climate grows warmer, more ICEBERGS drift south and melt, producing a layer of freshwater that floats above the seawater and inhibits the formation of cold, saline water that will sink to form the NADW. However, suppressing the formation of NADW may also suppress the Atlantic Conveyor, which carries warm water northward as the GULF STREAM and its branch the NORTH ATLANTIC DRIFT. A failure of the conveyor is likely to trigger a cold period, such as the Little Ice Age. This is the apparently paradoxical process by which GLOBAL WARMING could produce severely cold conditions in the Northern Hemisphere.

North Atlantic Drift (North Atlantic Current) A broad, shallow, warm surface current that is an extension of the GULF STREAM. It flows in a northeasterly direction from the middle of the North Atlantic Ocean and divides into two branches south of Iceland. The westerly branch flows northward and then turns south to join the EAST GREENLAND CURRENT. The other

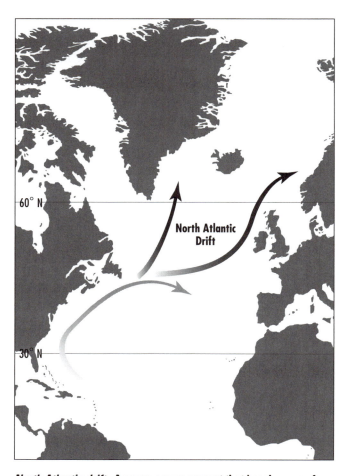

North Atlantic drift. A warm ocean current that breaks away from the Gulf Stream to form two streams, one flowing to the east of Iceland and past the coasts of northwestern Europe, the other flowing to the west of Iceland toward Greenland.

branch passes to the east of Iceland in a northeasterly direction, approaching close to the coasts of the British Isles and continuing to become the NORWEGIAN CURRENT. The North Atlantic Drift exerts a strong influence on the climate of the coastal regions of northwestern Europe, making it milder and wetter than it would be otherwise.

North Atlantic Oscillation (NAO) A periodic change that takes place on a scale of decades in the distribution of sea-level atmospheric pressure between Iceland and the Azores. The usual pattern is of low pressure over Iceland and high pressure over the Azores, and it is this distribution that drives the North Atlantic wind and storm systems. Its strength varies, however, and when pressure is lower than the average over Iceland it tends to be higher than average over the Azores, and vice versa.

The pattern is measured according to an NAO index. When pressure is lower than average over Iceland and higher than average over the Azores the index is high. When the Iceland low and Azores high are both weaker than average, the index is low. A high index produces cold winters in the northwestern Atlantic, mild winters in Europe, and dry weather in the Mediterranean region. The NAO index was high during much of the 1970s and especially high from the late 1980s to 1995. In 1996 the index fell to a very low value.

(There is more information at http://www.euronet.nl/users/e_wesker/nao.html.)

northeast storm   *See* NOR'EASTER.

North Equatorial Current   Two ocean currents, one in the North Atlantic and the other in the North Pacific, that flow from east to west parallel to and just north of the equator. The current flows within the upper 1,600 feet (500 m) of water at a speed of 0.6–2.5 mph (l–4 km h⁻¹) and is separated from the SOUTH EQUATORIAL CURRENT by the EQUATORIAL COUNTERCURRENT.

norther   A cold, strong winter wind that blows from the north across the southeastern United States, sometimes extending across the Gulf of Mexico and Central America into the Pacific. It is caused by COLD WAVES originating in the northwest that move POLAR

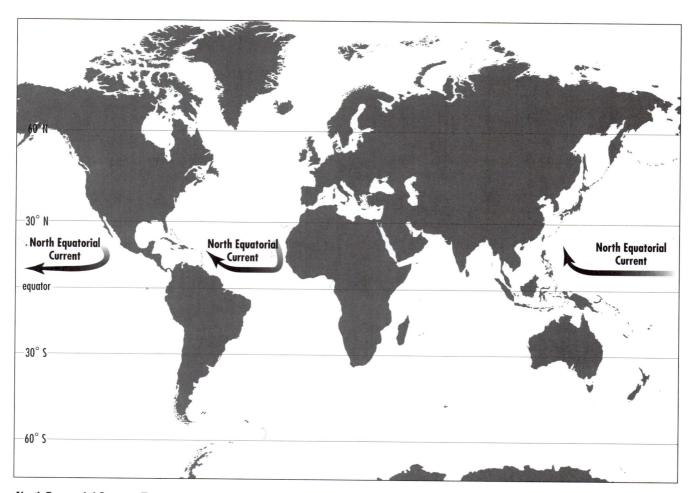

North Equatorial Current. **Two ocean currents that flow just north of the equator and parallel to it. The currents are deflected into more northerly paths, where they encounter continents.**

AIR southward. Near the gulf, where the weather is ordinarily warm, the sudden chill as a norther arrives is dramatic. The temperature falls to well below freezing. At San Antonio, Texas, the January mean temperature is 53° F (12° C), and it has been known to exceed 80° F (27° C), but during a norther it has fallen to 4° F (–15° C). At Houston, the temperature on one night was 75° F (24° C) at midnight. Then a norther arrived and at 9 A.M. the temperature was 22° F (–5° C). Sleet or snow may fall, but more commonly the sky remains clear. The wind continues to blow for a day or more.

northern circuit   The path that DEPRESSIONS usually follow as they cross the United States from west to east during the summer. The path takes them across the Great Lakes and Saint Lawrence River. Although depressions follow the circuit most often in summer, the depressions themselves are weaker then than they are at other times of year. In winter they usually follow the SOUTHERN CIRCUIT.

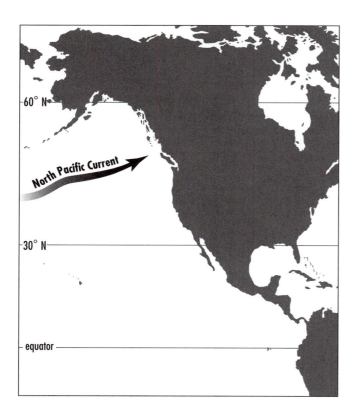

North Pacific Current. A current that carries warm water from Asia toward California.

Northern Rocky Mountain wind scale (NRM wind scale)   A scale of wind strength that has been adapted by the U.S. Forest Service for use in the forests of the northern Rocky Mountains. The forces and wind speeds are identical to those in the BEAUFORT WIND SCALE, but the visible effects described for winds of each force are those likely to be seen in that region.

north föhn   A FÖHN WIND that is generated by a movement of air from north to south across the Alps.

North Pacific Current   An ocean current that flows from west to east across the North Pacific Ocean, carrying warm water toward California. It is an extension of the KUROSHIO CURRENT.

North Temperate Zone   That part of the Earth that lies between the tropic of Cancer (*see* TROPICS) and the ARCTIC CIRCLE.

northwester   *See* NOR'WESTER.

Norway Current   *See* NORWEGIAN CURRENT.

Norwegian Current (Norway Current)   An ocean current that flows in a northeasterly direction, parallel to the northern coast of Norway. It is a continuation of the NORTH ATLANTIC DRIFT and carries relatively warm water into the Arctic Ocean.

nor'wester (northwester)   A wind or weather system that arrives from the northwest. In South Island, New Zealand, a nor'wester is a hot, dry wind that blows from the mountains. In northern India, nor'westers are storms that are caused by CONVECTION when dry, potentially cold air (*see* POTENTIAL TEMPERATURE) overruns warm, humid air. These conditions are especially common in the Ganges Delta in the weeks before the break of the summer MONSOON. In South Africa, a nor'wester is a DEPRESSION associated with an active front between MARITIME TROPICAL AIR and MARITIME POLAR AIR that carries storms, rain, overcast skies, variable winds, and cold weather to southwestern coasts in winter.

nowcasting   The issuing of local weather forecasts for the immediate future, up to two hours ahead. These

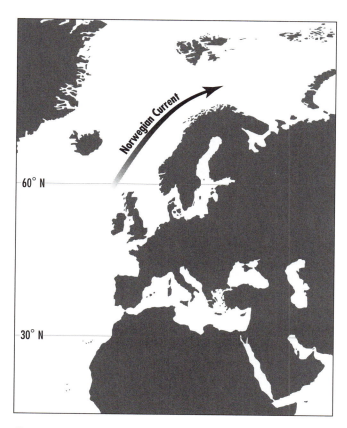

Norwegian Current. **A warm ocean current that flows parallel to the coast of Norway.**

forecasts give warning of approaching severe storms and TORNADOES.

NO$_x$ *See* NITROGEN OXIDES.

NRM wind scale   *See* NORTHERN ROCKY MOUNTAIN WIND SCALE.

Ns   *See* NIMBOSTRATUS.

NSIDC   *See* NATIONAL SNOW AND ICE DATA CENTER.

nuclear winter   A scenario that was proposed in 1983 by a team of scientists led by Richard P. Turco to warn politicians of the possible climatic consequences of a full-scale thermonuclear war. The team assumed that the war would be fought entirely in the Northern Hemisphere and that it would involve the detonation of approximately one-third of the global stock of nuclear weapons. At that time there were more than

50,000 such weapons, with a total explosive power equal to about 15,000 megatons of trinitrotoluene (TNT).

The initial flash of heat and IONIZING RADIATION from the explosions would ignite fires. Most of these would be extinguished by the blast wave that followed the radiation flash, but electrical sparks and smoldering remains would then reignite them. A cloud of black smoke would envelop the hemisphere. The smoke would absorb incoming solar radiation. This would warm the smoke and the air containing it, causing the cloud to rise, and a substantial amount of smoke would enter the STRATOSPHERE, where it would remain for many months or even years, because particles in the stratosphere cannot be washed to the surface by rain or snow.

As the smoke rose it would produce a region of low atmospheric pressure beneath the cloud. This would generate winds that would blow some of the smoke into the Southern Hemisphere. Early calculations suggested that the shading of the surface by the smoke cloud would reduce summer temperatures to levels typical of winter. This gave rise to the name *nuclear winter*. The fall in temperature would cause the area of ocean covered by ice to expand. The frozen sea would have a climate similar to that on dry land, with more extreme winter temperatures. This positive FEEDBACK would intensify the global cooling and prolong the "winter."

Precipitation over land would be reduced throughout the Northern Hemisphere and the Asian MONSOONS would fail. These climatic changes in addition to the cold would cause catastrophic harvest failures that would be followed by worldwide famine.

The study was meant as a warning, not a firm prediction, and it was based on a number of assumptions about the time of year the war took place, the distribution of targets, and the scale of the fires. Obviously, the calculations could not be tested.

The initial paper on nuclear winter led to further studies by a number of organizations. These confirmed the broad idea but found conditions would be more like those of late fall than of winter. This would still be severe enough to cause major harvest failures. It has also been calculated that 1 percent of the global stockpile of nuclear weapons would be sufficient to produce these effects if oil refineries were the primary targets.

Since the first study was published, tensions have been reduced between East and West. More than enough weapons remain, however, to produce a "nuclear fall" if they were to be used.

numerical forecasting   Weather folklore is based on the belief that weather patterns repeat themselves reliably. If this is true it means that a particular indicator, such as the color of the sky or shape of the clouds, can be depended on to be followed by weather of a particular kind. Folklore applies this idea crudely, but an element of modern weather forecasting is based on a similar idea. It assumes that weather in the future will closely resemble the kind of past weather that followed conditions like those obtaining at present. Unfortunately, the method is difficult to apply because the state of the atmosphere is rarely if ever identical on two occasions and because the atmosphere behaves chaotically (*see* CHAOS) so that quite small variations quickly develop into major differences.

Numerical forecasting aims to replace empirical methods with one that is firmly based on known physical laws. The forecaster begins with detailed measurements of the state of the atmosphere at many different places at regular intervals. These reveal the way conditions are changing. These changes are interpreted mathematically, by applying certain equations to them. These include the EQUATIONS OF MOTION, the laws of thermodynamics (*see* THERMODYNAMICS, LAWS OF), the equation of mass conservation, the EQUATION OF STATE, and equations of continuity. The equation of mass conservation relates changes in the DENSITY of air to the transport of its mass. The equations of continuity relate changes in the concentrations of the various constituents of the air, such as WATER VAPOR, CLOUD DROPLETS, and CARBON DIOXIDE, to their transport and to their sources and SINKS. Together these constitute the hydrodynamical equations. When used in conjunction with the temperature, pressure, humidity, and wind velocity at a particular time they make it possible to predict the state of the atmosphere at a future time.

The difficulty of applying this method is obvious: it calls for a truly prodigious number of separate calculations. VILHELM BJERKNES saw the possibility of developing such a method as long ago as 1904. He was unable to proceed very far with it because the detailed observations needed to supply the initial data were not available at the time. LEWIS FRY RICHARDSON made another attempt, which he published in 1922. This might have worked, but mathematicians have estimated that about 26,000 people would need to work full-time to perform by hand the calculations needed to predict the weather as it was occurring.

What Richardson needed was a computer, but it was not until 1953 that computers were sufficiently fast, powerful, and reliable to be used in weather forecasting. That was the year that the Joint Numerical Weather Prediction Unit (JNWP) was established in the United States. It began issuing forecasts in 1955.

Numerical methods are now used by most national weather services. Forecasters are supplied with PROGNOSTIC CHARTS that are generated by numerical MODELS. Although these charts picture the way a weather system may develop, they do not in themselves constitute the forecast. Experienced meteorologists use them as tools to help them identify emerging patterns and to recognize the significance of what is happening. The forecast that is finally produced results from the combination of the numerical forecast and the interpretive skill of the meteorologist.

Nusselt number   A DIMENSIONLESS NUMBER, related to the RAYLEIGH NUMBER, that is used in calculations of the transfer of heat in fluids. It is the ratio of heat that is transferred to the amount that would be transferred by pure conduction under ideal conditions. The number was discovered by the German physicist Ernst Kraft Wilhelm Nusselt (1882–1957).

oasis effect The cooling effect that produces a difference in local weather found where an area of moist ground (an oasis) is surrounded by dry ground (desert). It is due to cooling caused by evaporation. Over the moist ground, kept moist by irrigation or by a water table at or above ground level, the rate of evaporation exceeds the rate of precipitation and the warm air over the ground supplies the latent heat of vaporization. In the surrounding area, which is dry, evaporation and precipitation balance, but because the amount of precipitation is low, the rate of potential evaporation exceeds that of precipitation. Surplus heat is absorbed by the ground and warms the air in contact with the surface. This produces a large BOWEN RATIO over the dry ground and a negative Bowen ratio over the moist ground. Air is subsiding over the moist ground and rising over the dry ground. The oasis effect occurs not only at desert oases, but also over large, irrigated fields in areas with a semiarid climate. *See* ADVECTION.

oberwind A MOUNTAIN BREEZE that blows at night in the Salzkammergut region of Austria.

obscuration The condition that is reported in United States meteorological practice when the sky is completely hidden by a weather feature at ground level, such as FOG.

obscuring phenomenon Any atmospheric feature, other than clouds, that obscures a portion of the sky as seen from a weather station. The term is used in United States meteorological practice.

obstruction to vision An atmospheric feature that reduces horizontal VISIBILITY at ground level, such as FOG, smoke, or blowing snow. The term is used in United States meteorological practice.

occlusion The stage in the life cycle of a FRONTAL SYSTEM at which the advancing cold air has started to lift the air in the WARM SECTOR clear of the surface. On a weather map the COLD FRONT overrides the WARM FRONT, and it is shown as a front with alternating triangles and semicircles. Air on all sides of the warm sector is cold in relation to the warm-sector air, but in fact the air on one side of the warm sector is colder than that on the other side. Instead of classifying air as simply warm or cold, therefore, a third, category, cool, has to be introduced. It is then possible to classify an occlusion as warm if its passage causes a transition at the surface from cold to cool air, and as cold if it causes a change from cool to cold air.

occultation The passing of one celestial object in front of another, so that one of the objects is partly or completely hidden to an observer. For example, the Moon and Sun hide the stars by passing in front of them and planets hide their satellites in the same way. Planetary atmospheres can be studied during occultations. As the planet passes in front of a star, light from the star passes through the planet's atmosphere on its

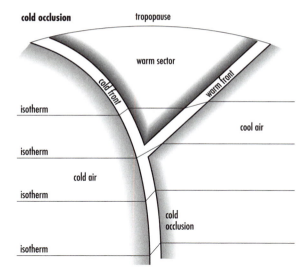

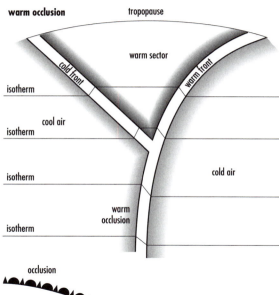

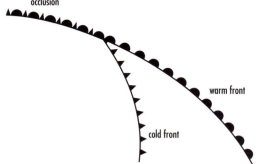

Occlusion. **The condition in which warm air is riding above cold air and all of it has risen above ground level. This is represented on a weather map by a line alternating the symbols for a warm and a cold front. If the air behind the occlusion is colder than the air ahead of it, it is a cold occlusion. If the air behind the occlusion is less cold than the air ahead of it, it is a warm occlusion.**

way to an observer on Earth. The spectrum of the light received at the observatory reveals the chemical elements present in the atmosphere through which the light passed.

occult deposition   The depositing of acid onto the surfaces of plants, buildings, and other objects that takes place when they are in direct contact with MIST or CLOUD DROPLETS containing dissolved acid. This is often a more serious form of pollution than acid delivered in rain, because the acid concentration is usually higher in very small droplets and these droplets tend to adhere to both upper and lower surfaces and to penetrate small fissures. Large raindrops deliver more dilute acid to exposed surfaces, leaving the undersides of leaves and other sheltered areas relatively dry. Occult deposition is one form of ACID RAIN.

oceanic climate   *See* MARITIME CLIMATE.

oceanicity   The extent to which the climate of a particular place resembles the most extreme type of MARITIME CLIMATE. Although climates can be classified as continental (*see* CONTINENTAL CLIMATES) or maritime, these types grade from one into the other. Except on ocean islands and some, but not all, coasts, a place with a maritime climate also experiences more continental conditions some of the time. Similarly, maritime influences extend a long way inland from the coasts of continents.

There is a need to refine the classification of continental and maritime climates in a way that reflects the gradations between them. The differences can be shown clearly on a HYTHERGRAPH, but only in a relative sense. The hythergraph compares the climates of two or three places (inclusion of more locations makes the graph too cluttered to be read easily) but gives no absolute value for their types of climate. This is possible if the essential climatic features of an area, together with its latitude, can be used to calculate a climatic index. Indices of CONTINENTALITY are widely used. These rate the continentality of a climate on a scale in which 0 indicates an extreme maritime climate and 100 an extreme continental climate. It is possible, therefore, to define the oceanicity of a climate as a low value for its index of continentality.

It is not entirely satisfactory to define something only by its difference from something else, however, and so several attempts have been made to devise an index of oceanicity. It is possible to calculate oceanicity as a percentage from the number of occasions in the year when the area being considered lies beneath MARITIME AIR compared with the total number of AIR MASSES that affect it during the same period. In this calculation

$$O = (M/N) \times 100$$

where O is the index of oceanicity, M is the number of maritime air masses affecting the area in a year, and N is the total number of air masses affecting the area in a year.

Two other methods are also widely used. Both of them base the calculation on the TEMPERATURE RANGE and one also takes account of precipitation. This method is given by

$$O = P \, d_t/100\Delta$$

where P is the average annual precipitation in millimeters, d_t is the number of days on which the temperature is in the range 0°–10° C, and Δ is the difference in temperature between the warmest and coldest months, in degrees Celsius.

The other method is given for the Northern Hemisphere by

$$O = 100 \, ((T_o - T_a)/A)$$

where T_o is the mean monthly daytime temperature for October, T_a is the mean monthly daytime temperature for April, and A is the average daytime temperature range. For places in the Southern Hemisphere, $T_a - T_o$ should be substituted for $T_o - T_a$. This index represents oceanicity as a percentage; 0 indicates an extreme continental climate and 100 indicates an extreme maritime climate.

Fiji, which is an oceanic island, obviously has a maritime climate. Its oceanicity index is 67. Thorshavn, in the Faeroe Islands, has an oceanicity index of 31. Omaha, Nebraska, has a continental climate. Its oceanicity index is 6.

oceanic stratosphere *See* COLD WATER SPHERE.

ocean weather station A weather station carried on a ship that is anchored in a specified location at sea. The station is staffed by observers equipped with instruments to measure both atmospheric and sea conditions, which are reported to a shore station at regular intervals. The requirements for an ocean weather station are specified by the WORLD METEOROLOGICAL ORGANIZATION.

octa *See* OKTA.

Odderade interstade An INTERSTADE that occurred in northern Europe during the Weichselian glacial (*see* DEVENSIAN GLACIAL). The evidence for it was first identified at Odderade, in Schleswig-Holstein, Germany. The interstade lasted from 70,000 to 60,000 years ago. Tundra vegetation, perhaps with a few cold-tolerant trees, grew throughout northern Europe during the interstade. When the interstade ended, ice-age conditions returned.

Odintsovo interstade An INTERSTADE that occurred in northern Europe during the Saalian glacial (*see* RISS GLACIAL). It was a time when plants typical of temperate, broad-leaved forest grew. Odintsovo, where the evidence for it was first identified, is a town lying to the west of Murmansk, in northern Russia close to the Finnish border.

Ofelia A TYPHOON that struck the Philippines, Taiwan, and China on June 23 and 24, 1990. It killed a total of 57 people.

offshore wind A wind that blows across the coast from the land in the direction of the sea.

ohm (Ω) The derived SYSTÈME INTERNATIONAL D'UNITÉS (SI) UNIT of electrical resistance, which is defined as the resistance between two points in an electric conductor when applying a constant potential difference of 1 VOLT between them produces a current of 1 AMPERE in the conductor. It is the oldest of all the electrical units and was adopted in 1838. The unit is named in honor of the German physicist Georg Simon Ohm (1789–1854).

okta (octa) A unit that is used to report the extent of cloud cover in eighths of the total sky. The CLOUD AMOUNT is measured by examining a reflection of the sky on a mirror marked out in a grid of 16 squares and counting the number of squares filled by cloud.

Older Dryas   A cold period that occurred in northern Europe from about 12,200 years ago to about 11,800 years ago, soon after the ice sheets had retreated at the end of the DEVENSIAN GLACIAL. It followed the BØLLING INTERSTADE and was first recognized from clays at a tile-making factory at ALLERØD, north of Copenhagen, Denmark. The clays are rich in remains of mountain avens (*Dryas octopetala*), an arctic–alpine plant.

Olga   A TYPHOON that struck Luzon, Philippines, in May 1976. It produced rains so heavy they caused widespread flooding in which 215 people died and at least 600,000 were rendered homeless. The floods caused $150 million of damage. Olga is also a category 1 typhoon (*see* SAFFIR/SIMPSON HURRICANE SCALE) that struck Japan and South Korea on August 2 and 3, 1999. In Japan it dropped 584 mm (23 inches) of rain on the area around Kochi and up to 25 mm (1 inch) of rain an hour on other places. One woman died when she was crushed by a landslide caused by the heavy rain. The heavy rain in South Korea caused 29 deaths and left 22 people missing. It destroyed more than 74,000 acres (30,000 ha) of farmland and more than 8,000 homes and left nearly 20,000 people homeless.

Oligocene   The epoch of geological time (*see* GEOLOGICAL TIME SCALE) that began about 35.4 million years ago and ended about 23.3 million years ago. It is the most recent epoch of the PALEOGENE period.

Olivia   A HURRICANE that struck Mazatlán, Mexico, on October 24, 1975. It killed 29 people.

OLS   *See* OPERATIONAL LINESCAN SYSTEM.

onshore wind   A wind that blows across the coast from the sea in the direction of the land.

opacus   A variety of clouds (*see* CLOUD CLASSIFICATION) in which the cloud forms a layer, sheet, or patch that is dense enough to hide the Sun or Moon completely. Opacus occurs with the cloud genera ALTOCUMULUS, ALTOSTRATUS, STRATOCUMULUS, and STRATUS. *Opacus* is the Latin word that means "opaque."

Opal   A HURRICANE, rated as category 4 on the SAFFIR/SIMPSON SCALE, that formed over the Yucatán Peninsula, Mexico, on September 27, 1995. A few days later, it weakened to a TROPICAL STORM, but by October 2 it had strengthened and was once again classed as a hurricane. By the time it reached Florida, on October 4, it had weakened to category 3. From Florida it moved into North Carolina, Georgia, and Alabama. Most of the damage it caused, estimated at more than $2 billion, was due to a storm surge that produced breaking waves and waves 12 feet (3.6 m) high. Opal killed 50 people in Guatemala and Mexico and 13 in the United States.

opaque sky cover   The proportion of the sky that is covered by cloud that completely hides anything that might be above it. The term is used in United States meteorological practice; the amount is usually expressed in tenths of the total sky.

operational linescan system (OLS)   The primary imaging system that is used on some of the satellites in the DEFENSE METEOROLOGICAL SATELLITE PROGRAM. Because the background to the images is dark it is possible to increase the gain on the OLS photomultiplier tube at night. This allows the OLS to detect LIGHTNING discharges, and the system has produced the longest set of data for lightning, dating from 1973. It also detects waste gas flares at oil wells and has revealed the large extent of this practice. The OLS scans the whole Earth once every day at visible and infrared WAVELENGTHS (*see* WAVE BAND) with a resolution of 1.74 and 0.37 miles (2.8 and 0.6 km).

operational weather limits   The minimum values for CEILING, VISIBILITY, and WIND at a particular airfield that permit aircraft to take off and land safely.

oppressive   *See* CLOSE.

optical air mass (airpath)   The length of the path through the atmosphere that is traveled by light from the Sun or any other celestial body. It is expressed as a multiple of the path length traveled by light when the source of the light is directly overhead (at the zenith).

optical depth (optical thickness)   A measure of the extent to which a cloud or layer of the atmosphere prevents the vertical passage of solar radiation. It can be applied to any layer of interest and is expressed as a dimensionless number, rather than a linear measure in feet or meters. It is calculated from the ratio of the amount of radiation falling on the upper boundary of the layer (I_t) to the amount emerging at the bottom (I_b), by the equation $I_t/I_b = \exp(-\tau/\cos Z)$, where τ is the optical depth of the layer and Z is the zenith angle, which is the angle between the solar radiation and the vertical. At sunrise and sunset $Z = 0$. A layer is considered to be optically thin if τ is less than about 0.2–0.5 and thick if τ is greater than 1.0.

optical rain gauge *See* RAIN GAUGE.

optical thickness *See* OPTICAL DEPTH.

orbit The path a particle or body follows around a central point or mass. As waves move across the sea or a lake individual water particles move up and forward and down and to the rear in approximately circular vertical orbits. The path the Earth follows around the Sun is its orbit, as are the paths satellites follow around the Earth or another larger body. WEATHER SATELLITES are placed in either GEOSTATIONARY ORBIT or POLAR ORBIT.

orbital forcing Changes in climate that are due to variations in the ORBIT of the Earth around the Sun. This alters the intensity of the solar energy the Earth receives. The onset and ending of GLACIAL PERIODS are thought to be due to orbital forcing. *See* MILANKOVICH CYCLES.

orbit period The time an orbiting body takes to complete one full journey around its orbital path. It is measured as the time that elapses between two successive passes of a fixed point. The Earth's orbit around the Sun has a period of one year, but the length of the year varies according to the way it is measured. A calendar year contains 365 days, with a leap year of 366 days in every fourth year. (Century years are leap years only if all four digits are exactly divisible by 4; 1900 was not a leap year, 2000 was, and 2100 will not be). A year counted from the interval between EQUINOXES, called a *tropical year*, contains 365.242 days. A sidereal year, measured by the positions of the fixed stars, has 365.256 days. An eclipse year, measured as the time that elapses between successive passages of the Sun through the points where the orbits of Earth and the

Moon intersect (called the Moon's *nodes*), has 346.620 days. A Gaussian year, measured with reference to the semimajor (smaller) of the two axes in the elliptical orbital path, has 365.257 days.

Orchid A TYPHOON that struck South Korea on September 11, 1980. It killed 7 people and more than 100 fishermen were lost at sea.

orographic Pertaining to mountains, and especially to their location and form. The word is derived from the Greek *oros*, "mountain," and *graphē*, "writing."

orographic cloud Cloud that forms above high ground as a result of OROGRAPHIC LIFTING. As air is forced to rise its flow becomes somewhat turbulent (*see* TURBULENT FLOW). This mixes the air, so its temperature and humidity are the same throughout the affected layer. If the air is fairly buoyant (*see* BUOYANCY) small CUMULUS clouds form, their positions indicating the

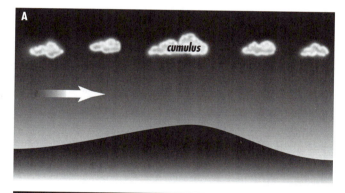

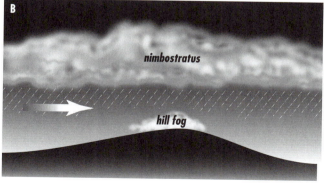

Orographic cloud. As air approaches the high ground (from the left in this drawing) and is forced to rise, cloud forms above the summit and to the upwind and downwind sides of it. The type of cloud depends on the relative humidity and buoyancy of the air.

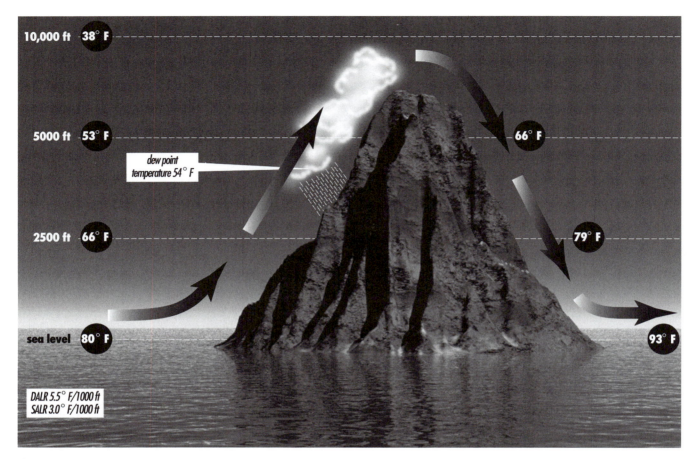

10,000 ft 38° F

5000 ft 53° F

dew point temperature 54° F

2500 ft 66° F

66° F

79° F

sea level 80° F

93° F

DALR 5.5° F/1000 ft
SALR 3.0° F/1000 ft

Orographic lifting. **As air rises to cross the mountain it cools at the dry adiabatic lapse rate (DALR) until it reaches its dew point temperature. Cloud then forms and rain falls. The air continues cooling at the saturated adiabatic lapse rate (SALR). On the lee side of the summit the descending air warms at the DALR, reaching sea level much warmer and drier than it was at sea level on the windward side.**

degree of lifting in different places. Because of the mixing, their bases all are at the same height. On the WINDWARD side of the high ground there are a few small clouds. The largest cloud forms above the crest of the hill, and downwind the clouds are smaller, but larger than those on the windward side. If the air is moist, it forms a thicker layer of NIMBOSTRATUS lying above the hill crest and extending to both sides, with a cap of STRATUS sitting as HILL FOG on the highest ground. LENTICULAR CLOUD is also a form of orographic cloud.

orographic lifting The forced raising of air as it crosses high ground. The effects are most pronounced when air crosses a mountain range that lies at right angles across its path. The Rocky Mountains are a good example. They are high and aligned north and south, and the prevailing air movement is from west to east. Air is constantly being forced to cross the mountains, and this action produces marked differences in the climates to the west and the east of them. Orographic lifting has several effects. The first is to increase precipitation on the WINDWARD slope.

As air rises it expands and cools in an ADIABATIC process. Unless it was already saturated, the air temperature decreases with height at the DRY ADIABATIC LAPSE RATE (if the air is saturated its temperature falls at the SATURATED ADIABATIC LAPSE RATE). When the rising air reaches the LIFTING CONDENSATION LEVEL water vapor starts to condense and cloud forms. Whether cloud forms and, if it does, whether it produces precipitation depend on the original humidity of the air. Air that has crossed the ocean, such as the air reaching the Rockies from the west, carries a large amount of moisture. Consequently, orographic lifting usually produces cloud and precipitation. A mountain range located deep in the interior of a continent has a much smaller effect, because the air reaching it is drier.

Tamarack, on the windward slopes of Mount Whitney, California, has the heaviest average snowfall in North America, 37.5 feet (11.4 m) a year; in January 1911, 32.5 feet (9.9 m) fell in one month. Other places sometimes receive more in a single season, however. In the winter of 1998–99, for example, 95 feet (29 m) of snow fell on Mount Baker, in Washington State. The people of Lloro, Colombia, probably experience the highest average rainfall in the world. Lloro, on the windward slopes of the Andes at an elevation of 520 feet (159 m), is estimated to have received 523.6 inches (13,300 mm) of rain a year over 29 years. Cherrapunji, at an elevation of 4,309 feet (1,313 m) in the mountains of northeastern India, receives the rainfall resulting from the orographic lifting of MONSOON winds. Its average annual rainfall is 425 inches (10,800 mm), but between August 1860 and July 1861 Cherrapunji received 1,041 inches (26,441 mm), which is approximately 86 feet (26 m) of rain.

Precipitation falls heavily on the windward side of the mountains, but the LEE side is dry, sometimes extremely dry. It lies in a rain shadow. Death Valley, California, lies in a rain shadow. It receives an average 1.6 inches (41 mm) of rain a year.

Rising, unsaturated air cools at the dry adiabatic lapse rate (DALR) of 5.5° F per 1,000 feet (10° C per kilometer). Suppose the air starts at a sea-level temperature of 80° F (27° C) and the DEW POINT TEMPERATURE is 54° F (12° C). The lifting condensation level, at which the air temperature falls to the dew point temperature, is at about 4,700 feet (1,430 m). Water vapor then starts condensing and, because of the release of LATENT HEAT, the air continues cooling at the saturated adiabatic lapse rate (SALR) of 3° F per 1,000 feet (5.5° C per kilometer). By the time it crosses the mountain peak, say at 10,000 feet (300 m), the air temperature is 38° F (3.3° C).

The air at the top of the mountain is saturated and water vapor is still condensing from it. Because its RELATIVE HUMIDITY (RH) is 100 percent, the dew point temperature and the actual air temperature must be the same. The dew point temperature at the mountain top is therefore 38° F (3.3° C). Air now starts descending on the lee side of the mountain. As it does so it is compressed and its temperature rises. It immediately rises above the dew point temperature, so its RH falls below saturation. Water vapor ceases to condense and the sinking air warms at the DALR. By the time it is once more at sea level its temperature has risen to 93° F (34° C) and it is extremely dry. This is why there is often a dry desert in the rain shadow on the lee side of a large mountain range.

It can also happen that air on the windward side of the mountain forms a pool that is partly trapped. Air at a higher level flows over the top of the pool and down the lee side. This causes a wind of the FÖHN type.

Hills and mountains are often windy places as a consequence of orographic lifting. The rising air pushes against the air above the mountain. This causes a series of vertical air movements that continue downwind from the peak. One consequence is that the wind speed is greatly reduced on the lower part of the slope on both sides of the barrier but greatly increased at the summit by the VENTURI EFFECT.

orographic occlusion   An OCCLUSION that forms when an advancing WARM FRONT reaches a mountain range that acts as a barrier to it. The front slows, causing the air in the WARM SECTOR to accumulate in a pool and allowing the COLD FRONT to undercut it more rapidly than it would have over level ground.

orographic rain   Rain that falls on the WINDWARD slope of a hill or mountain as a direct consequence of OROGRAPHIC LIFTING. The amount of orographic rain is calculated by subtracting the amount of rain a particular weather system would have been estimated to produce over level ground from the total amount that fell.

oscillatory wave   A wave that causes air or water to move about a point, but without advancing in the direction the wave is moving. Particles of air or water describe an approximately circular orbit, moving up and forward, then down and to the rear.

osmometer   See OSMOSIS.

osmosis   The process in which water molecules cross a membrane that is permeable to them and that separates two solutions of different concentrations. The molecules move from the solution of weaker concentra-

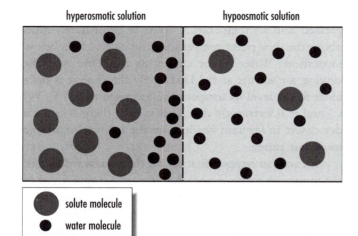

hyperosmotic solution   hypoosmotic solution

- solute molecule
- water molecule

Osmosis. **A selectively permeable membrane separates two solutions. Water molecules flow across the membrane from the hypoosmotic solution to the hyperosmotic solution. They continue to do so until the proportion of solute to water molecules is the same on both sides of the membrane.**

tion to the one of higher concentration. This process dilutes the stronger solution by the addition of water and strengthens the weaker solution by the removal of water. Water continues to cross the membrane until the concentrations of the two solutions are equal. The stronger solution is said to be *hyperosmotic* with respect to the weaker solution and the weaker solution is said to be *hypoosmotic* with respect to the stronger solution. Solutions of equal concentration are said to be *isoosmotic*. The membrane is said to be *selectively permeable*. Osmosis occurs regardless of the composition of the two solutions. It is only their relative concentrations that matters. Water flows by osmosis from seawater to a very concentrated sugar solution, because the seawater contains a lower concentration of solute molecules than does the sugar solution.

Osmosis exerts a pressure across the selectively permeable membrane. The osmotic pressure of a solution is directly proportional to its solute concentration, and it can be measured by an osmometer. In one type of osmometer, the specimen solution is contained in one section of a tube and separated by a selectively permeable membrane from distilled water in the other part of the tube. Osmosis causes water to flow into the solution, thereby increasing its volume. A piston exerts a pressure that is sufficient to prevent this expansion and

therefore to prevent water from crossing the membrane. The osmotic pressure is the pressure the piston needs to exert in order to maintain a constant volume in the solution.

Living cells can be damaged by the osmosis that occurs if the concentration of the internal and external solutions changes radically. Plant cells have walls that make them less susceptible to damage of this kind than animal cells, which lack walls. If a freshwater organism is immersed in seawater, so the seawater fills the intercellular spaces, water moves out of cells under osmotic pressure, causing cell dehydration. If a marine organism is immersed in freshwater the reverse happens. Water flows into cells under osmotic pressure. This action can cause animal cells to lyse (burst). Plant cells become turgid (rigid).

Freezing also causes injury by osmosis. When ice crystals form in the intercellular solution, its concentration increases as a result of the removal of water. This causes water to flow out of the adjacent cells, with consequent dehydration.

Osmosis is exploited industrially to remove salt from water that is too salty for drinking. Two processes are used: reverse osmosis and electrodialysis. In reverse osmosis, salt water is pushed against a selectively permeable membrane under a pressure that exceeds the osmotic pressure of the solution. Water flows across the membrane from the salt water. This is only partly successful, however, and although it can render brackish (slightly salty) water drinkable it does not purify seawater.

In electrodialysis two membranes are used and an electric current is passed through the solution. Common salt (NaCl) separates in solution into sodium IONS, carrying a positive charge (Na^+), and chlorine ions, carrying a negative charge (Cl^-). The electric current causes N^+ ions to move across one membrane and Cl^- ions to move across the other, leaving pure water in the area between the membranes. This technique, like reverse osmosis, is used only to treat brackish water, not seawater.

ouari   A southerly wind, similar to the KHAMSIN, that blows across Somalia, Africa.

outburst   A sudden, very heavy fall of PRECIPITATION from a CUMULIFORM cloud that is caused by the strong downcurrents in the cloud. Outbursts often accompany THUNDERSTORMS.

outflow boundary The location in a THUNDER-STORM where air that has been cooled by the EVAPORATION of moisture meets warm, moist air.

outlet glacier *See* VALLEY GLACIER.

outlook period *See* FORECAST PERIOD.

outwash plain *See* GLACIAL LIMIT.

overcast A layer of cloud that covers all or most of the sky. In reporting of weather conditions, the sky is said to be *overcast* when at least 95 percent or 7 OKTAS of the sky is covered by cloud. The sky is not overcast if it is covered by an OBSCURING PHENOMENON. It is then said to be *obscured*.

overrunning The situation in which warm air rides up a frontal surface, moving over the cold air that is beneath the front.

overseeding CLOUD SEEDING in which an excessive amount of nucleating material is delivered into the cloud. The aim of cloud seeding is to produce ICE CRYSTALS that will grow by the BERGERON–FINDEISEN MECHANISM until they are large enough to fall as PRECIPITATION. If the cloud is overseeded, a much larger number of ice crystals forms, but they are too small to fall and there are so many of them that the cloud is depleted of supercooled droplets (*see* SUPERCOOLING) and so the crystals cannot grow. The result is that whereas seeding may increase precipitation, overseeding inhibits it.

oxygen (O) The most abundant element in the Earth's crust, where it occurs combined with other elements (such as sulfates and carbonates) in the minerals from which rocks are composed, as well as in water. By weight, crustal rocks are 47 percent oxygen and the oceans are 88.8 percent oxygen. Its atomic number is 8, relative atomic mass 15.9994, melting point –353.92° F (–214.4° C), and boiling point –297.4° F (–183° C). It occurs in the atmosphere as a colorless, odorless gas (but is pale blue as a liquid) constituting 20.946 percent of the atmosphere by volume, with a concentration of 209,460 parts per million. In its most common form oxygen atoms bond in pairs as diatomic oxygen (O_2), but atoms can also bond in threes to form OZONE (O_3). Oxygen is extremely reactive and many oxidation reactions release energy. Aerobic respiration releases energy from oxidation reactions and combustion releases heat through chain reactions in which carbon and hydrogen are oxidized. Oxygen is slightly soluble in water and aquatic aerobic organisms, such as fish, use dissolved oxygen for respiration.

oxygen isotope ratios The two most common ISOTOPES of natural oxygen are ^{16}O and ^{18}O, and they occur in atmospheric oxygen in the ratio of about one part of ^{18}O to 499 parts of ^{16}O. Both isotopes bond equally well with hydrogen to form water, so for every 499 water molecules there is one molecule of $H_2{}^{18}O$. Chemically the two are identical, but the $H_2{}^{18}O$ is slightly heavier. Because $H_2{}^{16}O$ is lighter, it evaporates more readily than $H_2{}^{18}O$. Less energy is needed for molecules of $H_2{}^{16}O$ to break free of the liquid surface. Once in the air, gaseous water molecules condense to form liquid droplets, containing a higher proportion of the lighter isotope. The droplets fall to the surface once more as precipitation, so the relative proportions of the two isotopes remain constant in seawater.

During the onset of a GLACIAL PERIOD, however, snow falling over the expanding ICE SHEETS remains there. The ice sheets grow thicker, the snow is compacted into ice, and an increasing proportion of the world's water is removed from the oceans and stored on land. The snow is made from water containing more than 499 parts of ^{16}O to every 1 part ^{18}O, and since the water molecules evaporated primarily from the oceans, as ^{16}O is removed the proportion of ^{18}O in seawater increases.

Small animals called *foraminifers*, or *forams*, are abundant in the oceans. They are tiny, most of them less than 0.04 inch (1 mm) in diameter, but they live inside shells they make themselves and they have existed for at least the last 600 million years. Their shells are made from calcium carbonate ($CaCO_3$), which they synthesize from the water around them. The oxygen they use contains the two isotopes in the same proportion as in the water.

When they die, the soft parts of their bodies decompose and vanish, but their insoluble shells sink to the bottom of shallow water (at below about 2–3 miles [3–5 km]), called the *carbonate compensation depth,* $CaCO_3$ dissolves once more). There they form part of the sediment that eventually becomes sedimentary rock. Those rocks contain oxygen with a ratio of the ^{16}O and ^{18}O isotopes that reflects the ratio in the oceans at the time the $CaCO_3$ was made.

If the $CaCO_3$ is enriched in ^{18}O it indicates that ^{16}O was accumulating in snow and ice at the time and therefore that the ice sheets were expanding as the climate grew colder. The $CaCO_3$ can be dated, for example, by RADIOCARBON DATING, and so changes over time in the ratio of the oxygen isotopes can be interpreted as changes in the global climate.

Similarly, the isotope ratio is preserved in the ice sheets themselves. There they reflect changes in temperature. All the water molecules are enriched with ^{16}O, but by an amount that depends on the temperature. Although ^{16}O evaporates more readily than ^{18}O, ^{18}O also evaporates and the warmer the weather the more of it that is able to do so. Consequently, an increase in the proportion of ^{18}O in the ice indicates warmer weather, a decrease indicates colder weather, and this "thermometer" is so sensitive it can be used to detect the difference between winter and summer temperatures, a help in counting the annual layers in ICE CORES.

Analysis of these two oxygen isotopes is a major tool in the reconstruction of the history of the global climate.

oxygen isotopes Oxygen occurs in several ISOTOPES. Natural oxygen is mainly oxygen-16, usually written ^{16}O, in which the atomic nucleus contains 8 protons and 8 neutrons. This is the most abundant isotope, with a relative atomic mass of 15.9994, which is approximately 16, and it accounts for 99.76 percent of natural oxygen. The addition of one more neutron produces ^{17}O, accounting for 0.04 percent of natural oxygen, and adding two more neutrons produces ^{18}O, accounting for 0.2 percent of natural oxygen. There are also three, very short-lived radioactive isotopes, ^{14}O, ^{15}O, and ^{19}O. Information about past climates can be inferred from studies of the two most important isotopes, ^{16}O and ^{18}O (*see* OXYGEN ISOTOPE RATIOS).

Oyashio Current (Kamchatka Current) An ocean current that flows southward from the Bering Sea, past the Kuril Islands, to the northeast of Japan, where it

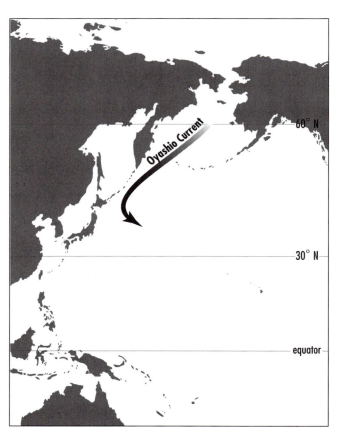

Oyashio Current. **A cold ocean current that flows southward from Alaska toward Japan.**

meets the KUROSHIO CURRENT. It flows at less than 1.5 mph (2.4 km h^{-1}) and carries cold water, with a low salinity of 33.7–34.0 parts per thousand (‰).

ozone A form of oxygen in which the molecule comprises three atoms (O_3) rather than the two of ordinary oxygen (O_2). It is a pale blue gas and a powerful oxidizing agent. In the troposphere, it forms by a series of reactions in which ULTRAVIOLET (UV) RADIATION reduces nitrogen dioxide (NO_2) to nitrous oxide (NO), releasing one oxygen atom (O), which reacts with oxygen (O_2) to form O_3. Ozone is involved in the formation of PHOTOCHEMICAL SMOG, can damage crops and other vegetation, and can accelerate the deterioration of rubber and other materials. It is also harmful to human health; some people are sensitive to ozone at concentrations of 0.001 parts per million (ppm). In many people, a concentration of 0.15 ppm causes irritation of the respiratory tract with associated discomfort, and at 0.17 ppm persons with asthma and other respiratory ailments are seriously affected. In most countries, standards that are set for air quality aim to limit human exposure to ozone. In the stratosphere, ozone absorbs UV radiation. *See* OZONE LAYER.

(More information on ozone as a pollutant can be found in Charles E. Kurchella and Margaret C. Hyland, *Environmental Science.* 2d ed. [Boston: Allyn and Bacon, 1986].)

ozone depletion potential A measure of the extent to which a particular chemical compound is likely to remove OZONE from the OZONE LAYER, compared with the extent to which CFC-11 freon-11 (chlorofluorocarbon-11) does so. CFC-11 is given a value of 1 and other compounds are evaluated on this scale. As well as CFCs, the list of compounds implicated in ozone depletion includes hydrofluorocarbons (HCFCs) and halons, which are compounds containing bromine.

ozone hole *See* OZONE LAYER.

ozone layer A region of the stratosphere between 20 km and 30 km (66,000 and 98,000 ft) above the ground where the concentration of OZONE (O_3) is usually higher than it is elsewhere, commonly reaching 10^{18}–10^{19} molecules per cubic meter, or 220–460 DOBSON UNITS (DU). The ozone forms in a two-step reaction. In the first step, a photon of ULTRAVIOLET (UV) sunlight, having a wavelength of 240 nanometers (nm) and 5.16 electron volts (eV) of energy, is absorbed by a molecule of oxygen (O_2). This energy is sufficient to break the bond between the two oxygen atoms, to form two single atoms. Some of these atoms then collide with oxygen molecules in the presence of a molecule of any other substance. The additional molecule is needed to absorb momentum from the oxygen atom and molecule, thus allowing them to bond together. The two steps in the reaction are

$$O_2 + photon \rightarrow O + O \ (1)$$

$$O_2 + O + M \rightarrow O_3 + M \ (2)$$

where the photon supplies the energy and M is the molecule of any other substance (usually nitrogen, N_2).

The short-wave UV radiation that dissociates oxygen molecules is known as *hard UV* or *UV-C. Soft UV,* or *UV-B,* has a longer wavelength. UV with a wavelength that includes some UV-C but more UV-B is absorbed by ozone and causes it to dissociate. This reaction is

$$O_3 + photon \rightarrow O_2 + O \ (3)$$

Ozone is also destroyed by reacting with single oxygen atoms:

$$O_3 + O \rightarrow 2O_2 \ (4)$$

Reaction 1, in which UV radiation is absorbed by oxygen molecules, occurs most strongly at very high altitudes above 80 km (50 miles), the height at which the density of the atmosphere is such as to cause a substantial proportion of the incoming UV-C to strike oxygen atoms. Ozone formation, by reaction 2, requires the collision of three molecules, and this is most likely at much lower altitudes, where molecular collisions are

Compound	Lifetime (years)	Ozone depletion potential
CFC-11	45	1.0
CFC-12	100	1.0
CFC-113	85	0.8
CFC-114	300	1.0
CFC-115	380	0.6
HCFC-22	12	0.05
Methyl chloroform	4.8	0.10
Carbon tetrachloride	35	1.06
Halon-1211	16	3.0
Halon-1301	65	10.0
Halon-2402	not known	6.0

frequent in the denser air. Air movements in the stratosphere cause enough mixing to allow some atomic oxygen to be carried to lower levels, where it can engage in reaction 2. Ozone levels are then maintained by reactions 2 and 3, with atomic oxygen being fed constantly from above. This balance causes the highest concentration of ozone to occur at about 30 km (19 miles) over the equator and about 18 km (11 miles) over the North and South Poles.

Reactions 1 and 3 depend on sunlight, so they cease at night, while reactions 2 and 4 continue. Reactions 1 and 3 proceed most strongly in low latitudes, where solar radiation is most intense. Ozone does not accumulate over the Tropics or disappear in winter from the polar stratosphere, because air currents at high level transport stratospheric ozone away from the equator and toward the Poles, constantly replenishing the supply.

In 1985, research by three British scientists, J. C. Farman, B. G. Gardiner, and J. D. Shanklin, revealed that the seasonal decrease in the concentration of stratospheric ozone over central Antarctica had become greater than in previous years. Their ground-based observation was quickly confirmed by examining satellite measurements. The decrease occurs naturally in latitudes above 60° S during late winter and early spring, from September to December, and its extent varies from year to year, but the studies indicated that some new factor was involved. Further research found the depletion of ozone was caused by halons (chemicals containing bromine) used in fire extinguishers, but mainly by chlorofluorocarbons (CFCs).

The region in which the stratospheric ozone concentration is reduced in late winter and early spring came to be known as the *ozone hole*. There is always some ozone present, so there is not literally a "hole," but in some years levels have fallen well below their usual values, to only a little more than 100 DU. In October 1993, the level fell to its lowest value, 91 DU (partly as a result of sulfate particles released from the eruption of Mount Pinatubo).

Stratospheric ozone depletion occurs within the POLAR STRATOSPHERIC CLOUDS that form inside the POLAR VORTEX.

(More information on the measurement of atmospheric ozone can be obtained from http://mdstud. chalmers.se/~md5mike/projekt/planet/technology.hl.)

ozonosphere An alternative name for the OZONE LAYER, which is a region of the stratosphere between 66,000 and 98,000 ft (20 and 30 km) above the ground where the concentration of OZONE (O_3) is higher than it is elsewhere.

P

Pa *See* PASCAL.

Pacific air MARITIME AIR that has crossed the Rocky Mountains and has been modified by its passage over the mountains. When it reached the coast the air was cool and moist. OROGRAPHIC LIFTING caused much of its WATER VAPOR to condense and fall as PRECIPITATION, and during its descent on the eastern side of the Rockies the air warmed in an ADIABATIC process. What was originally cool, moist air became warm, dry air.

Pacific- and Indian-Ocean Common Water (PIOCW) The deep waters of the Pacific and Indian Oceans, which are usually considered together, as a single mass of water. Their characteristics are very similar. Both have a temperature of 35.3° F (1.5° C) and a salinity of 34.7 parts per thousand (‰).

Pacific Decadal Oscillation (PDO) A change that occurs over a period of several decades in the ocean–atmosphere system in the Pacific basin. It affects the temperature of the lower atmosphere, passing through alternating warm and cold phases.

Pacific high An ANTICYCLONE that covers a large part of the subtropical North Pacific Ocean. There is a similar anticyclone over the South Pacific. The Pacific highs are SOURCE REGIONS for maritime AIR MASSES. The North Pacific high and North ATLANTIC HIGH together cover one-quarter of the Northern Hemisphere, and for six months of each year they cover almost 60 percent of it.

pack ice Large blocks of ice that cover the surface of the sea. In winter, pack ice covers about 50 percent of Antarctic water and about 90 percent of Arctic water. In summer, it covers about 10 percent of the sea in the Antarctic and about 80 percent in the Arctic. Ships can usually move through pack ice if it covers less than 75 percent of the surface and there is open water between blocks. The presence of pack ice can be detected from a distance by the appearance of ICE BLINK.

Paleocene The epoch of geological time (*see* GEOLOGICAL TIME SCALE) that began about 65 million years ago and ended about 56.5 million years ago. It is the earliest epoch of the PALEOGENE period.

paleoclimatology The scientific study of the climates that existed in the distant past. The *paleo-* in the name is derived from the Greek *palaios*, which means "ancient." By reconstructing past climates, paleoclimatologists provide a background against which the history of the Earth can be seen. Knowledge of paleoclimates helps geologists, botanists, zoologists, and ecologists to understand how particular regions of the world developed to the conditions in which we see them today. They are able to explain how it is that animals and plants once lived in places they would now find intolerable, such as the hippopotamuses and tropical lotuses that once lived in the center of London.

Paleoclimatologists also provide important information about climate itself. We know from their research that about 20,000 years ago—a very short time in geological terms—the Northern Hemisphere lay in the grip of the coldest part of an ice age (*see* WISCONSINIAN GLACIAL and DEVENSIAN GLACIAL). Sheets of ice thousands of feet thick covered much of North America and Eurasia, and where the bare ground was exposed it was deeply frozen. We also know that at other times the world has been much warmer than it is today. The paleoclimatological record shows the extremes of climate that are possible.

These studies are relevant to modern concerns. By describing past conditions they provide suggestions of what the world might be like if its climates were to become markedly warmer or colder and just how warm or cold they are capable of becoming. They provide warnings of the kind of surprises climate change may bring, such as the sudden drop in temperatures almost to ice-age levels that interrupted the rapid warming at the end of the most recent ice age (*see* YOUNGER DRYAS). This change is believed to have been due to one of several HEINRICH EVENTS, which suppress the formation of NORTH ATLANTIC DEEP WATER and partially or completely shut down the ATLANTIC CONVEYOR. There were also sudden and dramatic rises in temperature during ice ages, known as *DANSGAARD–OESCHGER EVENTS.*

Obviously, there are no written records of ancient climates. Paleoclimatologists must study indirect evidence in order to discover the causes that produced observable effects. Many lines of inquiry are open to them.

Wind, rain, and repeated freezing and thawing leave clear marks on rocks and on landscapes. Traces of sand dunes often survive for many thousands of years after the desert in which they formed has vanished. Rivers leave behind gravel from which their size and courses can be tracked. Soil that is made from wind-blown dust indicates a dry climate in which the winds are dusty, and the dust itself can often be traced to the rocks that were eroded to produce it. This reveals the direction of the PREVAILING WIND, which in turn reveals the kind of weather patterns that predominated at the time. That time can be calculated, for example, by the RADIOCARBON DATING of organic material present in the undisturbed soil.

Plants and animals tend to be associated with particular climates and both leave traces. Plants leave POLLEN or SPORES from which the type of plant, and sometimes even the species, can be identified (*see* POLLEN ANALYSIS). Animals leave identifiable teeth and fragments of bone, and beetles, some species of which tolerate only a narrow range of temperature, leave behind their wing cases, or elytra (*see* BEETLE ANALYSIS). Groups of plants and animals, called *assemblages,* give a clear indication of the weather conditions at the time and place where they lived together.

Purely physical changes also leave clues in the form of changes in OXYGEN ISOTOPE RATIOS. These can be related to rising or falling temperature. Dust trapped in ice sheets and extracted from ICE CORES indicates whether climates were generally dry or wet, because the air contains more dust in dry weather than it does in wet weather.

Paleoclimatologists are building a fairly detailed history of the world's climate that extends about 100,000 years into the past. Still further back in time they have "snapshot" pictures of conditions at particular times. Obviously the weather would have varied from year to year, but the snapshots probably represent the general kind of weather that was typical over long periods.

Paleogene The period of geological time (*see* GEOLOGICAL TIME SCALE) that began about 65 million years ago and ended about 23.3 million years ago. It includes the PALEOCENE, EOCENE, and OLIGOCENE epochs.

paleoglaciology The study of GLACIERS and ICE SHEETS that existed in prehistoric times from the evidence they have left in surface rocks.

Palmer Drought Severity Index (PDSI) An index for classifying droughts that was introduced in 1965 by W. C. Palmer, a meteorologist at the U.S. Weather Bureau ("Meteorological Drought," *U.S. Department of Commerce Research Paper No. 45,* Washington, D.C.). It measures the extent to which the water supply departs from what is normal for a particular place and it is widely used in the United States for monitoring droughts. It is especially valuable in agriculture. The index is calculated from the amount of precipitation, the temperature, and the amount of water available in the soil. Its disadvantages are that the values it uses are based on data from Iowa and western Kansas and are arbitrary; the index is sensitive to the water-retaining capacities of soils so it is difficult to apply over a region containing soils of varying types; snow and frozen ground are not taken into account; and no allowance is made for the lag that occurs between precipitation and its runoff. The index centers on 0 and runs from 4.00 to –4.00.

PDSI Classification

4.00 or more	Extremely wet
3.00–3.99	Very wet
2.00–2.99	Moderately wet
1.00–1.99	Slightly wet
0.50–0.99	Incipient wet spell
0.49–-0.49	Near normal
–0.50–-0.99	Incipient dry spell
–1.00–-1.99	Mild drought
–2.00–-2.99	Moderate drought
–3.00–-3.99	Severe drought
–4.00 or less	Extreme drought

(More information about the Palmer Drought Severity Index can be found at http://nris.state.mt.us/wis/indices.html.)

palynology The study of POLLEN grains, SPORES, and the shells of some aquatic organisms in order to classify them and discover their distributions. Although the terms *palynology* and *POLLEN ANALYSIS* are often used synonymously, there is a difference between them. Palynology developed from pollen analysis and is used in many fields, including archaeology, petroleum exploration, and PALEOCLIMATOLOGY.

pampero A violent squall, with winds of gale force, that blows from the southwest across the pampas of Argentina and Uruguay, north of the River Plate. The squalls form along a COLD FRONT and are accompanied by CUMULUS or CUMULONIMBUS cloud, sometimes with thunderstorms, and clouds of dust. The squall usually lasts only a few minutes, but the winds continue for several hours. The pampero is similar to the SOUTHERLY BURSTER of New South Wales, Australia, and the NORTHER of the United States.

PAN *See* PEROXYACETYL NITRATE.

panas oetara A strong, warm, dry northerly wind that blows in Indonesia in February.

pancake ice Patches of ice, floating on the surface of the sea, that are fairly thin and approximately circular in shape. They are produced when FRAZIL ICE covers a large area and the movement of the sea breaks it into pieces. These constantly collide with one another and their shape results from the collisions. During its formation, salt water becomes trapped between ice crystals.

pan coefficient The ratio of the amount of water that evaporates from a unit area of the exposed surface of a large body of water to the amount that evaporates from an EVAPORATION PAN of similar area.

pannus An ACCESSORY CLOUD that comprises ragged patches of cloud attached to or beneath another cloud. The word *pannus* is Latin for "shred." Pannus is most often seen with CUMULONIMBUS, CUMULUS, ALTOSTRATUS, and NIMBOSTRATUS.

papagayo *See* NORTE.

parcel method A way of testing the STABILITY OF AIR in which the consequences of displacing particular PARCELS OF AIR are calculated. It is assumed that only those parcels are affected, but in fact their stability is similar to that of the larger body of air around them.

parcel of air (air parcel) A volume of air that can be considered separately from the air surrounding it and from which it is assumed to be physically isolated. The concept is theoretical, since no volume of air can be completely isolated from the air around it, but it is useful in calculating the behavior of the atmosphere. It is used in calculating ADIABATIC changes and calculations involving the EQUATION OF STATE.

parhelion *See* SUN DOG.

partial obscuration The condition of the sky when part of it, up to 90 percent or 7 OKTAS, is hidden by an OBSCURING PHENOMENON on the surface. The term is used in United States weather reporting.

partial pressure In a mixture of gases, the share of the total pressure that can be attributed to one of the constituent gases. If the atmospheric pressure is 1,000 mb and oxygen accounts for 21 percent of the mass of the air, then the partial pressure for oxygen is 210 mb.

particulate matter The fine particles that are suspended in the atmosphere. Some enter the atmosphere as a consequence of natural events, such as volcanic eruptions, desert winds, and forest and grass fires ignited by LIGHTNING. POLLEN grains and the SPORES of fungi and bacteria also form part of the atmospheric particulate matter. Other particles result from human activities, especially plowing of dry soil, and deliberate setting of fires, as well as creation of soot from the burning of coal and oil. Collectively, atmospheric particles are known as AEROSOLS. Small particles can be harmful to health when they are inhaled, because they are able to penetrate deep into the lungs. Particles that are less than 25 μm in diameter, known as *PM25*, are believed to cause a number of respiratory illnesses, and even smaller particles, 10 μm in size (PM10), are also suspected of causing harm.

partly cloudy The condition of the sky when clouds are present but for most of the time cover less than the whole sky. In reporting of weather conditions, the term is used rather more precisely to indicate that over a period of 24 hours the average cloud cover has been between 10 percent and 50 percent (1–4 OKTAS).

pascal (Pa) The derived SYSTÈME INTERNATIONAL D'UNITÉS (SI) UNIT of pressure, which is equal to 1 NEWTON per square METER. The unit is named in honor of the French physicist, mathematician, and philosopher BLAISE PASCAL (1623–62). 10 Mpa = 1 BAR; 0.145 kPa = 1 pounds force (lbf) in^{-2}.

Pascal, Blaise (1623–1662) French *Mathematician, physicist, and theologian* A French mathematician, physicist, and theologian who discovered that the atmosphere has an upper limit and that atmospheric pressure decreases with altitude, Pascal was born at Clermont-Ferrand, in the Auvergne region of France. The family moved to Paris in 1631. His father, a mathematician and government official, supervised Blaise's education. Blaise was soon revealed as a mathematical prodigy. In 1640, not yet 17 years of age, he wrote *Essai pour les coniques,* about the geometry of conic

Blaise Pascal. **The French mathematician, physicist, and theologian who proved that atmospheric pressure decreases with altitude.** *(John Frederick Lewis Collection, Print and Picture Collection, The Free Library of Philadelphia)*

sections, an essay that made René Descartes envious. Between 1642 and 1644, he designed and made a calculating machine, based on cogwheels, to help his father in the many calculations his work required. This machine was the ancestor of the modern cash register, and the computer programming language Pascal is named after its inventor. Pascal made and sold about 50 of them and several are still in existence.

Pascal learned of and became interested in the work of EVANGELISTA TORRICELLI. He repeated the Torricelli experiment but used red wine and water instead of mercury. Wine is even less dense than water, so Pascal's barometers had tubes 39 feet (12 m) tall and were fixed to the masts of ships!

He also reasoned that if air has weight, then it must cover the surface of the Earth rather like an ocean, with an upper surface. This implied that the weight of air measured by a barometer must decrease with height, because the greater the height of the instrument above the surface the smaller must be the mass of air weighing down from above.

In 1646, he returned to Clermont-Ferrand to test this idea on the Puy-de-Dôme, an extinct volcano, 4,806 feet (1,465 m) high, not far from the town. Pascal was never physically strong, and by 1646 he suffered from severe indigestion and insomnia. There was no question that he could make the strenuous ascent of the Puy-de-Dôme—the climb is very steep—and so he enlisted the help of his brother-in-law, Florin Périer. Périer carried a barometer up the mountain, at intervals recording the pressure it registered. This showed a steady decrease in pressure with increasing height. Similar records were made from a second barometer, left at the foot of the mountain. These showed no change in pressure throughout the day. The experiment confirmed Torricelli's discovery that air has weight. This achievement is recognized in the name of the derived SYSTÈME INTERNATIONAL D'UNITÉS (SI) unit of pressure or stress, the pascal.

Pascal made many more contributions to science and mathematics before his death, in Paris, probably of meningitis associated with stomach cancer.

passive front   *See* INACTIVE FRONT.

passive instrument   An instrument that measures radiation falling on it (so it is passive), rather than sending out a signal that returns to it.

past weather   The weather that has prevailed over the period, usually of six hours, since a weather station last submitted a report. The past weather is reported as a series of numbers between 0 and 9:

0 Cloud covered half of the sky or less throughout the period.
1 Cloud covered more than half the sky for part of the period and less than half for the remainder of the period.
2 Cloud covered more than half the sky throughout the period.
3 Sandstorm, dust storm, or blowing snow.
4 Visibility was less than 0.6 mile (1 km) as a result of fog, ice fog, or haze.
5 Drizzle.
6 Rain.

7 Snow or a mixture of rain and snow (in Britain, called *sleet*).
8 Showers.
9 Thunderstorms, with or without precipitation.

Pat   A TYPHOON, rated category 4 on the SAFFIR/SIMPSON SCALE, that struck Kyushu, Japan, on August 30, 1985. It generated winds of up to 124 mph (200 km h^{-1}) and killed 15 people.

path length   The distance that incoming solar radiation must travel between the top of the atmosphere and the land or sea surface. The path length has a value of 1 when the Sun is directly overhead, or 90° above the horizon, and it increases as the angle of the Sun decreases. The path length can therefore be calculated as 1/cos *A,* where *A* is equal to 90° minus the angle of the Sun. If the Sun is at 45°, the path length is 1.4 and if the Sun is at 30° the path length is 2 (90 − 30 = 60; 1/cos 60 = 2). Solar radiation travels twice as far through the air when the Sun is 30° above the horizon as it does when the Sun is directly overhead.

Paul   A HURRICANE, rated category 4 on the SAFFIR/SIMPSON SCALE, that struck Sinaloa, Mexico, on September 30, 1982. It generated winds of 120 mph (193 km h^{-1}) and left 50,000 people homeless.

Pauline   A HURRICANE that struck southern Mexico October 8 to 10, 1997. It generated winds of up to 115 mph (185 km h^{-1}) and waves up to 30 feet (9 m) high and caused extensive damage in the city of Acapulco and in coastal villages in the states of Oaxaca and Guerrero. The hurricane killed 217 people and rendered 20,000 homeless.

PDO   *See* PACIFIC DECADAL OSCILLATION.

PDSI   *See* PALMER DROUGHT SEVERITY INDEX.

peak gust   The highest wind speed that is recorded at a weather station during a period of observation, commonly of 24 hours. Despite the name, a peak gust may be a sustained wind rather than a GUST.

pearl-necklace lightning (beaded lightning, chain lightning)   A rare type of LIGHTNING, in which the luminosity of the discharge varies along its length, so the lightning appears as a chain of lights, or a string of pearls.

pea souper A type of SMOG in which visibility was reduced to less than 30 feet (10 m)—sometimes to very much less. The name refers to the yellowish color of the smog when it contained more smoke than fog. Smog made window curtains and laundry hanging outdoors filthy and deposited soot particles on the hands, faces, and clothes of people outdoors in it. It was visible as a haze even indoors. Traveling was difficult. Street lights remained lit throughout the day and road vehicles used their lights, but at times drivers were unable to see the edge of the road and became badly disoriented. Pedestrians fared only slightly better. In the thickest smogs it was difficult for them to see the ground beneath their feet. Smog of this type was especially associated with London, where it occurred on a few days in most winters, but there were similar smogs in most other industrial cities. Legislation severely curtailing the burning of coal in domestic fires greatly improved urban air quality, and pea soupers no longer occur.

pedestal rock *See* MUSHROOM ROCK.

Peggy A TYPHOON that struck the northern Philippines on July 9, 1986. It caused floods, landslides, and mudslides, producing extensive damage to property. More than 70 people were killed. It then moved to southeastern China, where it arrived on July 11 and caused widespread flooding. More than 170 people were killed and at least 1,250 injured, and more than 250,000 homes were destroyed.

Peléean eruption *See* VOLCANO.

penetrative convection The type of CONVECTION that occurs when the ground or water surface is warmer than the air immediately above it. This situation most often develops when cool air moves across a warm surface, and it is especially common in low latitudes. The air is warmed from below, producing CONDITIONAL INSTABILITY, which leads to the formation of CUMULIFORM clouds. Unlike cellular convection (*see* CONVECTION CELL), which produces LAMINAR FLOW and regular cloud patterns, penetrative convection produces TURBULENT FLOW and is chaotic (*see* CHAOS.).

Penman formula A mathematical formula for measuring the rate of EVAPORATION from an open surface that was published in 1948 by the British climatologist H. L. Penman ("Natural evaporation from open water, bare soil and grass," *Proceedings of the Royal Society,* **193**, pages 120–145). The advantage of the Penman formula is that it expresses evaporation losses in terms of the duration of sunshine, the mean air temperature, the mean humidity, and the mean wind speed. The duration of sunshine is related to the amount of radiation received at the surface, and the other factors limit the amount of heat and moisture that are lost from the surface. These four factors are regularly measured at weather stations and so the formula is easy to apply. It is also widely used in the construction of climate MODELS.

(If you wish to follow the mathematical steps by which Penman arrived at his formula, these are set out at www-das.uwyo.edu/~geerts/cwx/notes/chap04/penman.hl.)

pennant A triangular symbol, resembling a pennant flag, that is used on a SYNOPTIC CHART to indicate wind speeds greater than 48 KNOTS (55 mph [89 km h^{-1}]). It is drawn so that the shaft of the pennant indicates the wind direction and the pointed tip of the pennant points toward the region of low pressure.

pentad A period of five days. Meteorologists and climatologists often use the pentad in preference to the

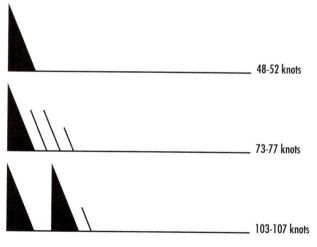

Pennant. **Three examples of the symbol used to indicate higher wind speeds on a synoptic chart.**

week, because there is an exact number of pentads in a 365-day year (73).

penumbra The less deeply shaded area that lies near the edge of a shadow. The central, totally shaded area is the *umbra*. Although they refer to any shadow, the terms are most often applied to ECLIPSES of the Moon or Sun and to SUNSPOTS. If the source of light is a point, the shadow is sharply defined and there is no penumbra. If the source covers a large area, such as a substantial portion of the sky, then some of the light illuminates the area around the umbra, producing a penumbra.

perched aquifer An AQUIFER that lies above a lower aquifer. At the base of the sequence, a layer of impermeable material lies above the bedrock and there is a layer of permeable material above that (*see* PERMEABILITY). Water drains into the permeable layer from outside the area and saturates a band of material above the impermeable layer. This flows as GROUNDWATER. This is the lower aquifer. There are a second impermeable layer above the unsaturated material and permeable material above that all the way to the surface. Water drains downward through the permeable material but is held at the impermeable barrier, where it accumulates to form a second, higher aquifer. This is the perched aquifer.

percolation The downward movement of water through wet soil until it reaches and joins the GROUNDWATER. The movement of water into dry soil is called

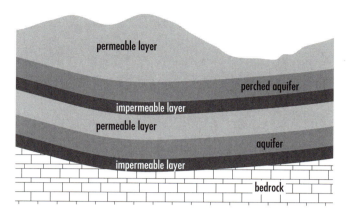

Perched aquifer. **A perched aquifer lies above a layer of impermeable material, with a second sequence of permeable and impermeable layers below. Water enters the lower sequence from outside. Consequently there are two aquifers, the upper one perched above the lower one.**

water infiltration. It is the percolation of water that removes dissolved minerals from the upper layers of soil. The rate at which water infiltrates dry soil or percolates through wet soil depends on the PERMEABILITY of the soil.

pergelisol *See* PERMAFROST.

perhumid climate In the THORNTHWAITE CLIMATE CLASSIFICATION, a climate in which the monthly MOISTURE INDEX is greater than 100 and the monthly POTENTIAL EVAPOTRANSPIRATION is greater than 44.9 inches (114 cm). It is designated *A*. In terms of THERMAL EFFICIENCY, this climate is megathermal (*A'*).

periglacial climate The climate that prevails near the edge of an ICE SHEET. Its most distinctive feature is the frequency of cold, dry winds that blow outward from the semipermanent ANTICYCLONE over the ice sheet. This means the periglacial climate is cold and dry. There is a PERMAFROST layer beneath the surface, but the ACTIVE LAYER thaws during the summer. During the brief thaw the soil of the active layer loosens and large rocks tend to start sliding downhill, but before they can travel far, the ground freezes once more. Repeated cycles of freezing and thawing gradually arrange the rocks in patterns that can be recognized long after the climate has ceased to be periglacial.

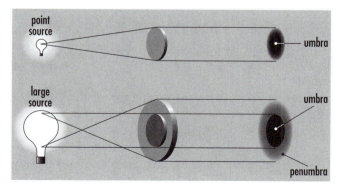

Penumbra. **The dark center of a shadow is the umbra. An area adjacent to the umbra that is only partly shaded is called the *penumbra*.**

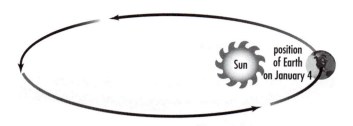

Perihelion. **The point in its elliptical solar orbit at which a body is closest to the Sun.**

These patterns provide one of the clues that are used by paleoclimatologists (*see* PALEOCLIMATOLOGY).

perihelion   The point in the eccentric solar orbit of a planet or other body when it is closest to the Sun. At present, Earth is at perihelion on about January 4 each year, but the dates of APHELION and perihelion change over a cycle of about 21,000 years (*see* MILANKOVICH CYCLES and PRECESSION OF THE EQUINOXES). The Earth receives 7 percent less solar radiation at aphelion than it does at perihelion.

period   The amount of time that elapses between two events, and in particular the time that elapses between repetitions of the same event. The period of a wave is the time interval between two wave crests (or troughs) passing a fixed point. The FREQUENCY of a wave is given by $1/T$, where T is the period, and the speed (c) of a wave is given by $c = \lambda T$, where λ is the WAVELENGTH. When air or water moves within a confined space, such as a lake or coastal bay, the resulting waves may oscillate with a period that is determined by the configuration of the boundaries containing them. This is known as the *natural period* for that place and it is equal to the reciprocal of the natural frequency.

perlucidus   A variety of cloud (*see* CLOUD CLASSIFICATION) that forms an extensive layer, sheet, or patch that includes open spaces. These may be very small but are quite distinct. Blue sky can be seen through them, and when they are in the correct position the Sun or Moon can shine through them. Perlucidus occurs with the cloud genera ALTOCUMULUS and STRATOCUMULUS. The name *perlucidus* is derived from the Latin words *per*, which means "through," and *lucidus*, which means "bright" (from the verb *lucēre*, "to shine").

permafrost (pergelisol)   Ground that remains frozen throughout the year. In order for permafrost to form the temperature must remain below freezing for at least two consecutive winters and the whole of the intervening summer. Permafrost covers approximately 26 percent of the land area of the Earth, including more than half of the land area of Canada.

The depth of the permafrost layer varies. In Canada it is about 6.5 feet (2 m) along the southern edge of the permafrost zone and up to 1,000 feet (300 m) thick in the far north. On the North Slope of Alaska the permafrost is 2,000 feet (600 m) thick in places, and in parts of Siberia it is up to 4,600 feet (1,400 m) thick.

Soil is a poor conductor of heat and changes in the average air temperature take many years to penetrate. After about 100 years the change affects material about 500 feet (150 m) below the surface, and it is 1,000 years before it is felt at 1,640 feet (500 m). Material deep below the surface is so well insulated from temperature changes above the surface that scientists working out past climatic conditions sometimes use it. They take long, vertical cores of soil from the ground and measure the temperature at intervals along them. Some of the Canadian, Alaskan, and Siberian permafrost has remained frozen for several thousand years, since the end of the most recent (WISCONSINIAN or DEVENSIAN) GLACIAL PERIOD.

In the northern part of the Arctic, adjacent to the Arctic Circle and to the north of it, the permafrost is continuous. This means that all of the ground is frozen all of the time. Farther south, and covering a much larger land area, the permafrost is discontinuous. It occurs in patches and is found where the average temperature locally is lower than that for the surrounding area, on north-facing slopes that are never exposed to direct sunshine, and on land that is wet for most of the time.

Except for the northernmost tip of the Antarctic Peninsula, all of Antarctica lies within the Antarctic Circle. Ground that is not covered by the ICE SHEET is permanently frozen. Apart from the continent and its offshore islands, there are no other regions of permafrost in the Southern Hemisphere, because there is no large expanse of land in a sufficiently high latitude. Even the southern islands, such as the Falkland Islands (Las Malvinas), have no permafrost.

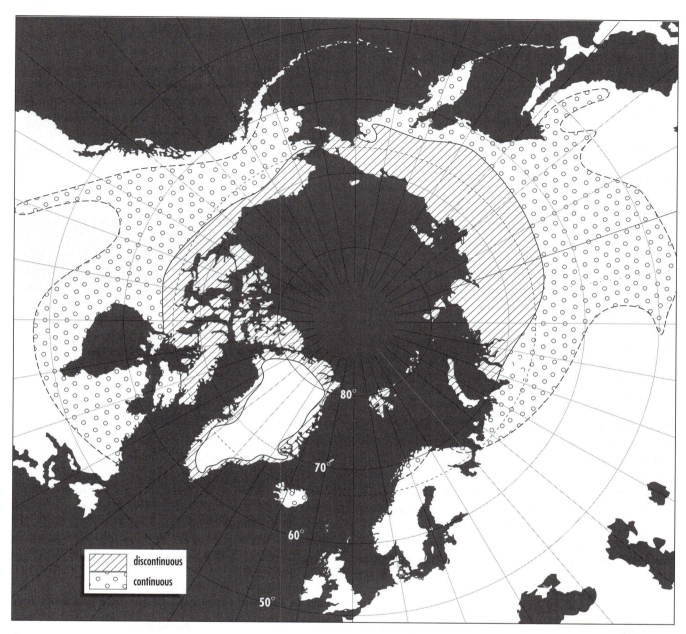

Permafrost. Together, the regions of continuous and discontinuous permafrost in the Northern Hemisphere cover more than half of Canada and much of Siberia.

Permafrost may include areas that remain unfrozen. These are called *talik*. Where summer temperatures remain above freezing for several weeks or months the upper part of the soil thaws. This produces a thin ACTIVE LAYER in which a few plants can grow. Over the years as plants die and shed leaves, plant material accumulates to form a mat of dead vegetation that insulates the ground below, preventing the summer warmth from penetrating. The existence of an active layer therefore tends to perpetuate the permafrost layer beneath it.

A permafrost layer impedes drainage, and, because it is hard as rock, it presents an impenetrable barrier to plant roots.

permanent drought   One of the types into which DROUGHTS are formally classified. It is the type of drought that characterizes deserts. There are no permanent streams or rivers; precipitation rarely occurs, although it is often heavy when it does; and crops can be grown only on irrigated land.

permanent wilting percentage   *See* WILTING.

permanent wilting point   *See* WILTING.

permeability (hydraulic conductivity)   The ability of a soil, sediment, or rock to allow water or air to move through it. This is measured as the volume of water that flows through a unit cross-sectional area in a specified time. Soil permeability is categorized as slow, moderate, or rapid according to the rate of flow and is reported as the distance traveled in one hour. If the water is moving vertically downward through the soil the process is called *percolation*. It is reported as the time taken for the water to move a specified distance. The classification system comprises seven classes for both permeability and percolation.

Permeability and the percolation rate depend on the structure of the material through which air or water moves. Sand does not hold water well. Pour water onto a sandy beach on a fine day and the sand is dry again within at most a few minutes. Clay becomes sodden, because water does not pass through it easily. The difference is due to the size and shape of the particles from which the soil is made.

The spaces between soil particles are called *pores,* and if all the particles are spherical and the same size,

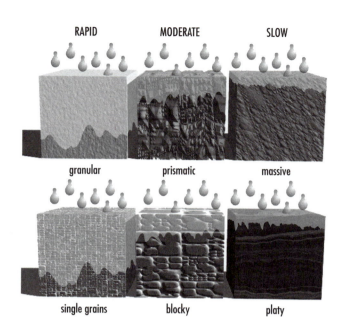

Permeability. **The rate at which water flows through a soil depends on the size and arrangement of the soil particles.**

the pore space amounts to about 45 percent of the total volume filled by the particles. This proportion is the same whatever the size of the particles themselves. In a soil, however, the particles are not all of the same size or shape.

If the particles are large, the pore spaces between them are also large. This arrangement allows water and air to move through the soil rapidly. Gravel and sand do not retain water very long because they consist of large particles. Sand grains have a granular texture and

Class	Permeability		Percolation	
	inches	cm	minutes	
	per hour		per inch	per centimeter
1. very slow	less than 0.05	less than 0.1	more than 1200	more than 470
2. slow	0.05–0.20	0.1–0.5	300–1,200	118–470
3. moderately slow	0.20–0.80	0.5–2.0	75–300	30–118
4. moderate	0.80–2.50	2.0–6.4	24–75	9–30
5. moderately rapid	2.50–5.00	6.4–12.7	12–24	5–9
6. rapid	5.00–10.00	12.7–25.4	6–12	2–5
7. very rapid	more than 10.00	more than 25.4	less than 6	less than 2

gravel consists of single stones, or grains. Water permeates and percolates rapidly through them.

If the particles are very small, the amount of pore space is the same, but the individual pores are small. Water tends to be held to the surfaces of the grains by molecular attraction, and the distance between one surface and the next is so small that there is little room for water to flow past the water adhering to the particles. Some very small particles tend to pack together to form a solid mass. This squeezes out the pore spaces and produces a massive texture. Clay particles pack together, but in layers. The particles themselves are flat, like flakes, but microscopically small. They arrange themselves with their flat sides adjacent. This produces a platy structure that prevents water passing. Water permeates and percolates slowly through soils of these types.

Between the two extremes there are soils made from particles that arrange themselves into prismatic columns. Water is able to move between the columns. Other soils consist of big, irregular particles. These form a blocky structure with channels along which water can move. Water permeates and percolates moderately quickly through soils of these types.

Permeability affects the way soils respond to the weather. Areas where the permeability is rapid are not susceptible to flooding, because water leaves them quickly. Dry weather can parch the ground just as quickly, however, so there is a relatively high risk of DROUGHT. Where permeability is slow, drought is less of a risk, but flooding is more likely. Also, soils that retain water tend to be cold in early spring, because of the high HEAT CAPACITY of water. This can delay the sowing of spring crops.

(You can read more about soils and permeability in Michael Allaby, *Elements: Earth* [New York: Facts On File, 1993].)

peroxyacetyl nitrate (PAN) A chemical compound ($CH_3CO·O_2NO_2$) that forms by a complicated series of reactions involving the oxidation of hydrocarbons, especially the unburned hydrocarbons in vehicle exhausts. PAN is fairly stable in the cold air of the upper TROPOSPHERE, but in warm air it decomposes to release nitrogen dioxide (NO_2) and the highly reactive peroxyacetyl radical $CH_3CO·O_2$. This process contributes to the atmospheric content of NITROGEN OXIDES (NO_x), which are implicated in the formation of PHOTOCHEMICAL SMOG and ACID RAIN.

PAN decomposes and reforms by a reversible reaction, depending on the air temperature:

$$CH_3CO·O_2NO_2 \leftrightarrow CH_3CO·O_2 + NO_2$$

PAN is constantly forming, decomposing, and reforming, but in warm air the reaction favors the release of NO_2 and peroxyacetyl radical. Some of the PAN survives to be carried aloft by CONVECTION currents, however, and in colder air the same reaction favors the PAN. Consequently, this essentially urban pollutant can be dispersed over a wide area.

perpetual frost climate *See* ICE CAP CLIMATE.

persistence The length of time during which a particular feature of the weather remains unchanged. In the case of the wind, persistence is calculated as the ratio of the mean wind vector (*see* VECTOR QUANTITY) to the average wind speed (ignoring the direction).

persistence forecast A weather forecast that predicts a continuation of present conditions for several hours ahead. Such a forecast might state that the rain that is falling will continue to do so, or that the present fine weather will remain unchanged. Such a forecast cannot predict changes in the direction or speed with which weather systems move or the formation or dissipation of FRONTAL SYSTEMS. Consequently, a persistence forecast usually remains valid for no more than 12 hours, seldom for as long as a full day, and often fails in as little as six hours.

Peru Current (Humboldt Current) An ocean current that flows northward from the WEST WIND DRIFT, past the western coast of South America, to join the SOUTH EQUATORIAL CURRENT. It carries cold water and is broad and slow-moving. The current is noted for the many UPWELLINGS along its course that move nutrients near to the surface.

Peruvian dew *See* GARÚA.

PGF *See* PRESSURE GRADIENT FORCE.

pH *See* ACIDITY.

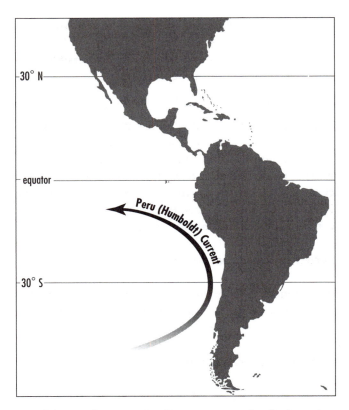

Peru (Humboldt) Current. **A cold ocean current that flows north-ward from the Southern Ocean, parallel to the western coast of South America.**

phase   1. A part of a system that is of the same composition throughout and that is clearly distinct from all other parts of the system. A block of ice is a system consisting of one phase, as is a mass of liquid water or water vapor. A mixture of water and ice constitutes a two-phase system, and a mixture of ice, liquid water, and water vapor constitutes a three-phase system. A solution of salt or sugar in water is a single-phase system. When water changes between gas, liquid, and solid, it is said to change phase. At a pressure of 6.03 mb and a temperature of 32.018° F (0.01° C) water can exist as gas, liquid, and solid simultaneously. This is known as its *triple point* (*see* BOILING). 2. The stage that a regularly repeating motion has reached is also called a *phase*. Usually the phase of a periodic motion (*see* PERIOD) is described by comparison with another motion having the same WAVELENGTH. If the peaks and troughs of two or more waves coincide, the waves are said to be in phase. If the peaks of one wave coincide with the troughs of the other, the two are said to

be out of phase. If the peaks and troughs are between the two extremes of being in phase or out of phase, the distance between them is called the *phase difference*.

phenology   The scientific study of periodic events in the lives of plants and animals that are related to the climate. Such events include the dates on which deciduous trees come into leaf; crops germinate, flower, ripen, and are harvested; and migratory birds arrive and depart. Phenological events comprise all the familiar signs of the changing seasons. The dates of particular events at places some distance apart can be used to compile a phenological gradient marking the geographical movement of the seasons across a continent or, on a much smaller scale, to measure the effects of such influences as exposure or shade within a garden. Phenological data collected in Europe over a period of 30 years from the INTERNATIONAL PHENOLOGICAL GARDENS have shown that during this time spring events have advanced to occur 6 days earlier and autumn events occur 4.8 days later. The dates of European

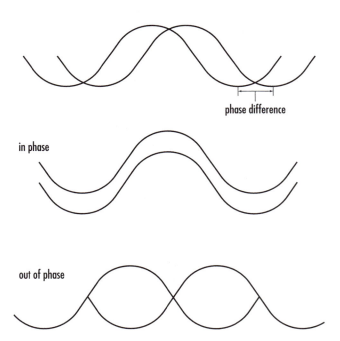

Phase. **The distance between the troughs and peaks of two waves of the same wavelength is the phase difference between them. The waves are in phase when their peaks and troughs coincide and out of phase when the peaks of one coincide with the troughs of the other.**

grape harvests have been of particular value to historians tracing the history of climate.

(The use of grape harvests features strongly in Emmanuel Le Roy Ladurie, *Times of Feast, Times of Famine: A History of Climate since the Year 1000* [New York: Doubleday and Co., 1971].)

photochemical smog   A form of AIR POLLUTION that occurs when ULTRAVIOLET RADIATION in strong sunlight acts upon hydrocarbon compounds emitted by vehicle exhausts. It is quite different from SMOG of the London type.

Compounds such as aldehydes (compounds containing the –COOH group joined directly to another carbon atom), ketones (compounds containing the C·CO·C group), and formaldehyde (or methanal, HCHO) give photochemical smog a characteristic odor, and nitrogen dioxide (NO_2) and solid particles cause a brownish haze. OZONE (O_3), aldehydes, and PEROXY-ACETYL NITRATE (PAN) cause irritation to the eyes and throats of persons exposed to smog, and plants are damaged by nitrogen oxides (NO_x), O_3, PAN, and ethene ($CH_2=CH_2$, also called *ethylene*). The formation of photochemical smog begins with the photolytic cycle, in which NO_2 is broken down and reformed. Atomic oxygen (O) is produced in the first stage of the

photolytic cycle. This is highly reactive and oxidizes hydrocarbons (H_c) to hydrocarbon free radicals (H_cO^*). Free radicals are atoms or molecules that have unpaired electrons, as a result of which they are extremely reactive. These react further to reform NO_2, allowing the cycle to continue, and to yield O_2 and the hydrocarbon ingredients of smog. Using oxygen from the photolytic cycle, the reactions are

$$O + H_c \rightarrow H_cO^* \ (1)$$

$$H_cO^* + O_2 \rightarrow H_cO_3^* \ (2)$$

$$H_cO_3^* + NO \rightarrow H_cO_2^* + NO_2 \ (3)$$

$$H_cO_3^* + H_c \rightarrow \text{aldehydes, ketones, etc. } (4)$$

$$H_cO_3^* + O_2 \rightarrow O_3 + H_cO_2^* \ (5)$$

$$H_cO_x^* + NO_2 \rightarrow \text{PAN } (6)$$

photochemistry   The branch of chemistry that studies the chemical effects of electromagnetic radiation. Many chemical reactions take place only when radiation supplies the energy that is needed to drive them. OZONE is formed in the STRATOSPHERE by the action of ULTRAVIOLET RADIATION on oxygen (*see* PHOTODISSOCIATION). PHOTOCHEMICAL SMOG is the product of a series of reactions that take place in strong sunlight. These are photochemical processes, but the best known and most important photochemical reactions are those in PHOTOSYNTHESIS.

photodissociation   The splitting of a molecule into smaller molecules or single atoms, using light as a source of energy. In the upper stratosphere oxygen (O_2) molecules absorb solar ultraviolet radiation (UV) at wavelengths below about 0.3 μm and are photodissociated into single oxygen atoms (O + O). UV radiation at less than 0.23 μm causes the photodissociation of chlorofluorocarbon (CFC) compounds in the stratosphere. In the troposphere, UV radiation at 0.37–0.42 μm photodissociates nitrogen dioxide (NO_2) into nitrogen oxide (NO) and oxygen (O). Photodissociation occurs when an atom or molecule absorbs a photon, which is a unit (quantum) of electromagnetic radiation, possessing precisely the energy needed to allow it to break free of the group to which it is attached.

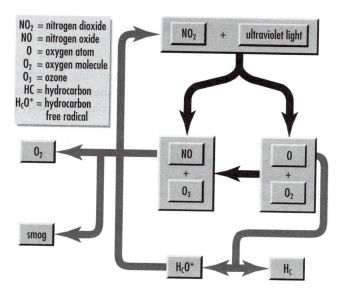

Photochemical smog. Pollution occurs when some of the atomic oxygen produced by the photolytic cycle combines with unburned hydrocarbons to form hydrocarbon free radicals.

photoionization   The IONIZATION of an atom that occurs when it absorbs a photon of electromagnetic radiation with a wavelength of less than about 0.1 μm, which is in the ULTRAVIOLET part of the spectrum. Photoionization occurs in the region of the upper atmosphere known as the *IONOSPHERE,* where the density of the air is such that collisions between solar radiation and gas molecules are common. PHOTODISSOCIATION, the first step in photoionization, separates gas molecules into their constituent atoms. Individual atoms are then ionized to produce separate charged atoms and free electrons. This process imparts an electrical charge to the ionosphere, and it also absorbs photons of ultraviolet light.

photolytic cycle   A naturally occurring sequence of chemical reactions in the course of which ULTRAVIOLET RADIATION supplies the energy for the PHOTODISSOCIATION of nitrogen dioxide (NO_2) in the TROPOSPHERE. The NO_2 then reforms. The reactions are

$$NO_2 + UV \rightarrow NO + O \quad (1)$$

$$O + O_2 \rightarrow O_3 \quad (2)$$

$$O_3 + NO \rightarrow NO_2 + O_2 \quad (3)$$

If hydrocarbons from vehicle exhausts are also present in the air the natural cycle is disrupted and a range of other compounds are produced, causing PHOTOCHEMICAL SMOG.

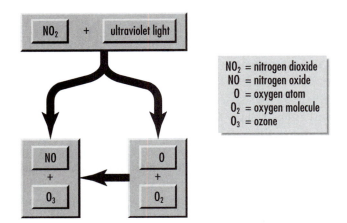

Photolytic cycle. **Ultraviolet light splits nitrogen dioxide into nitrogen oxide and atomic oxygen. Oxygen atoms then combine with oxygen molecules to form ozone.**

photoperiod   The number of hours of daylight that occur during a 24-hour period. Except at the equator, this varies with the season. Many plants respond physiologically to changes in the photoperiod, in the process of photoperiodism.

The most common responses are those affecting the time of flowering. Some plants do not flower if the daily cycle includes long periods of darkness. These are known as *long-day plants* (but are really short-night plants) and they flower in late spring and early summer, when the days are lengthening and the nights are growing shorter. Lettuce, wheat, and barley are long-day plants. Strawberries and chrysanthemums are among the plants that flower between late summer and early spring, when days are short and nights are long. They are called *short-day plants*. Not all plants are affected by the photoperiod. Tomatoes and cucumbers are among the plants in which the time of flowering is not determined by the duration of daylight. These are called *day-neutral plants*.

photoperiodism   *See* PHOTOPERIOD.

photopolarimeter–radiometer (PPR)   An instrument that is used in remote sensing. It supplies data from which the temperature and cloud formation in the atmosphere of a planet or satellite can be determined, as well as some surface detail. The PPR measures the intensity and polarization of sunlight in the visible part of the spectrum (*see* WAVE BAND).

photosphere   The visible surface of the Sun or any other star, in the case of the Sun comprising a layer of gas about 300 miles (500 km) thick that is opaque to radiation at the base but transparent at higher levels. It is the layer from which the solar radiation is emitted and it is visible because it emits light. The temperature of the photosphere is about 6,000 K at the base and falls to about 4,000 K at the top. Above the photosphere lies the CHROMOSPHERE.

photosynthesis   The series of chemical reactions by which green plants (as well as certain bacteria and cyanobacteria) synthesize (construct) sugars, using carbon dioxide and water as the raw materials. The first stage in the process depends on light as a source of energy and is called the *light-dependent* or *light stage*. The second stage also takes place in light, but it does not use light energy and so it is called the *light-independent* or *dark stage.*

The overall set of reactions can be summarized as

$$6CO_2 + 6H_2O + \text{[light energy]} \rightarrow C_6H_{12}O_6 + 6O_2\uparrow$$

The oxygen is released into the air, indicated by the arrow pointing upward. $C_6H_{12}O_6$ is a simple sugar from which more complex sugars and starches can be made.

The process begins with chlorophyll, the green pigment that is held in bodies called *chloroplasts* in the cells of leaves and some stems. Chlorophyll is a very complex substance with the property of absorbing light energy. When a photon of light possessing exactly the right amount of energy strikes a chlorophyll molecule, one electron in the chlorophyll molecule absorbs that energy and becomes excited. It jumps from its ground state to its excited state and escapes from the molecule but is immediately captured by a neighboring molecule. An electron is then passed from molecule to molecule along an electron-transport chain until it is used to split water into hydrogen and oxygen:

$$H_2O \rightarrow H^+ + OH^-$$

Having lost an electron, the chlorophyll molecule carries a positive charge. The hydroxyl ion (OH) produced by the PHOTODISSOCIATION of water carries a negative charge in the form of an extra electron. It passes this to the chlorophyll. Both hydroxyl and chlorophyll are then neutral and hydroxyls combine to form water:

$$4OH \rightarrow 2H_2O + O_2\uparrow$$

The oxygen is released into the air. Free hydrogen atoms attach themselves to molecules of nicotinamide adenine dinucleotide (NADP), converting it to NADPH. This completes the light stage.

NADP loses its hydrogen again during the dark stage, and the NADP then returns to the light stage. The dark stage begins when a molecule of CO_2 becomes attached to one of ribulose biphosphate (RuBP) in the presence of an enzyme, RuBP carboxylase (rubisco). The captured carbon then enters a sequence of reactions that end with the construction of sugar molecules and also of the RuBP, with which the process began. Because the RuBP is reconstructed so it can be used again, the reactions form a cycle. It is

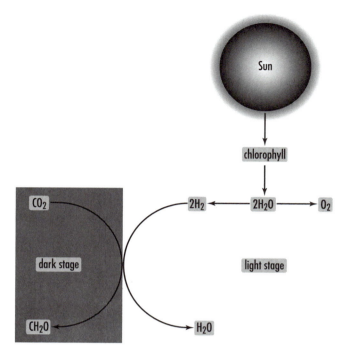

Photosynthesis. **The series of chemical reactions by which plants use the energy of sunlight to manufacture sugars has two stages. Water is broken down into hydrogen and oxygen in the light stage, and in the dark stage carbon from carbon dioxide is used to make sugar.**

known as the *Calvin cycle,* because its details were discovered by the American biochemist Melvin Calvin (1911–97). Calvin was awarded the 1961 Nobel Prize in Chemistry for this work.

There are variations on the photosynthetic pathways. If the first product of the dark stage, made when CO_2 joins RuBP, is 3-phosphoglycerate, the pathway is called *C3,* because 3-phosphoglycerate has three carbon atoms. Most plants, including all trees, are C3 plants.

Other plants use the C4 pathway, in which CO_2 combines with phosphoenol pyruvic acid (PEP) rather than RuBP, and the first product is oxaloacetic acid, which has four carbon atoms. This pathway uses more energy than the C3 pathway, but it produces more sugar for a given leaf area, making C4 plants grow faster than C3 plants. C4 plants can also tolerate higher light intensities and lower CO_2 concentrations than C3 plants. Most C4 plants are either grasses or plants that grow in desert. The grasses include sugarcane and corn (maize).

Some desert plants, including cacti and pineapple, have evolved a third pathway. Plants exchange gases through their STOMATA, but water vapor also passes through the stomata if they are open in bright sunshine, when the rate of EVAPORATION is high. To minimize the loss of water, these plants keep their stomata closed during the day and open them for gas exchange at night. CO_2 enters at night, combines with PEP to produce oxaloacetic acid, and this is then converted into malic acid. The carbon is stored as malic acid until the next day, when it is broken down, releasing CO_2 which enters the Calvin cycle.

This method of photosynthesis was first observed in plants of the family Crassulaceae, and it is known as *crassulacean acid metabolism* (CAM). The Crassulaceae are succulent herbs and small shrubs that are found mainly in warm, dry regions. The family includes the stonecrops and houseleeks.

phreatic water   See GROUNDWATER.

phreatic zone   See GROUNDWATER.

Phyllis   A TYPHOON that struck the Japanese island of Shikoku in August 1975. It killed 68 people. A week later the island was struck by Typhoon RITA.

physical climatology   The branch of CLIMATOLOGY that deals with exchanges of mass and energy in the TROPOSPHERE, that is to say, with the physical processes that produce and regulate climate. This is contrasted with DYNAMIC CLIMATOLOGY, which is concerned with motion and dynamic processes, and SYNOPTIC CLIMATOLOGY, which relates the circulation of the atmosphere to the differences in climates. Physical climatologists study the amount of solar radiation received at a particular place and the processes by which it is converted into other forms of energy, such as KINETIC ENERGY and chemical energy, and finally returned to space as INFRARED RADIATION. Many of these processes involve the CONDENSATION, EVAPORATION, DEPOSITION, and SUBLIMATION of water, and it is the absorption of solar energy that drives the HYDROLOGICAL CYCLE. All of these responses of water to the gain and loss of energy are central to physical climatology. The study of URBAN CLIMATES also forms an increasingly important part of physical climatology.

physical meteorology   The branch of METEOROLOGY that is concerned with the physical processes involved in producing the day-to-day weather. These include such phenomena as evaporation and CONDENSATION and the formation of clouds and FOG, the mechanisms by which CLOUD DROPLETS grow and cause PRECIPITATION, the separation of electrical charge that leads to THUNDERSTORMS, and the development of SUPERCELLS, MESOCYCLONES, and TORNADOES.

Physical Oceanography Distributed Active Archive Center (PO.DAAC)   The branch of the Data Information System of the EARTH OBSERVING SYSTEM that is responsible for storing and distributing information about the physical state of the ocean. Most of the data held at the PO.DAAC were obtained from satellites. They are technical and intended for research and educational use. The data are available to anyone free of charge, but must not then be sold. The center is part of the National Aeronautics and Space Administration (NASA) and is located at the Jet Propulsion Laboratory, California Institute of Technology.

(You can learn more about the PO.DAAC and its products at podaac.jpl.nasa.gov/in.)

physisorbent   See ADSORPTION.

phytoclimatology   The scientific study of the climatic conditions in the air between and adjacent to growing plants and on the surfaces of plants.

pibal   See PILOT BALLOON.

PICASSO-CENA   A joint U.S.–French satellite that is due to be launched in 2003. It will be commanded and monitored from a center in France and its data will be transmitted to the National Aeronautics and Space Administration (NASA) Langley Research Center at Hampton, Virginia. The satellite will be placed in a Sun-synchronous POLAR ORBIT at a height of 438 miles (705 km) and will fly in formation with one of the satellites of the EARTH OBSERVING SYSTEM. It will use a two-wavelength laser to measure AEROSOLS, thus providing high-vertical resolution profiles of aerosol properties. The acronym stands for *Pathfinder Instruments for Cloud and Aerosol Spaceborne Observations—Climatologie Etendue des Nuages et des Aerosols*.

(More information can be found at http://graphix2.larc.nasa.gov/gs/project/picasso/implement.html.)

piezoresistance barometer *See* BAROMETER.

pigs seeing the wind An old English country belief holds that pigs can see the wind. When gales are imminent they become very restless, running around their sties and scattering their bedding.

pileus An accessory cloud (*see* CLOUD CLASSIFICATION) that extends horizontally for only a short distance but forms a smooth, thin covering above or attached to the top of a CUMULIFORM cloud, like a cap or hood. It is most often seen while the main cloud is developing.

Pileus is the Latin name for a felt cap.

pilot balloon (pibal) A weather balloon (*see* BALLOON SOUNDING) that is filled with a measured amount of hydrogen to ensure that it ascends at a predetermined rate. As it rises, the balloon is tracked by a THEODOLITE. At intervals, the altitude of the balloon is calculated from its known rate of ascent and its AZIMUTH angle is read from the THEODOLITE. From this information the WIND VELOCITY is calculated for each height. If the sky is obscured by cloud, the height of the CLOUD BASE can be measured.

pilot report A description of the current weather conditions that is radioed to air traffic control or to a meteorological center by the pilot or other crew member of an aircraft. The report may consist of nothing more than the height of the CLOUD BASE or CLOUD TOP or the presence of CLEAR AIR TURBULENCE. A full pilot report should contain, in this order, the extent or location of the reported conditions; the time they were observed; a description of the conditions; the altitude of the conditions; and, in the case of a report of clear air turbulence or ICING, the type of aircraft.

Pinus longaeva *See* BRISTLECONE PINE.

PIOCW *See* PACIFIC- AND INDIAN-OCEAN COMMON WATER.

pitot–static tube *See* PITOT TUBE.

pitot tube (**pitot–static tube**) A device that is used to sample air in order to measure the speed with which the air is moving (or the speed with which the pitot tube is moving in relation to the air surrounding it) and the atmospheric pressure. Pitot tubes are used at weather stations, but their most widespread use is on aircraft. Every aircraft carries a pitot tube to supply air to its ALTIMETER, airspeed indicator, and vertical speed indicator (which shows the rate at which the aircraft is climbing or descending).

It comprises two thin-walled tubes, one enclosing the other, and mounted so that the ends of the tubes face into the airstream. Strictly, it is only the tube used to measure the speed of the airflow that should be called a pitot tube. This was invented by the French physicist Henri Pitot (1695–1771). The second tube is a static tube. The end of the pitot tube that faces into the airstream is open. The end of the static tube is closed, but there is a belt of small holes around its circumference a distance from the forward end of the tube that is equal to not less than five times the diameter of the tube. Usually the pitot tube is housed inside the static tube.

Air enters the pitot tube through the open end and enters the static tube through the belt of small holes. Pipes conduct the sampled air from both tubes to the instruments.

When air enters the pitot tube it is brought to rest, exerting a pressure of $1/2\ \rho v^2$, where ρ is the density of the air and v is its speed. The air pressure in the static tube is exactly equal to the external atmospheric pressure.

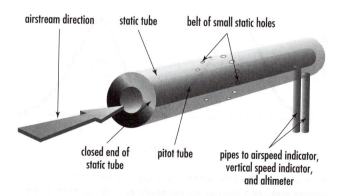

airstream direction   static tube   belt of small static holes

closed end of static tube   pitot tube   pipes to airspeed indicator, vertical speed indicator, and altimeter

Pitot tube. The outer, static tube encloses the inner, pitot tube. Air from both tubes is fed to the instruments that indicate wind speed or, in an aircraft, airspeed, altitude, and rate of climb or descent (vertical speed).

The WIND SPEED or airspeed is calculated from the difference in pressure between the two tubes. In an aircraft this is shown on the airspeed indicator. Changes in the pressure in the static tube indicate that the tube is changing its altitude and are used to calculate the vertical speed, which is shown on the vertical speed indicator. Comparing the pressure in the static tube with the sea-level pressure gives the height above sea level and is shown as altitude on the ALTIMETER.

pixel   *See* DIGITAL IMAGE.

plage   A bright area on the CHROMOSPHERE of the Sun that is associated with increased emission of radiation in the X-RAY, extreme short-wave ULTRAVIOLET, and radio wavelengths.

Planck's law   A description of the relationship between the wavelength of electromagnetic radiation and the temperature of the body emitting it that was first stated in 1900 by the German physicist Max Planck (1858–1947). The law states that the intensity of radiation emitted at a given wavelength is determined by the temperature of the emitting body. This can be written as

$$E_\lambda = c_1/[\lambda^5(\exp(c_2/\lambda T) - 1)]$$

where E_λ is the amount of energy (expressed in watts per square meter per micrometer of wavelength); λ is the wavelength (in micrometers); T is the temperature in kelvins; c_1 is the first radiation constant, with the value 3.74×10^{16} W m^{-2}; and c_2 is the second radiation constant, with the value 1.44×10^{-2} m K.

plane of the ecliptic   An imaginary disk, the circumference of which is defined by the path the Earth follows in its orbit about the Sun. Each day of the year, the noonday Sun is at a slightly different position in the sky from the one it was in on the preceding day. If its position is plotted for every day of the year on a picture of the landscape that shows a clear view in the direction of the equator, its varying positions appear to follow a path across the sky. This path marks the ecliptic and the disk it encloses is the plane of the ecliptic. If the rotational axis of the Earth were normal (at right angles) to the plane of the ecliptic the position of the noonday Sun would be the same on every day of the year and, therefore, there would be no SEASONS. In fact, however, the axis is tilted. At present, its angle to the ecliptic is 66.5°, so it is tilted 23.5° from the normal (90 − 66.5 = 23.5), but this angle changes in the course of a cycle with a period of about 41,000 years (*see* MILANKOVICH CYCLES). Latitudes 23.5° N and S mark the TROPICS of Cancer and Capricorn and latitudes 66.5° N and S mark the ARCTIC and ANTARCTIC CIRCLES.

planetary boundary layer (atmospheric boundary layer, surface boundary layer)   The lowest part of the atmosphere, where the movement of the air is strongly influenced by the land or sea surface. Friction causes eddies, making the flow of air turbulent (*see* TURBULENT FLOW). The depth of the planetary boundary layer varies from place to place and from time to time, but it is usually less than about 1,700 feet (519 m). The air above the planetary boundary layer constitutes the FREE ATMOSPHERE.

planetary vorticity (*f***)**   The VORTICITY about a vertical axis that a mass of fluid moving on the surface of the Earth possesses by virtue of the rotation of the Earth. Its magnitude is equal to that of the CORIOLIS EFFECT, so it is zero at the equator and at its maximum at the North and South Poles. Planetary vorticity is positive (causing the fluid to flow counterclockwise) in the Northern Hemisphere and negative (causing the fluid to flow clockwise) in the Southern Hemisphere.

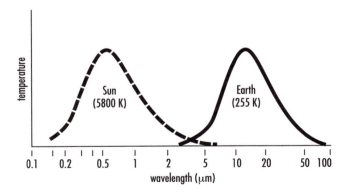

Planck's law. **The intensity of the radiation emitted at different wavelengths by a body at the temperature of the surface of the Sun compared with that emitted by a body at the temperature of the surface of the Earth. Obviously, the Sun emits far more radiation than does the Earth, but for ease of comparison the graphs assume the same amount for both bodies.**

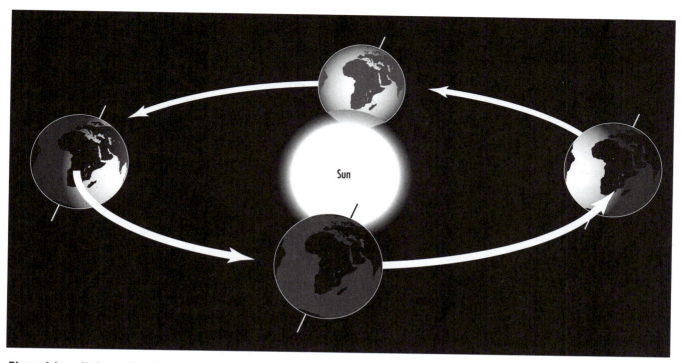

Plane of the ecliptic. An imaginary disk, the edge of which marks the Earth's orbital path around the Sun.

planetary waves *See* ROSSBY WAVES.

planetary wind Any wind that has a speed and direction caused wholly by the interaction of solar radiation and the rotation of the Earth.

plane wave A WAVE FRONT that is not curved because of its distance from the source. As waves spread outward, like ripples on a pond, their circumferences increase in size until a point is reached at which any short section of the wave front is effectively straight. It is then a plane wave.

plasma One of the four states of matter (the others are gas, liquid, and solid) in which a gas is ionized (*see* IONIZATION). The KINETIC ENERGY of particles in a plasma exceeds the energy of attraction (POTENTIAL ENERGY) between particles that are close together. Electrons move rapidly among the particles, neutralizing any net charge, so that each charged particle is surrounded by a cloud of particles with an opposite charge and the electric forces within the plasma are low. These clouds overlap, and consequently each par-

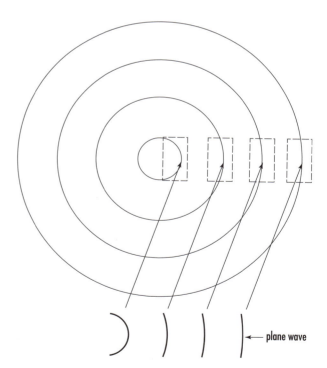

Plane wave. A short arc taken from the outermost circle appears almost straight because of the length of the circumference. This is a plane wave.

ticle is linked to many others. Particles rarely collide. Plasmas occur naturally in the atmospheres of stars, including the Sun, and constant bombardment by the charged particles of the SOLAR WIND continuously creates a plasma in the region of space immediately surrounding Earth. The MAGNETOSPHERE consists entirely of plasma.

plate tectonics The theory that describes the surface of the Earth as a number of solid sections that are able to move in relation to one another, thereby causing the deformation of rocks and the production of new structures. The sections are called *plates,* and *tectonics* (from the Greek *tektonikos,* meaning "carpenter") is a geological term referring to rock structures and the forces that produce them.

The theory of plate tectonics explains the presence of features that must have formed under conditions very different from those of today. Limestone rocks, for example, are abundant in many parts of the world. Limestone forms only by the heating and compression of sediments on the floor of a shallow sea. Consequently, limestone regions must once have been covered by sea. Coal measures can form only in mud beneath shallow coastal waters in the Tropics. Areas where coal is found, such as the northern United States, northern Europe, Russia, and China, must once have lain close to the equator. Other places, such as parts of Devon in southwestern England, have rocks of a type that forms only in hot, dry deserts.

The theory developed slowly over a long period. Geographers had speculated about the fact that the shape of the continent of Africa looks as though it would fit snugly against Central and South America but assumed this was mere coincidence. Then, in 1879, Sir George Darwin (1845–1912, a son of Charles Darwin) suggested that the Moon might have formed by breaking away from the Earth. Geologists believed the

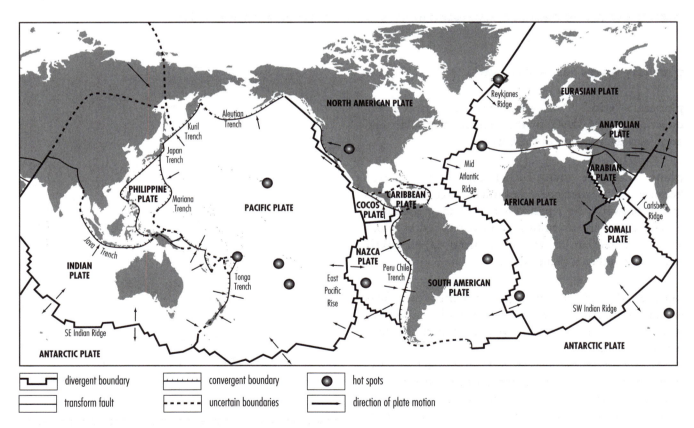

Plate tectonics. **The major plates into which the Earth's crust is divided, with ridges at which plates are separating and new rock is being added and trenches at which one plate is being subducted beneath its neighbor.**

mantle, beneath the solid rocks of the Earth's crust, is liquid, and in 1882 and 1889 the Reverend Osmond Fisher (1817–1914) said that the Pacific Ocean might fill a basin caused by the removal of the rocks that formed the Moon. Osmond thought the continents on each side of the Pacific might have moved together to close the gap and that this movement might have caused a split that widened to form the Atlantic. The most comprehensive proposal for the motion of continents was made some years later by ALFRED WEGENER, but by then the mantle was believed to be solid and his idea was dismissed.

Support for Wegener's idea began to grow in the middle 1940s, when scientists first acquired the technological means to study the floor of the deep oceans. The most important development was the discovery that rocks on each side of the ridges running across all the major oceans formed distinct bands of reversed magnetic polarity. Scientists knew that from time to time the Earth's magnetic field reverses its polarity, so that north becomes south and south north. Mineral grains in molten rock align themselves with the magnetic field and as the rock solidifies their magnetic orientation becomes fixed. The magnetic bands in the rocks of the ocean floor matched each other on each side of the ridges. This suggested that over a long period new, molten rock had emerged from the ridges and solidified and that the ocean floor had moved away from the ridges to accommodate the new rock. In 1963 an American naval oceanographer, Robert Sinclair Dietz (1914–95), called this *seafloor spreading*. It was in 1967 that Professor Dan McKenzie (born 1942) of the University of Cambridge drew all the different strands of evidence together and proposed the theory of plate tectonics.

The theory proposes that the Earth's crust is formed from a number of major and minor plates. Geologists differ in the way they classify some of the plates, but most now consider that there are eight major plates: the African, Eurasian, Pacific, Indian, North American, South American, Antarctic, and Nazca plates. The Cocos, Caribbean, Somali, Arabian, Philippine, and Scotia plates are smaller and are classed as lesser plates. In addition, there are minor plates such as the Juan de Fuca and Gorda plates, as well as microplates and fragments of former plates. The plates are made from solid rock, and they move because of convection currents in the hot rock of the mantle. In

some places where the oceanic crust is thin there are hot spots, where volcanoes are especially active. Earthquakes and volcanism are also common in the vicinity of plate margins.

Plate margins are of several types. At divergent, or constructive, margins plates are moving apart and new crustal rock is being added. At convergent, or destructive, margins plates are moving toward each other and the denser oceanic rock is being drawn beneath the lighter continental rock. This process is called *subduction*. At other margins the plates are moving parallel to each other but in opposite directions, a process that produces a type of rock fracture called *transform faulting*.

Plate movements cause the continents to change their positions. About 200 million years ago, for example, there was just one "supercontinent" called *Pangaea* (from the Greek *pan*, meaning "all," and *gi*, meaning "Earth") and one ocean, called *Panthalassa* (*thalassa* means "sea"). An arm of the sea, called *Tethys*, penetrated deeply into Pangaea, partly separating it into a northern part, called *Laurasia*, that contained all the present northern continents, and a southern part, called *Gondwana*, containing the southern continents. About 180 million years ago Pangaea began to break apart. North America broke away first from Africa and about 150 million years ago from Europe.

(You can learn more about plate tectonics at vulcan.wr.usgs.gov/Glossary/PlateTectonics/description_plate_tectonics.html.)

Pleistocene The geological epoch during which the most recent sequence of GLACIATIONS occurred began about 1.64 million years ago and ended about 10,000 years ago, when the WISCONSINIAN GLACIAL ended. The Pleistocene epoch was followed by the HOLOCENE EPOCH, in which we are living today. Together these two epochs constitute the Quaternary subera or the Cenozoic era of geological time (*see* GEOLOGICAL TIME SCALE). Technically, the commencement of the Pleistocene is dated from the first appearance in sedimentary rocks of fossils of a bottom-dwelling foraminiferan (foraminifera are animals resembling amoebae, but with shells) called *Hyalinea baltica*.

When the Pleistocene was first identified, its climate was thought to have been uniformly cold throughout. It was equated with the "Great Ice Age," the existence of which was discovered by LOUIS AGAS-

SIZ. In fact, the climate was much more complex. About 3 million years ago, the continents of North and South America joined, closing a seaway that had previously separated them. This created conditions that allowed glaciations to develop in the Northern Hemisphere. The first of these occurred about 2.36 million years ago (during the Pliocene epoch of the Tertiary subera). During the Pleistocene glaciations occurred at intervals of about 100,000 years, and their onset and ending were driven by the astronomical events of the MILANKOVICH CYCLES. During each glaciation ICE SHEETS extended approximately to a line running from Seattle to New York and from London to Berlin and Moscow. Sea levels fell to about 395 feet (120 m) lower than they are today, because of the large amount of water held as ice.

Between glaciations there were INTERGLACIALS. These were episodes of warmer climates lasting an average of 10,000 years. During some interglacials, including the SANGAMONIAN INTERGLACIAL, which was the one prior to the present Flandrian interglacial, temperatures were markedly higher than those of today. About 528,000 cubic miles (2.2 million km³) of ice melted from the West Antarctica ice sheet during the Sangamonian, raising the sea level to about 20 feet (6 m) above its present level, but the East Antarctica and Greenland ice sheets remained intact.

Pleistogene The period of geological time (*see* GEOLOGICAL TIME SCALE) that began about 1.64 million years ago and that continues to the present day. It constitutes the whole of the QUATERNARY subera.

Plinian eruption *See* VOLCANO.

Pliocene The epoch of geological time (*see* GEOLOGICAL TIME SCALE) that began about 5.2 million years ago and that ended about 1.64 million years ago. It constitutes the most recent epoch of the NEOGENE period.

plume rise The height to which a CHIMNEY PLUME rises after leaving the top of the smokestack. The effective stack height is equal to the sum of the height of the stack and the plume rise. Plume rise depends on the height, internal diameter and shape, and diameter at the mouth of the stack; on the temperature and exit velocity of the plume; and on the prevailing wind speed and LAPSE RATE.

plum rains *See* MAI-U.

pluvial A prolonged period of increased precipitation that affects a large region. It is caused by increased evaporation from the ocean and is associated with generally warmer conditions. Pluvial periods are separated by drier interpluvial periods.

pluvial lake A lake that formed during a time of increased rainfall and later disappeared. Lakes that formed during the PLEISTOCENE glacial advances and of which only traces now remain are the best known examples. The lakes formed during warmer episodes, when rainfall increased, glaciers retreated, and rivers flowed. In North America, lakes in the Great Basin expanded to form large inland seas, such as LAKE BONNEVILLE and Lake Lahontan, and extensive lakes also formed in East Africa. All of these lakes reduced greatly in size during the interpluvial periods that coincided with the INTERGLACIALS and separated the pluvials.

PMO *See* PORT METEOROLOGICAL OFFICER.

PO.DAAC *See* PHYSICAL OCEANOGRAPHY DISTRIBUTED ACTIVE ARCHIVE CENTER.

point discharge An upward flow of IONS carrying positive charge from tall objects such as trees and buildings that is induced by the negative charge at the base of a cloud that is producing a thunderstorm. This is the more important of the two processes that produce a negative charge at the surface of the Earth (the other is LIGHTNING) and in this way replenish the charge that is lost as positive ions are conducted through the air downward from the IONOSPHERE. In the absence of point discharges and, to a lesser extent, lightning strokes, within about 15 minutes the surface would acquire a positive charge equal to that in the ionosphere and the Earth's electrical field would break down. Occasionally a point discharge is strong enough to be visible as a glow, known as *Saint Elmo's fire*, around the structure from which it flows. This is sometimes seen near the top of a ship's mast.

point rainfall The amount of rain that falls over a specified period into a particular RAIN GAUGE, or the amount that is estimated to have fallen during that period at a particular place. The point rainfall often refers to the amount of rain that fell during a single storm, or during a period of unusually wet or dry weather.

Poisson's equation An equation from which it is possible to calculate the temperature of a PARCEL OF AIR at any height, provided the air is on a DRY ADIABAT. The equation is

$$(T \div \Phi) \times c_p \div R = p \div p_0$$

where T is the actual temperature, Φ is the POTENTIAL TEMPERATURE, R is the specific gas constant of air, c_p is the specific heat at constant pressure, p is the pressure at the position of the air parcel, and p_0 is the surface pressure. The equation was devised by the French mathematician Siméon Denis Poisson (1781–1840).

polacke A cold, dry, KATABATIC WIND that blows in winter over northern Bohemia, in the Czech Republic. It descends from the Sudeten (Polish, Sudety) Mountains, carrying air from Poland.

polar air Cold air that originates in the high-pressure regions of Siberia, northern Canada, and the Southern Ocean (*see* SOURCE REGION). In winter, a CONTINENTAL polar (cP) AIR MASS covers all of Eurasia north of the Himalayas, with the exception of western Europe and North America from the far north of Canada (where cP air gives way to continental ARCTIC AIR) to the south of the Great Lakes. The air is stable and produces COLD WAVES. As it passes over the lakes it is modified to cPk air, producing LAKE-EFFECT SNOW. There is no cP air mass over the Antarctic in winter or in summer. MARITIME polar (mP) air forms in both winter and summer over the North Atlantic and North Pacific Oceans and over the northern part (to the north of the mA air) of the Southern Ocean. In North America, mP air from the North Pacific produces mild, humid conditions at all times of year, often with SHOWERS in winter. The air is more stable in summer and produces low STRATUS cloud and FOG near coasts.

polar-air depression A type of NONFRONTAL DEPRESSION that occurs only in the Northern Hemi-

sphere. It forms when unstable arctic or polar MARITIME AIR moves southward along the eastern side of a large RIDGE extending along a north–south line.

polar automatic weather station An AUTOMATIC WEATHER STATION that is designed to operate in extremely cold climates. The instruments are mounted on a sled with pontoons on each side to provide additional support.

polar cell That part of the atmospheric circulation in which air subsides over high latitudes, flows away from the Pole at low level, then rises in middle latitudes, where it meets the FERREL CELL at the POLAR FRONT. It then flows back toward the Poles, completing the vertical cell. Subsiding air produces high surface pressure and DIVERGENCE. The flow of air away from the poles

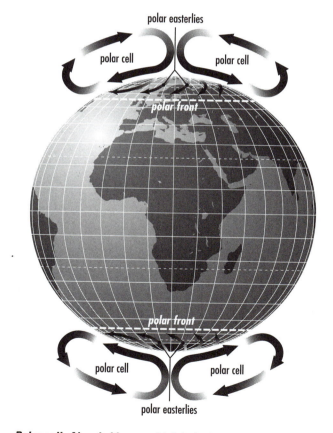

Polar cell. Air subsides over high latitudes and diverges at the surface, flowing away from the Poles and producing the polar easterlies. Where the air encounters the polar front it rises and flows back toward the Poles.

is subject to the CORIOLIS EFFECT. This deflects the air in a westerly direction, producing a belt of easterly winds, the polar easterlies. Together, the Ferrel cell, HADLEY CELL, and polar cell constitute the THREE-CELL MODEL of meridional atmospheric circulation.

polar climate A high latitude climate in which the mean monthly temperature remains below freezing throughout the year. Lichens and mosses may grow sparsely where the land surface is free of ice and snow, but otherwise there is no vegetation. There is no land at the North Pole. There, some thawing of the sea ice occurs in summer, producing slushy areas and stretches of open water.

During summer the hours of daylight are long. Everywhere inside the ARCTIC CIRCLE and ANTARCTIC CIRCLE there is at least one day when the Sun does not sink below the horizon. At the Poles themselves this period of "midnight Sun" lasts about six months. Despite this, the Sun remains fairly low in the sky and the sunlight is not very intense. During winter the days are very short. At Murmansk, in northern Russia, the Sun remains above the horizon for 70 days during the summer but is below the horizon from November 26 until January 20. Nevertheless, SCATTERING and REFRACTION allow some sunlight to reach the surface when the Sun is less than 18° below the horizon. This dim light that results is called *astronomical twilight*. There are also moonlight, starlight, and AURORAS to provide light, so the darkness is seldom total.

At Eismitte, the station at an elevation of 9,941 feet (3,032 m) near the center of the GREENLAND ICE SHEET that was established by ALFRED WEGENER, the mean annual temperature is –22° F (–30° C). The coldest month is February, when the mean temperature is –53° F (–47° C), and the warmest month is June, when the mean temperature is 4° F (–15.5° C). At Little America, close to sea level in Antarctica, the mean annual temperature is –14° F (–25.5° C). July is the coldest month, with a mean temperature of –39° F (–39.4° C), and the warmest month is February, when the mean temperature is 7° F (–13.9° C).

Together, the amount of solar radiation that is reflected by the snow and ice over the Arctic Ocean and the amount of heat emitted by the sea as INFRARED RADIATION exceed the amount reaching the surface from the Sun by 60 percent. Temperatures over the ocean are much higher than this would suggest, however, because ocean currents carry in warm water. The water then releases heat by radiation and CONVECTION. Water from the GULF STREAM keeps the Norwegian and Barents Seas open throughout the winter. Even in the coldest areas of the Arctic Ocean, the sea-surface temperature does not fall below 29° F (1.7° C).

In winter, the sea ice covers an area of about 4.5 million square miles (11.7 million km²). The ice drifts from east to west and this movement causes it to crack, leaving open spaces, called *leads,* even at the North Pole. In summer, the ice melts from the coasts of the surrounding continents until it occupies about 3 million square miles (4.8 million km²). For about two months in summer the ocean is sometimes mainly open, with patches of drifting ice.

There are three principal types of polar climate, known as *POLAR WET, POLAR WET-AND-DRY,* and *POLAR DRY* (see ICECAP CLIMATE, MARINE SUBARCTIC CLIMATE, and TUNDRA CLIMATE).

polar desert An area inside the ARCTIC or ANTARCTIC CIRCLE where the annual precipitation is very low, although most of the surface is covered with snow and ice. The low precipitation is due mainly to two factors. The first is that air is subsiding and diverging at the surface (see POLAR CELL). This air previously ascended to the height of the TROPOPAUSE along the POLAR FRONT. As it rose its temperature fell and it lost most of its moisture, so by the time it subsides again it is very dry. Its low-level DIVERGENCE prevents moister air from entering. The second factor is the low air temperature over the Arctic and Antarctic regions, which prevents the air from holding much moisture.

(You can learn more about the polar deserts in Michael Allaby, *Ecosystem: Deserts* [New York: Facts On File, 2001].)

polar dry climate The climate that is found over the continent of Antarctica away from the coast. It is produced by continental polar (cP) air (see CONTINENTAL AIR and POLAR AIR) and is type EF in the KÖPPEN CLIMATE CLASSIFICATION and EF' in the THORNTHWAITE CLIMATE CLASSIFICATION. Over most of Antarctica more energy is lost from the surface by reflection (see ALBEDO) and INFRARED RADIATION (see BLACKBODY RADIATION) than is received from the Sun. This is possible because warmth that is carried into the region by ocean currents and the circulation of the atmosphere is lost to space. Consequently, Antarctica is the part of the world where the loss of surplus heat helps maintain a constant global mean temperature.

There are several reasons for the extremely low temperatures over Antarctica. It is a continent and therefore does not benefit from the moderating influence of MARITIME AIR (*see* CONTINENTALITY). Also, its elevation is generally high. At the South Pole the surface of the ICE SHEET is 10,000 feet (3,000 m) above sea level, and VOSTOK STATION is at an elevation of 11,401 feet (3,475 m). Finally, the South Pole receives 7 percent less solar radiation during its winter than the North Pole receives during its winter. This is because the elliptical orbit of the Earth places the South Pole 3 million miles (4.8 million km) farther from the Sun at the June SOLSTICE than the North Pole is at the December solstice. The coldest temperatures occur a few days after the Sun has risen above the horizon.

At the Amundsen–Scott Station, at the South Pole, the mean annual temperature is –56.8° F (–49.4° C). August is the coldest month, with a mean temperature of –75.9° F (–60.0° C). The warmest month is January, when the mean temperature is –18.7° F (–28.2° C). PRECIPITATION is very low, probably not exceeding an annual mean of 2 inches (50 mm), and the RELATIVE HUMIDITY can fall to 1 percent. Winds are strong, especially in winter. They are KATABATIC WINDS produced by air moving gravitationally away from the higher elevations often with enough force to overcome the PRESSURE GRADIENT. The winds are strongest when the pressure gradient coincides with the topographic gradient. Then they routinely exceed hurricane force (75 mph [121 km h^{-1}]).

polar easterlies   *See* POLAR CELL.

polar-easterlies index   A scale of values that allows the strengths of the polar easterlies (*see* POLAR CELL) to be compared. The index is calculated from difference in the average sea-level air pressure at latitudes 55° N and 70° N. The index is expressed as the east-to-west component of the GEOSTROPHIC WIND in meters per second.

polar front   The front that is located in middle latitudes and marks the boundary between polar air on the poleward side and tropical air on the equatorial side. This is also the boundary between the direct polar cell and indirect midlatitude cell in the THREE-CELL MODEL of the GENERAL CIRCULATION of the atmosphere. Prevailing winds are easterly in the polar air and westerly in the tropical air and air masses travel in the same direction as the winds. The polar front extends from the surface to the TROPOPAUSE, where it generates the POLAR FRONT JET STREAM. *See also* POLAR FRONT THEORY.

polar front jet stream   The JET STREAM that is associated with the POLAR FRONT. There is a steep temperature gradient across the polar front, where TROPICAL AIR and POLAR AIR meet. The gradient is at a maximum at the TROPOPAUSE, where the jet stream occurs in the tropical air on the side of the front nearer to the equator. The wind at the core of the jet stream can reach 95 mph (150 km h^{-1}) in summer and 185 mph (300 km h^{-1}) in winter.

polar front theory   An explanation for the way CYCLONES form (*see* CYCLOGENESIS) and cross the middle latitudes that was devised by VILHELM BJERKNES and his colleagues at the BERGEN GEOPHYSICAL INSTITUTE, in Norway, and described by JACOB BJERKNES in the article "On the Structure of Moving Cyclones," published in 1919. Using data obtained from balloon observations, the Bergen team proposed the existence of the POLAR FRONT. At the surface, they found that the polar front is often broken into sections that are separated by regions in which the temperature gradient is much shallower than it is in the FRONTAL ZONE. Waves develop in the sections of frontal zone, and the team coined the names *WARM FRONT* and *COLD FRONT* to distinguish the two types of front that enclose the FRONTAL WAVE. In 1923, Bjerknes and Halvor Solberg published another article, "The Life Cycle of Cyclones and the Polar Front Theory of Atmospheric Circulation" (in *Geofisiske Publikasjoner* 3, no. 1), in which they drew together their ideas about the polar front and the dynamics of cyclone formation. The combined theories described the formation and movement of midlatitude frontal cyclones in a context of the transport of heat away from the equator and the GENERAL CIRCULATION of the atmosphere. The polar front theory soon became, and has remained, a central feature of the science of meteorology, although several of its details have been revised in the light of more recent discoveries. It is now known to be relevant to the formation of frontal cyclones along other fronts, as well as along the polar front.

polar glacier   *See* COLD GLACIER.

polar high   The persistent region of high surface atmospheric pressure that covers the Arctic Basin and Antarctica. In winter it consists of continental arctic (cA) air (*see* CONTINENTAL AIR and ARCTIC AIR) over both polar regions. In summer the cA air continues to cover Antarctica, but maritime arctic (mA) air (*see* MARITIME AIR) covers the Arctic. The polar highs are the source of the polar easterlies (*see* POLAR CELL).

polar hurricane   *See* POLAR LOW.

polar ice   A layer of ice that forms a complete covering on a large area of the surface of the sea and that is more than one year old. Polar ice is less saline than WINTER ICE and YOUNG ICE, because in summer, when it partly melts, the spaces in which salt water are trapped open, allowing salt to drain away. Salinity is lowered further as a consequence of the low thermal conductivity of ice. This insulates the water below the ice, preventing its temperature from falling and so decreasing the rate at which ice accumulates on the underside of the surface layer. At the same time, snow falling on the surface remains there, diluting the salt content of the ice as a whole.

polar low (**polar hurricane**)   A small, intense CYCLONE that forms during winter in the cold AIR MASS on the side of the POLAR FRONT that is nearer the pole. Polar lows usually produce heavy hail and snow, and winds of up to gale force.

They are 125–500 miles (200–800 km) in diameter and develop when large amounts of very cold air spill out from the ice-covered continents across a markedly warmer sea. A small, upper-level TROUGH may also need to be present to trigger the disturbance that grows into the cyclone.

Once it has formed, a polar low is similar to a TROPICAL CYCLONE in many respects, although it is much smaller. Like a tropical cyclone, it is circular, generates strong winds, and has a cloud-free center surrounded by towering CUMULIFORM cloud sustained by CONVECTION and extending to the TROPOPAUSE. Air flows outward from the cyclone at high level, producing CIRRUS cloud. A polar low dissipates quickly when it crosses land. In other respects it is unlike a tropical cyclone. From its first appearance a polar low reaches its full strength within 24 hours or less. It travels at up to 30 knots (34.5 mph [55.5 km h⁻¹]), which is much faster than a tropical cyclone, and lasts no longer than 48 hours before it reaches land and dies.

Polar lows occur in many parts of the North Pacific and North Atlantic Oceans and also over the Tasman Sea, close to New Zealand. So far as is known, they rarely form over the Southern Ocean. They are most common in the Greenland, Norwegian, and Barents Seas, but sometimes they also form on the western side of Greenland and in the Beaufort Sea, to the north of Alaska.

polar mesospheric cloud   A name for NOCTILUCENT CLOUD that is preferred scientifically because it is more descriptive. Rather than suggesting a cloud that is visible at night, the preferred name indicates that it is seen only in polar latitudes and that it occurs in the upper MESOSPHERE.

polar molecule   A molecule in which the electromagnetic charge is separated, so that one end of the molecule carries a positive charge and the other end carries a negative charge, although the molecule as a whole is neutral. Because it carries charge at its ends, the molecule is a dipole. The water molecule is polar. This is because its two hydrogen atoms share their single electrons with the oxygen atom. Lines drawn from the two hydrogen atoms to the center of the oxygen atom meet at an angle of 104.5°, so both hydrogen

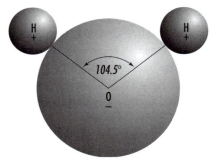

Polar molecules. **Water molecules (H_2O) are polar because their hydrogen atoms share electrons with the oxygen atom, and both hydrogens, bearing positive charge, are on the same side of the oxygen, which carries negative charge.**

atoms are positioned on the same side of the oxygen atom. This gives the oxygen side of the molecule a negative charge and the hydrogen side a positive charge, and it is this characteristic that allows water molecules to form HYDROGEN BONDS and liquid water to act as a very efficient solvent.

polar night vortex   *See* POLAR VORTEX.

polar orbit   A satellite orbit that passes close to the North and South Poles at an altitude of about 534 miles (860 km), which is one-seventh the radius of the Earth. A polar orbit may be fixed in relation to the position of the Sun (SUN-SYNCHRONOUS ORBIT), but forming an angle with the meridians, so the entire surface of the Earth is scanned in successive passes.

polar outbreak   An extension of POLAR AIR into lower latitudes. Polar outbreaks are often produced during the later stages of the INDEX CYCLE, when the JET STREAM follows a deeply undulating path. A

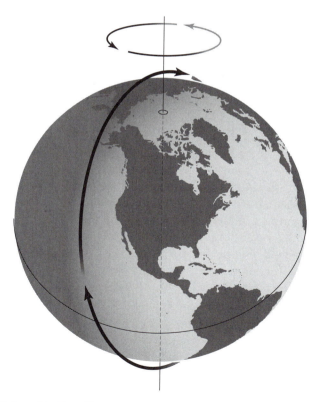

Polar orbit. **An orbit that carries a satellite close to or across the North and South Poles.**

TROUGH in the middle TROPOSPHERE usually extends over eastern North America in both summer and winter. It is possibly a LEE TROUGH resulting from the effect of the Rocky Mountains on the high-level westerly winds. The flow of air around the trough tends to carry polar air southward, especially when the trough is strong. When it is weak the westerly flow of air is stronger and polar outbreaks are less likely.

polar stratospheric clouds (PSCs)   Clouds that form in winter over Antarctica and, less commonly, over the Arctic. They occur in the STRATOSPHERE at a height of 9–15.5 miles (15–25 km). Usually they are too thin to be visible, but when the Sun is about 5° below the horizon they can sometimes be seen as NACREOUS CLOUDS. What is known about them has been discovered by means of LIDAR.

There are two principal types of PSC, known as *type 1* and *type 2*, and there are two or possibly three varieties of type 1 PSCs, types 1a, 1b, and 1c. All type 1 PSCs form at about 9 miles (15 km) altitude in air that is just above the frost point. At this height the frost point is about 195 K (–109° F [–78° C]). As the temperature falls below the frost point type 1 clouds begin to form rapidly, with very small ICE CRYSTALS acting as FREEZING NUCLEI. The source of the water vapor to produce ice crystals is not known, but it may result from the oxidation of methane (CH_4) to carbon dioxide (CO_2) and water (H_2O). The type 1 particles then grow rapidly as nitric acid (HNO_3) and more water vapor condenses onto them. The resulting solution may be either liquid or solid, depending on the conditions around it.

Type 1a clouds are believed to consist of irregularly shaped, liquid particles about 0.004 inch (0.1 mm) in diameter made from approximately one molecule of HNO_3 to three molecules of H_2O.

Type 1b clouds are made from much smaller (about 0.00004 inch [0.001 mm]) liquid particles that are spherical and probably made from a mixture of sulfuric acid (H_2SO_4) and HNO_3 dissolved in water. There may also be type 1c clouds, made from solid crystals of HNO_3 and water.

Type 2 PSCs are made from ice crystals. They form at lower temperatures and, therefore, at a greater altitude, most commonly at around 15.5 miles (25 km), where the temperature is about 188 K (–121° F [–85° C]). Type 2 PSCs occur over Antarctica but are very rarely observed over the Arctic, where winter temperatures seldom fall low enough for them to form.

The chemical reactions involved in the removal of OZONE from the OZONE LAYER take place on the surface of PSC ice crystals. Chlorine (Cl) is present in the stratosphere in two forms that are fairly inert chemically. These are hydrochloric acid (HCl) and chlorine nitrate ($ClONO_2$). On the surface of PSC particles they are converted into the much more reactive forms Cl_2 and HOCl by the reactions

$$HCl + ClONO_2 \rightarrow HNO_3 + Cl_2$$

$$ClONO_2 + H_2O \rightarrow HNO_3 + HOCl$$

In both reactions the HNO_3 remains inside the cloud particles. HOCl then reacts further to release free atomic chlorine (Cl).

Chlorine then reacts to remove ozone (O_3) in a series of steps.

$$Cl + O_3 \rightarrow ClO + O_2$$

This reaction takes place twice, to yield two molecules of ClO.

$$ClO + ClO + M \rightarrow Cl_2O_2 + M$$

$$Cl_2O_2 + h\upsilon \rightarrow Cl + ClO_2$$

$$ClO_2 + M \rightarrow Cl + O_2 + M$$

where M is any air molecule and $h\upsilon$ is a quantum of solar energy of near-ultraviolet wavelength.

Nitrogen oxides (NO_x) are also removed by reactions on the surface of PSC particles. The process is called *denoxification* and it is important because nitrogen dioxide (NO_2) removes chlorine oxide (ClO), which is a key ingredient in the ozone-depletion process, by the reaction

$$ClO + NO_2 + M \rightarrow ClONO_2 + M$$

NO_2 changes back and forth into gaseous N_2O_5:

$$4NO_2 + O_2 \leftrightarrow 2N_2O_5$$

Denoxification then removes gaseous nitrogen oxides by the reactions

$$N_2O_5 + H_2O \rightarrow 2HNO_3$$

$$N_2O_5 + HCl \rightarrow ClNO_2 + HNO_3$$

Nitric acid is also removed, because as the PSC particles grow bigger their weight increases and they start to settle out of the stratosphere.

(You can find out more about polar stratospheric clouds from www.atm.ch.cam.ac.uk/tour/psc.html and www.awi-potsdam.de/www-pot/atmo/psc/psc.html.)

polar trough   A TROUGH in the upper TROPOSPHERE that extends toward the equator far enough to reach the TROPICS. The part of this high-level cold air that is closest to the equator sometimes becomes separated from the main part of the trough, to form a CUT-OFF LOW with a cold center. Air flows in a CYCLONIC direction around the center, and if this flow extends to the surface it triggers a SUBTROPICAL CYCLONE.

polar vortex   The large-scale circulation that dominates the middle and upper TROPOSPHERE in high latitudes and that is centered in polar regions of both hemispheres. Air circulates in a CYCLONIC direction around the vortex. In the Northern Hemisphere the vortex has two centers, one near Baffin Island and the other over northeastern Siberia.

A vortex also forms in winter in the STRATOSPHERE, with a strongly GEOSTROPHIC WIND circulating around it. This is sometimes called the *polar night vortex* and it is within this vortex that POLAR STRATOSPHERIC clouds form. These are the site of the chemical reactions by which OZONE is destroyed in the OZONE LAYER.

polar wet-and-dry climate   The climate that is typical of the coastlines surrounding the Arctic Ocean, along the northern coasts of Canada, Alaska, and Eurasia, and around Greenland, Iceland, and the smaller northern islands. This climate also occurs over the islands in the Southern Ocean. About 5 percent of the total land area of the Earth experiences this type of climate. It is produced by maritime polar (mP) and continental polar (cP) air (*see* MARITIME AIR, CONTINENTAL AIR, and POLAR AIR) and supports tundra vegetation, comprising lichens, mosses, sedges, grasses, herbs, shrubs, and low-growing trees.

Winters are cold, summers cool, and most of the PRECIPITATION falls in summer, so the summer is cloudy and wet and the winter is dry, with generally clear skies. The RELATIVE HUMIDITY averages 40–80 percent in summer and 40–60 percent in winter. It is an ET climate in the KÖPPEN CLIMATE CLASSIFICATION and E' in the THORNTHWAITE CLIMATE CLASSIFICATION. The mean annual temperature is below freezing, but temperatures rise above freezing in summer. During the summer, which lasts for an average of two to three months but in some places as long as five months, mean temperatures seldom exceed 41° F (5° C), although on some days the temperature may rise to about 80° F (27° C). Winds are usually light and variable in direction and storms are rare. Subsiding air on the high-latitude side of the POLAR CELL produces high surface pressure and fairly still air, especially in winter.

polar wet climate The climate that is typical of the Southern Ocean surrounding Antarctica. It is produced by maritime polar (mP) air (*see* MARITIME AIR and POLAR AIR) and is type Em in the KÖPPEN CLIMATE CLASSIFICATION and type AE' in the THORNTHWAITE CLIMATE CLASSIFICATION. The RELATIVE HUMIDITY is always high, with a yearly average of 50 percent, and the sky is usually cloudy. Cloud cover averages 80 percent during the winter and is rather less in summer. The amount of PRECIPITATION varies from place to place, depending on the latitude, but there is at least a 25 percent chance of precipitation on any day and in any one place there is little seasonal variation. Precipitation amounts range from 14.6 inches (370 mm) to 115 inches (2,920 mm). Summer temperatures can rise to 50° F (10° C). In winter the temperature averages 19°–32° F (–7 to 0° C). Storms are common, especially in winter. The gales associated with these storms led sailors to call the latitudes in which they occur the *roaring forties, furious fifties,* and *shrieking sixties.*

Pole of Inaccessibility The point in the Arctic Ocean that is farthest from land. It lies between the North Pole and Wrangel Island, off the coast of eastern Siberia, and is sometimes taken to be the center of the Arctic.

pollen The male reproductive cells of seed plants. Seed plants are plants that reproduce by producing seeds, rather than SPORES. The group includes the coniferous plants (gymnosperms) and the flowering plants (angiosperms). Pollen consists of individual pollen grains. These are produced in vast numbers. Many insect-pollinated plants produce pollen grains that are sticky or barbed, to make them adhere to the pollinator. Pollen from wind-pollinated plants is usually smooth. Most pollen lives for a very short time (a few hours in the case of grasses), but the protective outer coating, called the EXINE, is very tough and survives thousands of years under favorable conditions. The science of identifying and classifying pollen grains is called *PALYNOLOGY* and the interpretation of pollen and spores that are found in sediments is called POLLEN ANALYSIS; the two terms are often used synonymously.

pollen analysis The reconstruction of past climates and environments through the study of POLLEN grains and plant SPORES that are recovered from sediments. Pollen and spores bear surface markings, which allow the plants that produced them to be identified at least to the family level and sometimes to the genus or even species. The identification and classification of pollen and spores constituted PALYNOLOGY and pollen analysis is the interpretation of the record they leave; the two terms are often used synonymously.

A pollen grain is contained in a coat with two layers. The inside layer, called the *intine,* is soft; the outer layer, the EXINE, is very tough and often sculptured. The pollen grains of some plants have pores in the exine through which the intine protrudes and this adds to the markings on the exine. These sculptured shapes and markings are visible under a powerful microscope (usually at a magnification of at least ×300 and more often ×400 or ×1,000). They allow palynologists to identify the family of the plant that produced the pollen (for example, the birch family, Betulaceae). In some cases it is possible to go further and identify the genus (for example, an alder, *Alnus,* which is a genus of trees belonging to the birch family) or even the species (for example, green alder, *A. crispa*).

Once the pollen has been identified, its presence can be interpreted. The Betulaceae are a family that grows in cool or cold climates, so its pollen indicates that at one time the area had a climate like that of northern Canada or Eurasia, regardless of what the climate is like today. Alders grow near water, and so their pollen indicates that the ground was wet. It might have been a riverbank or the shore of a lake.

An interpretation as simple as this would be unreliable. It reads too much into the small amount of evidence. Pollen can travel, stuck to the skins or coats of

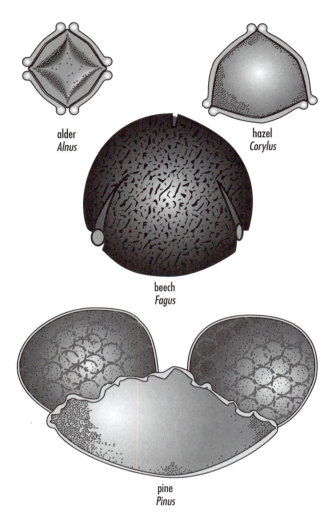

alder
Alnus

hazel
Corylus

beech
Fagus

pine
Pinus

Pollen analysis. **Pollen grains from four different tree genera, all drawn to the same scale. Note the air sacs on the pine pollen.**

animals or blown by the wind, so its presence in a particular place does not necessarily mean that that is where the plant actually grew. Conifers are wind-pollinated and produce pollen grains that have two air sacs to increase their buoyancy. This pollen can travel very long distances.

Plant species are never alone, however. There is always a community of them, and so where pollen is found it represents several species. That makes interpretation much more reliable, because of the improbability that pollen from a group of species would be brought together by chance. It is much more likely that the plants grew close to each other. They then

constituted a life assemblage. After they all died their pollen became a death assemblage. Its composition is different from that of the life assemblage, because the amount of pollen produced varies greatly from one species to another, the pollen grains themselves are distributed differently, and not all pollen survives equally well. The scientist analyzing the pollen allows for these factors.

The pollen is preserved by being buried in soil or lake sediment. It is best preserved under anaerobic, acid conditions, such as those of a peat bog or the bed of a lake. Samples are recovered by drilling vertically through the sediment to extract a core. Organic material found at carefully marked depths can then be dated by RADIOCARBON DATING. This gives a date for the pollen found at those depths. Several samples are needed from each site and it is usually necessary to count at least 200 pollen grains in each sample.

Pollen analysis began early in the 20th century. The first scientist to apply it was the Swedish geologist Lennart von Post.

(You can learn more about the history of pollen analysis in the United States from www.geo.arizona.edu/palynology/plns1295.html.)

pollen-assemblage zone *See* POLLEN ZONE.

pollen diagram A diagram that is used in paleoclimatology to illustrate the pattern of vegetation that occupied a site at particular times in the past. It is compiled from counts of the absolute or relative pollen frequency for certain species, genera, or families of plants. The pollen counts are made from samples of soil that are taken from different depths. Often the samples are obtained from a cliff face or exposed soil profile in a ditch or other cutting. In the diagram, pollen from each plant is shown as a vertical bar, plotted against depth with the ground surface at the top. The bar varies in thickness according to the amount of that pollen present at each level.

The resulting representation of vegetation types reveals the climatic conditions at different levels, because plants are sensitive to climatic change. If birch (*Betula*) and pine (*Pinus*) species predominate, for example, the climate was cold and similar to that of northern Canada and Siberia today. The appearance of

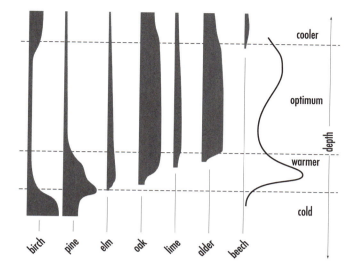

Pollen diagram. **A drawing that shows how the amount of pollen of particular plant genera changes with depth through a section cut vertically through the soil. This indicates the relative abundance of each genus at different times, from which climatic changes can be inferred.**

elm (*Ulmus*) and oak (*Quercus*) species at a higher level, accompanied by a decline in birch and pine, indicates that the climate was growing warmer. Species of basswoods, also called *lime* and *linden* (*Tilia*), grow only in warm temperate climates so their presence indicates a period of warmth. A decline in the abundance of these trees, accompanied by an increase in birch and the appearance of beech (*Fagus*) species, indicates cooler conditions.

pollen zone (pollen-assemblage zone) An assemblage of POLLEN grains and SPORES that is considered to be characteristic of a particular climate that was once the climate of a large region. The concept was introduced in 1940 by the English botanist Sir Harry Godwin, and although the zones he proposed are now known to overlap locally, they are still used.

Godwin proposed eight zones, identifying them with roman numerals. They extend from the latter part of the DEVENSIAN GLACIAL until the present day. The eight zones are summarized in the table:

pollution Any direct or indirect alteration of the properties of any part of the environment in such a way as to present an immediate or potential risk to the health, safety, or well-being of any living species. The alteration may be chemical, thermal, biological, or radioactive. Pollution is usually associated with human activities, but it can also occur naturally. Volcanic eruptions release large quantities of pollutants, for example, and many plants emit ISOPRENES and terpenes that contribute to the formation of OZONE and PHOTOCHEMICAL SMOG.

Chemical pollution occurs when substances are released that are toxic, such as tetraethyl lead, or that are harmless in themselves but engage in reactions that yield toxic products, such as PEROXYACETYL NITRATE. Thermal pollution occurs when gases or liquids are released at a markedly different (almost always higher) temperature than that of the medium into which they discharge. This can change the way air moves.

Years ago	Zone	Plants	Climate
14,000	I	creeping willow	Older Dryas; cold
12,000	II	birch	Allerød; milder
	III	creeping willow	Younger Dryas; cold
10,000	IV	birch, pine	Pre-Boreal; dry
	V	hazel, birch	Boreal; cool, dry
8,800	VIa	hazel, pine	Boreal; cool, dry
8,000	VIb	hazel, pine	Boreal; warmer, dry
	VIc	hazel, pine	Boreal; warmer, dry
7,500	VIIa	alder, oak, elm, lime	Atlantic; warm, moist
5,000	VIIb	alder, oak, lime	Sub-Boreal; cooler, drier
2,800	VIII	alder, birch, oak, beech	Sub-Atlantic; cool, wet

Discharges of warm water can harm aquatic animals by reducing the capacity of the water to hold dissolved oxygen. Biological pollution involves the release of harmful bacteria, viruses, or fungal spores. These can be carried long distances in the air. Radioactive pollution is caused by the release of radioactive substances that may emit IONIZING RADIATION. *See* AIR POLLUTION.

Pollution Standards Index (PSI) An internationally agreed scale that provides a measure of the AIR QUALITY at a particular place. The PSI compares the national air-quality standard with the amount present in the air of the pollutant that occurs at the highest concentration. If that amount is equal to that specified in the national standard the air is given a PSI value of 100. If the amount is less than the national standard, the air has a PSI value of less than 100 and its quality is considered to be moderate or good. If the amount is greater than the national standard, the PSI value is more than 100 and the air quality is poor. It is then graded from unhealthy to hazardous. A value of 200 is considered "very unhealthy," above 300 is "hazardous," and above 400 is "very hazardous." During the forest fires that swept Indonesia in 1997–98, the *Borneo Bulletin* reported that on April 12 the PSI reached 500.

Polly A TROPICAL STORM that caused a STORM SURGE with waves 20 feet (6 m) high on August 30 and 31, 1992, at Tianjin, China. It killed 165 people along the southeastern coast and rendered more than 5 million homeless.

polyn'ya (pl. *polynyi*) An area of open water that is surrounded by SEA ICE.

polynyi *See* POLYN'YA.

polytropic atmosphere A hypothetical atmosphere that is used in climate MODELS. It is in HYDROSTATIC EQUILIBRIUM and its temperature decreases with height at a constant LAPSE RATE.

ponding *See* FROST HOLLOW.

ponente A westerly wind that blows along the Mediterranean coast of France and in Corsica.

poniente A westerly wind that blows through the Strait of Gibraltar.

poor man's weatherglass *See* SCARLET PIMPERNEL.

poriaz (poriza) A strong, northeasterly wind that blows across the Black Sea in the region of the Bosporus, which is the strait that links the Black Sea with the Sea of Marmara and forms part of the boundary between Europe and Asia.

poriza *See* PORIAZ.

port meteorological officer (PMO) An official of a national meteorological service who is appointed to supervise the VOLUNTARY OBSERVING SHIPS (VOS) scheme. PMOs are based at ports and spend much of their time visiting participating ships. They are responsible for enrolling vessels into the VOS scheme, supervising the supply and installation of the necessary instruments and other equipment, collecting the meteorological logbooks from ships returning to port, and generally ensuring that the weather reports from ships at sea meet the required standard.

(You can learn more about PMOs at www.vos. noaa.gov/pmo11.html.)

positive axis A line that is drawn through the point of maximum curvature in the STREAMLINE of an EASTERLY WAVE. The term is most often used in connection with EQUATORIAL WAVES. The axis may be positive or negative. A positive axis indicates a trough in the Northern Hemisphere and a ridge in the Southern Hemisphere.

Postglacial *See* HOLOCENE EPOCH.

postglacial climatic revertence A period during which the climate became cooler and wetter than it had been previously. The change began abruptly about 2,500 years ago, and the cool, wet conditions have continued to the present day, marking the SUBATLANTIC PERIOD. The onset of the deterioration was marked by a decline in the number of lime trees (*Tilia* species) in England and Wales. Lime trees demand warm conditions. There was also a decline in the number of pine trees (*Pinus* species) in northern Britain, indicating that summers were cool.

(You can learn more about the history of the post-glacial climate from "The Flandrian: The Case for an Interglacial Cycle" at www.envf.port.ac.uk/geog/teaching/quatgern/q8b.htm.)

potato blight Two fungal diseases of potatoes that are most likely to cause damage under certain weather conditions. The less serious of the two is early blight, caused by *Alternaria solani*. It occurs in hot, dry weather. If the temperature remains below about 81° F (27° C) early blight produces brown marks on the leaves of the potato plant, but at higher temperatures it may destroy the foliage. Late blight is caused by *Phytophthora infestans* and occurs when the weather is cool and wet, with daytime temperatures between about 50° F and 78° F (10°–25° C). It can rapidly destroy the foliage. The tubers then start rotting and soon turn into an inedible brown pulp. It was late blight that caused the failure of the potato crop over Britain in 1845 and 1846 and the potato famine in Ireland.

potential energy The energy that is stored in a body by virtue of its position or state. Energy is stored in a ball that is stationary at the top of an incline and is converted into KINETIC ENERGY if the ball should start rolling. Gravitational, chemical, nuclear, and electrical energy are all forms of potential energy.

potential evaporation *See* EVAPORATIVE POWER.

potential evapotranspiration (PE) A concept that was introduced by CHARLES W. THORNTHWAITE into the second (1948) revision of his system of climate classification (see THORNTHWAITE CLIMATE CLASSIFICATION). It is the amount of water that would leave the ground surface by EVAPORATION and TRANSPIRATION if an unlimited supply of water were available. This is equivalent to the amount that would evaporate from an open water surface. It is calculated in centimeters from the mean monthly temperature, in degrees Celsius, corrected for the day length. Provided the monthly temperatures are known the value for *PE* can be read from tables. The equation for calculating them is

$$PE = 1.6(10t/I)^a$$

where t is the mean monthly temperature, I is the sum for 12 months of $(t/5)^{1.514}$, and a is an additional complex function of I.

potential instability *See* CONVECTIVE INSTABILITY.

potential temperature The temperature a volume of a fluid would have if the pressure under which it is held were adjusted to sea-level pressure, of 1,000 mb (100 kPa), and its temperature were to change in an ADIABATIC process. Potential temperature depends only on the actual temperature and pressure of the fluid. It is conventionally represented by the Greek letter phi (Φ), which is the *phi* in TEPHIGRAM.

In meteorology, the concept of potential temperature is used to calculate the STABILITY OF AIR—the likelihood that it will move vertically with consequent condensation or evaporation of moisture. It also explains why cold air does not sink from the upper atmosphere to the surface.

Air temperature decreases with height and according to the GAS LAWS, as the temperature of a mass of gas decreases so does its volume. If a given mass contracts, its density increases. Why is it, then, that the very cold air near the TROPOPAUSE remains there? Suppose the air is fairly dry, with no clouds in the sky, and the temperature near to ground level is 80° F (27° C). Up near the tropopause, at a height of 33,000 feet (10 km), suppose the air temperature is –65° F (–54° C). This is very much colder than the air temperature near the ground. Convert the actual temperature to the potential temperature, however, and the reason the cold air does not sink becomes evident. Increase the pressure to its sea-level value and, as the air is compressed, it warms at the DRY ADIABATIC LAPSE RATE (DALR) for air, of 5.4° F per 1,000 feet (9.8° C km^{-1}).

$$Φ = (DALR × A) + t_t$$

where A is the altitude of the cold air and t_t is its temperature. Therefore,

$$Φ = (5.4 × 33) − 65$$

$$Φ = 113.2° F$$

The potential temperature of the air at the tropopause (113.2° F [45° C]) is much higher than the actual temperature of air at ground level, which is 80° F (27° C). That is why the cold air remains aloft.

The potential temperature of any parcel of air is said to be *conserved*. This means it does not change as a consequence of the vertical movement of the air and knowledge of the potential temperature makes it possible to study the thermodynamic characteristics of the air and to represent them in diagrams. The concept is therefore of great importance to meteorologists and is widely used in weather forecasting.

potential temperature gradient   The difference between the ADIABATIC lapse rate and the ENVIRONMENTAL LAPSE RATE (ELR). If the two are the same, the potential temperature gradient is zero and a rising PARCEL OF AIR has a constant BUOYANCY. If the adiabatic lapse rate is the larger, the gradient is negative and the buoyancy of a rising parcel of air increases with height. If the ELR is the larger, the gradient is positive and the parcel of air becomes less buoyant with height, eventually reaching a level at which its buoyancy is zero.

power-law profile   A mathematical expression that describes the variation of the wind with height. Many attempts have been made to devise such a formula, but they tend to fail when the air is very stable. The most successful is probably

$$u = (u*/k)[(\ln\{z/z_o + b/4L'\})(z - z_o)]$$

where u is the wind speed, $u*$ is the FRICTION VELOCITY, k is the VON KÁRMÁN CONSTANT, ln means the natural logarithm, z is the height, z_o is the roughness length (see AERODYNAMIC ROUGHNESS), b is a coefficient, and L' is the length of the gradient, and the gradient is assumed to remain constant with height.

Poza Rica incident   An industrial accident in 1950 at Poza Rica, Mexico, that caused serious AIR POLLUTION. Equipment failed at the sulfur-recovery unit of an oil refinery, leading to the release of large amounts of hydrogen sulfide (H_2S), which was trapped beneath an INVERSION. Hydrogen sulfide is poisonous and evil-smelling (it smells like rotten eggs). At low concentrations it causes headaches and at high concentrations it is lethal. A total of about 320 people were made ill at Poza Rica and 22 people died.

PPR   See PHOTOPOLARIMETER–RADIOMETER.

praecipitatio   A supplementary feature of clouds (see CLOUD CLASSIFICATION) that consists of PRECIPITATION falling from the cloud and appearing to reach the ground. The name is the Latin for *I fall headlong*. Praecipitatio is most often seen with CUMULUS, CUMU-LONIMBUS, STRATOCUMULUS, NIMBOSTRATUS, STRATUS, and ALTOSTRATUS clouds.

prairie climate   *See* GRASSLANDS CLIMATE.

Pre-Boreal period   The first of the five ages into which the HOLOCENE EPOCH is divided on the basis of POLLEN ANALYSIS. It lasted from about 10,300 years ago until about 9,600 years ago and was a time when forests were expanding rapidly. In central North America there were spruce forests in the north and broad-leaved forest farther south. Birch and pine forests replaced tundra in Europe.

precession of the equinoxes   A change in the dates at which the Earth reaches APHELION and PERIHELION and therefore in the position of the Earth in its orbit at the EQUINOXES and SOLSTICES. Precession is a property of gyroscopes. When a force is applied to its rotational axis, the axis moves at right angles to the force in the direction of rotation. Because it is spinning, the Earth behaves as a gyroscope and forces on its axis (more strictly on its equatorial bulge) are exerted by the gravitational attraction of the Moon, Sun, and, to a lesser extent, Jupiter. This force causes the axis to wobble, like that of a toy gyroscope or spinning top (see MILANKOVICH CYCLES). If the equator is extended to the edge of the universe, the circle it forms is called the *celestial equator*. This is at an angle to the PLANE OF THE ECLIPTIC, because of the tilt in the Earth's rotational axis. The equinoxes occur when the Earth reaches the two positions in its orbit at which the celestial equator intersects the ecliptic. The axial wobble causes the celestial equator to change its position, and this change causes the orbital positions of the equinoxes to change. The result is that the orbital positions of the Earth at the equinoxes move westward by 50.27" (arcseconds) every year and complete a circuit of the ecliptic in about 25,800 years (360° ÷ 50.27" = 25,800). At present, Earth is at perihelion in early January and at aphelion in early June. In about 12,000 years, it will be at perihelion in June and at aphelion in January.

precipitable water vapor   The total amount of water vapor that is present in a column of air above a point on the surface of the Earth, or in a column of air within a layer of the atmosphere that is defined by the atmospheric pressure at its base and top. It is measured as the mass of water vapor in a unit area (such as pounds per square yard or kilograms per square meter). It is the amount of water that would fall as precipitation if it were to condense, but it is also the amount of water vapor that will react, as vapor, with outgoing radiation, thereby affecting the rate of atmospheric heating.

precipitation Water that falls from the sky to the surface in either liquid or solid form. The word *precipitation* is from the Latin *praecipitatio,* which means "I fall headlong."

When the RELATIVE HUMIDITY of the air exceeds 100 percent (*see* SUPERSATURATION), WATER VAPOR condenses in the presence of CLOUD CONDENSATION NUCLEI to form CLOUD DROPLETS or freezes in the presence of FREEZING NUCLEI to form ICE CRYSTALS. Depending on the temperature, the droplets or crystals then grow either by collision (*see* COLLISION THEORY) or by the BERGERON–FINDEISEN MECHANISM. They fall from the cloud when their weight exceeds the ability of vertical air currents to support them.

Water droplets, ice crystals, HAILSTONES, and SNOWFLAKES that fall from clouds are called *HYDROMETEORS.* This type of precipitation includes DRIZZLE, RAIN, FREEZING RAIN, HAIL, GRAUPEL, SLEET, SNOW, and ICE PELLETS.

Not all precipitation falls from clouds. The term is also applied to DEW, WHITE DEW, HOAR FROST, RIME FROST, GLAZE, FOG, and FREEZING FOG. Precipitation that falls from a cloud but evaporates before reaching the ground is called *VIRGA.*

precipitation area The area on a SYNOPTIC CHART over which PRECIPITATION is falling. On TV weather maps it is often shown by shading.

precipitation ceiling The vertical VISIBILITY that is measured looking upward into PRECIPITATION. This measure is used when precipitation obscures the CLOUD BASE.

precipitation cell An area indicated by RADAR within which PRECIPITATION is fairly continuous.

precipitation current A downward flow of electric charge that is caused by the fall of charged HYDROMETEORS.

precipitation echo The image on a RADAR screen that is caused by the reflection of the radar transmission by PRECIPITATION.

precipitation-efficiency index A value that indicates the amount of water that is available for plant growth. It was devised by CHARLES W. THORNTHWAITE and forms one of the bases of the THORNTHWAITE CLIMATE CLASSIFICATION. Precipitation efficiency is calculated from measurements of the temperature, the precipitation, and the amount of water that evaporates from an exposed water surface in the course of one month. This is given by $115(r/t - 10)^{10/9}$, where r is the mean monthly rainfall in inches and t is the mean monthly temperature in degrees Fahrenheit. This calculation is made for each month, and the sum of the indexes for 12 months is the precipitation-efficiency index.

precipitation fog *See* FRONTAL FOG.

precipitation-generating element A small region inside a cloud where ICE CRYSTALS are growing more rapidly than they are elsewhere in the cloud at the expense of a concentration of supercooled (*see* SUPERCOOLING) water droplets. When they exceed a certain size the ice crystals fall through the cloud, generating precipitation by the BERGERON–FINDEISEN MECHANISM.

precipitation intensity The amount of PRECIPITATION that falls to the surface within a specified period. It is measured in inches or millimeters per hour or day.

precipitation inversion (rainfall inversion) An INVERSION that inhibits PRECIPITATION. Such inversions sometimes develop in mountain areas, but they are most common in the TROPICS, where they are associated with the HADLEY CELLS. Subsiding air on the high-latitude sides of the Hadley cells warms by an ADIABATIC process and some of it forms warm pockets of dry air at a height of 50,000–65,000 feet (1.5–2.0 km) over the eastern sides of all the tropical oceans. These pockets restrict the vertical movement of air rising by CONVECTION and this reduces the amount of CONDENSATION and consequent precipitation.

precipitation physics The branch of PHYSICAL METEOROLOGY that is concerned with the physical processes that are involved in the formation of CLOUD DROPLETS, ICE CRYSTALS, and the resulting PRECIPITATION.

precipitation station   A weather station where only the amount and type of PRECIPITATION are measured and recorded.

prefrontal surge   The descent of dry air over an OCCLUSION and ahead of the upper-level front. The air that descends over an occluded front from near the top of the TROPOSPHERE is called a *dry slot;* it is a prefrontal surge only if it is ahead of the front. The dry air lies above relatively warm, moist air. This situation can cause INSTABILITY, resulting in the formation of CUMULONIMBUS clouds and sometimes THUNDERSTORMS.

present weather   The current weather conditions that are included in a report from a weather station. These are represented by a series of two-digit numbers, from 00 to 99. Most of the categories are listed below. Categories 30–39 and 40–49 are not listed in detail to prevent repetition. The descriptions are also simplified from the official versions, which are worded very carefully to prevent ambiguity and are therefore not easy to understand. When the present weather is reported, the number that is used is the highest that is applicable to the conditions. In the descriptions that follow *freezing* means freezing on impact.

00 No cloud developing during the past hour
01 Cloud dissolving during the past hour
02 Cloud generally unchanged during the past hour
03 Cloud developing during the past hour
04 Visibility reduced by smoke
05 Haze
06 Dust widespread
07 Dust or sand raised by local wind, but not by dust storms, sandstorms, or whirls (devils)
08 Dust or sand whirls seen in the past hour, but no dust storms or sandstorms
09 Dust storm or sandstorm seen nearby during the past hour
10 Mist
11 Shallow, patchy fog or ice fog
12 Shallow continuous fog or ice fog
13 Lightning but no thunder
14 Precipitation seen, but not reaching the surface
15 Precipitation seen reaching the surface in the distance

16 Precipitation seen reaching the surface nearby, but not at the station
17 Thunderstorm but no precipitation seen
18 Squalls at the time of observation or during the past hour
19 Funnel cloud seen at the time of observation or during the past hour
20 Precipitation, fog, or thunderstorm during the past hour but not at the time of observation
21 Drizzle (not freezing) or snow grains, but not in showers
22 Rain (not freezing), but not in showers
23 Rain and snow or ice pellets, but not in showers
24 Freezing drizzle or freezing rain
25 Rain showers
26 Showers of rain and snow (British sleet) or snow
27 Showers of hail and rain or hail
28 Fog or ice fog in the past hour
29 Thunderstorm
30–39 Dust storms, sandstorms, drifting snow, or blowing snow
40–49 Fog or ice fog at the time of observation
50 Drizzle (not freezing) that is intermittent and slight at the time of observation
51 Drizzle (not freezing) that is continuous at the time of observation
52 Drizzle (not freezing) that is intermittent and moderate at the time of observation
53 Drizzle (not freezing) that is continuous and moderate at the time of observation
54 Drizzle (not freezing) that is intermittent and heavy at the time observation
55 Drizzle (not freezing) that is continuous and heavy at the time of observation
56 Slight freezing drizzle
57 Moderate or heavy freezing drizzle
58 Slight drizzle and rain
59 Moderate or heavy drizzle and rain
60–69 Same as 50–59, but with rain instead of drizzle and in 58 and 59 snow instead of rain
70 Snowflakes, intermittent and slight at the time of observation
71–75 Same as 51–55, but with snow instead of drizzle
76 Ice prisms with or without fog
77 Snow grains with or without fog
78 Isolated, starlike, snow crystals with or without fog
79 Ice pellets
80 Rain showers, slight
81 Rain showers, moderate or heavy
82 Rain showers, violent
83 Rain and snow showers, slight

84 Rain and snow showers, moderate or heavy
85 Snow showers, slight
86 Snow showers, moderate or heavy
87 Slight showers of snow pellets, encased in ice or not, with or without rain or rain and snow (British sleet) showers
88 Moderate showers of snow pellets, encased in ice or not, with or without rain or rain and snow (British sleet) showers
89 Slight hail showers, without thunder, with or without rain or rain and snow (British sleet)
90 Moderate or heavy hail showers, without thunder, with or without rain or rain and snow (British sleet)
91 Slight rain
92 Moderate or heavy rain
93 Slight snow, or rain and snow (British sleet), or hail
94 Moderate or heavy snow, or rain and snow (British sleet), or hail
95 Slight or moderate storm with rain and/or snow, but no hail
96 Slight or moderate storm with hail
97 Heavy storm with rain and/or snow, but no hail
98 Storm with sandstorm or dust storm
99 Heavy storm with hail

pressure altitude The height above sea level at which the AIR PRESSURE in a STANDARD ATMOSPHERE would be the same as the pressure measured at the surface in a particular place (*compare* DENSITY ALTITUDE). The use of pressure altitudes allows atmospheric pressure to be expressed in terms of altitude. For example, in summer the pressure altitude at the VOSTOK STATION in Antarctica is about 10 percent higher than the true altitude, and in winter it is about 15 percent higher. The difference between the true altitude and pressure altitude is called the *pressure-altitude variation.*

pressure-altitude variation *See* PRESSURE ALTITUDE.

pressure anemometer *See* ANEMOMETER.

pressure center The center of an area of low (CYCLONE) or high (ANTICYCLONE) pressure as it appears on a SYNOPTIC CHART or weather map.

pressure-change chart (pressure-tendency chart) A chart that shows the BAROMETRIC TENDENCY. This is the change in AIR PRESSURE that has occurred over a specified period across a surface at a constant height.

pressure-fall   center   (isallobaric   low,   katabaric center) The place where the AIR PRESSURE has fallen further than it has anywhere else over a specified period.

pressure gradient (isobaric slope) The rate at which atmospheric pressure changes over a horizontal distance. The ISOBARS that join points of equal pressure on a surface resemble the contour lines on a physical map, and a line drawn at right angles to them shows the direction of the gradient, or slope, that inclines from a region of high pressure to one of low pressure. The distance between isobars indicates the steepness of the gradient, just as does the distance between contour lines. That distance is proportional to the difference in pressure between the high and low regions, just as the distance between contour lines is proportional to the elevation of high and low areas of ground. Air moves because there is a pressure gradient, but it does not move directly down the gradient (*see also* GRADIENT WIND).

pressure gradient force (PGF) The force that accelerates air horizontally across the surface of the Earth. It is produced by the PRESSURE GRADIENT, and its magnitude is proportional to the steepness of the gradient. The PGF has both vertical and horizontal components, but the vertical component, which tends to make the air rise, is balanced by the force of gravity and therefore can be ignored.

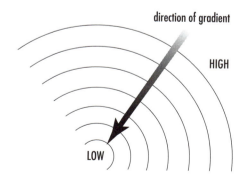

Pressure gradient. **The rate at which pressure changes in air between centers of high and low pressure varies according to the distance between the centers. This constitutes a gradient, like the slope of a hillside with the high pressure at the top of the hill and low pressure in the valley. The isobars resemble contours**

The pressure gradient force always acts in the same direction as the pressure gradient. Consequently, that is the direction in which it tends to move the air—by the most direct route from a region of high pressure to a region of low pressure until the two are equal and the pressure gradient disappears. This is not what happens in fact, because of the CORIOLIS EFFECT, but it is the direction in which the PGF acts.

A force that continues to act causes the body on which it acts to accelerate, so the PGF is a force of acceleration. Its magnitude can be calculated provided the relevant factors are known. These are the air density, the distance between the two points over which the PGF is being calculated, and the difference between the pressures at those two points. These values must be in compatible SYSTÈME INTERNATIONAL D'UNITÉS (SI) units. Distances must be converted to meters and pressures from MILLIBARS to pascals (1 mb = 100 Pa). One pascal is the pressure that imparts an acceleration of 1 meter per second per second, per kilogram, per square meter. The equation is:

$$F_{PG} = (1 \div d) \times (\Delta_p \div \Delta_n)$$

where F_{PG} is the PGF, d is the density of the air (in kilograms per cubic meter), Δ_p is the difference in pressure (in pascals), and Δ_n is the distance between the two places (in meters).

Suppose the sea-level air pressure is 1,004 mb at Boston and 980 mb at New York, about 200 miles away.

1,004 mb = 100,400 Pa;
980 mb = 98,000 Pa;
Δ_n (distance of 200 miles) = 321,800 m;
d (density of air) at sea level = approximately 1 kg m^{-3};
Δ_p (pressure difference) = 100,400 – 98,000 Pa = 2,400 Pa

Applying the equation:

$$F_{PG} = (1 \div 1) \times (2,400 \div 321,800) = 0.00746 \text{ m s}^{-2}$$

This is a very small acceleration, but it applies to only 1 kilogram of air, the mass of air that occupies 1 cubic meter, and it is an acceleration, not a speed. Although the sea-level air density is 1 kg m^3, so the first term in the equation is 1 ÷ 1, the PGF is usually measured at some height above sea level, where the density of air is lower, and the first term does not cancel.

pressure jump line   *See* GUST FRONT.

pressure law   *See* GAS LAWS.

pressure melting   The melting of ice that is subjected to a strong pressure. As the pressure increases, the freezing temperature of water decreases by approximately 1° F for every 95.5-lb in^{-2} increase in pressure (1° C for every 140 bars). The weight of ice in a TEMPERATE GLACIER exerts sufficient pressure to melt a thin layer of ice at the base. The temperature at which ice begins to melt under a given pressure is known as the *pressure melting point*.

pressure pattern   The distribution of AIR PRESSURE as it is shown by the ISOBARS on a SYNOPTIC CHART. The patterns made by the isobars indicate the location and intensity of CYCLONES, ANTICYCLONES, RIDGES, and TROUGHS and the surface area that is affected by them.

pressure-plate anemometer   *See* ANEMOMETER.

pressure-rise center (anabaric center, isallobaric high)   The place where the AIR PRESSURE has risen further than it has anywhere else over a specified period.

pressure system   An atmospheric feature that is characterized by AIR PRESSURE. The term is usually applied to a CYCLONE or ANTICYCLONE, but it may also refer to a RIDGE or TROUGH.

pressure tendency   *See* BAROMETRIC TENDENCY.

pressure-tendency chart   *See* PRESSURE-CHANGE CHART.

prester   A very hot, burning WHIRLWIND that appears accompanied by lightning in the Mediterranean region and especially in Greece. The same name is also applied to a waterspout accompanied by lightning.

prevailing visibility   The greatest horizontal VISIBILITY that extends over at least one-half of the horizon around an observation point. It is the visibility that is reported on a STATION MODEL.

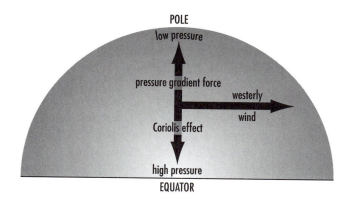

Prevailing westerlies. **Pressure at any height decreases with distance from the equator. This produces a pressure gradient force directed toward the Pole. As soon as air moves in response to this force the Coriolis effect deflects it to the east. When the two forces balance, the resultant wind blows from west to east.**

prevailing westerlies (westerlies) The westerly winds that predominate in middle latitudes. These winds are strongest at about latitudes 35°–40° N and S, on the side of the SUBTROPICAL HIGHS closer to the Poles. Here they also become more pronounced with height, all the way up to the TROPOPAUSE.

prevailing wind The direction from which the wind experienced at a particular place blows more frequently than it blows from any other direction (*see* WIND DIRECTION). In middle latitudes, for example, the prevailing winds are westerlies: that is, they blow from the west. Winds in the Tropics usually blow from the east, so there the prevailing winds are easterlies. A WIND ROSE is compiled to determine the direction of the prevailing wind.

primary circulation That part of the GENERAL CIRCULATION of the atmosphere that comprises large-scale, persistent features. Although details of these features change from time to time, their overall character remains constant over large areas of the world. The primary circulation is driven by the latitudinal differences in the amount of solar radiation that is received at the surface. These differences drive the tropical HADLEY CELLS, FERREL CELL, and POLAR CELL (*see* THREE-CELL MODEL). They produce the INTERTROPICAL CONVERGENCE ZONE and EQUATORIAL TROUGH, as well as the SUBTROPICAL HIGHS. These features and the

winds associated with them constitute primary circulation. More ephemeral features constitute the SECONDARY CIRCULATION.

primary cyclone (primary low) An area of low atmospheric pressure (a CYCLONE) that has one or more smaller (secondary) cyclones within its circulation.

primary front The front that is the first to form in a FRONTAL SYSTEM. One or more secondary fronts then develop from it along the FRONTAL WAVE (*see* FRONTOGENESIS.)

primary low *See* PRIMARY CYCLONE.

primary pollutant A substance that is released into the environment, where it causes immediate POLLUTION. The most widespread and serious primary pollutants are PARTICULATE MATTER, SULFUR DIOXIDE, NITROGEN OXIDES (NO_x), and unburned hydrocarbons. NO_x are both primary and SECONDARY POLLUTANTS, because as well as being released in vehicle exhausts and from certain industrial processes, they are formed by chemical reactions that take place in the air and involve PEROXYACETYL NITRATES. Chlorofluorocarbons (CFCs) are also considered primary pollutants, because of their role in depleting the OZONE LAYER and their GLOBAL WARMING POTENTIAL, as are HALONS and the other GREENHOUSE GASES.

probability forecast A weather forecast that states the expected likelihood of a particular type of weather, usually PRECIPITATION. The forecast might say there is a "60 percent chance of rain" or "a 0.6 chance of rain." Both mean the same thing, because probabilities can be expressed as a decimal between 0 (no chance at all) and 1 (absolute certainty). The statement means that during the FORECAST PERIOD there is a 60 percent chance that it will rain where you are and a 40 percent chance that it will not. It does not mean there is a 60 percent chance that rain will fall somewhere in the forecast area and a 40 percent chance that the entire area will remain dry. *Precipitation* is interpreted to mean 0.01 inch (0.25 mm) of rain or its equivalent at any particular place covered by the forecast at any time during the forecast period.

Calculating the probability begins by determining whether or not precipitation-bearing clouds will enter the area during the specified period. If it is fairly certain that they will (for example, because they are close to the edge of the area and advancing toward it) then there might be a 90 percent (0.9) probability of precipitation. The clouds may not pass over the entire area, however. Their size and predicted track might mean they will cross about 70 percent (0.7) of the area. The chance of precipitation in any particular place within the area is therefore $0.9 \times 0.7 = 0.63$, which is approximately 60 percent. Wherever you are, this means there is a 60 percent chance that it will rain, but there is a 90 percent chance that it will rain somewhere in the forecast area.

probable maximum flood *See* PROBABLE MAXIMUM PRECIPITATION.

probable maximum precipitation An estimate of the greatest amount of PRECIPITATION that could conceivably fall on a given drainage area over a given period. It is calculated from records of the worst storms known in the area. The amount of precipitation is then converted into the amount of water that will flow through streams and rivers as a result. This figure is used to calculate the probable maximum flood, a figure that engineers use when designing dams. Although this calculation provides an estimate of the magnitude of the most severe flood, it does not calculate the probability that such a flood will occur within any stated period (such as 50 or 100 years).

PROFS *See* PROGRAM FOR REGIONAL OBSERVING AND FORECASTING SYSTEMS.

prognostic chart A SYNOPTIC CHART that shows the patterns of pressure, the height of pressure surfaces, temperature, wind speed and direction, or other features of the weather as these are expected to appear at some specified time in the future. The position of fronts may also be drawn. It is a forecast chart. If the forecast is for more than about two days ahead the prognostic chart will show the average conditions expected, which are calculated from the range of predicted possibilities.

Program for Regional Observing and Forecasting Services (PROFS) A program that began in 1980–81 under the auspices of the NATIONAL OCEANIC AND ATMOSPHERIC ADMINISTRATION (NOAA). Its aim was to test and apply new scientific knowledge and technological innovations in order to improve operational weather services. The program was later renamed the *FORECAST SYSTEMS LABORATORY*.

progressive wave A wave or group of waves that move in relation to the surface of the Earth. Waves that remain stationary in relation to the surface are called *standing waves* (*see* LEE WAVE, WAVE CHARACTERISTICS).

proxy data DATA that do not refer to the climate directly but can be interpreted to yield information about climate. TREE RINGS (*see* DENDROCLIMATOLOGY), ICE CORES, POLLEN (*see* POLLEN ANALYSIS), and animal remains (*see* BEETLE ANALYSIS) are among the sources of proxy data. Proxy data are the only data available for the reconstruction of prehistoric climates (*see* PALEOCLIMATOLOGY).

PSCs *See* POLAR STRATOSPHERIC CLOUDS.

pseudoadiabat A line that is drawn on a THERMODYNAMIC DIAGRAM to show the LAPSE RATE of air that is rising past the LIFTING CONDENSATION LEVEL. The pseudoadiabatic lapse rate is almost the same as the SATURATED ADIABATIC LAPSE RATE.

pseudoadiabatic chart *See* STÜVE CHART.

pseudofront A boundary, resembling a small front, that develops between air that is cooled by rain falling through a large CUMULONIMBUS cloud and the warmer air adjacent to it.

PSI *See* POLLUTION STANDARDS INDEX.

psychrometer A HYGROMETER that uses two thermometers mounted parallel to each other and a short distance apart. One, the DRY-BULB THERMOMETER, measures the air temperature; the other, the WET-BULB THERMOMETER, indirectly measures the rate of EVAPORATION. The bulb of this thermometer is wrapped in a wick, usually made from muslin, partly immersed in a reservoir of distilled water. Water is drawn into the wick from where it evaporates. The LATENT HEAT of vaporization is taken from the thermometer bulb, thus depressing its temperature. The rate at which

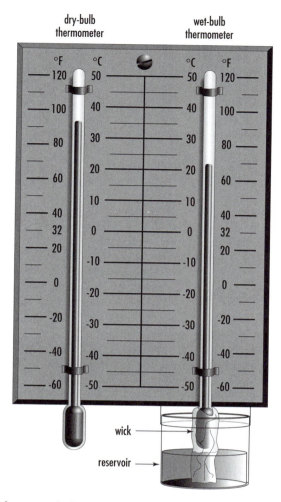

dry-bulb thermometer

wet-bulb thermometer

wick

reservoir

Psychrometer. **An instrument for measuring wet-bulb and dry-bulb temperatures. Two identical thermometers are mounted side by side on a board. The wet bulb is wrapped in a wick that is partly immersed in a reservoir of distilled water.**

water evaporates, and therefore the amount of latent heat supplied by the bulb, varies according to the RELATIVE HUMIDITY of the air. It is necessary to provide a flow of air over the thermometers and to ensure that the wick remains wet at all times, but is not allowed to become sodden. Neither relative humidity nor the DEW POINT TEMPERATURE can be read directly from the instrument, but both can be calculated from its readings.

psychrometric equation An equation that is used to calculate the SPECIFIC HUMIDITY of air from the WET-BULB TEMPERATURE and DRY-BULB TEMPERATURE registered by a PSYCHROMETER.

$$q(T) = q_s(T_w) - \lambda(T - T_w)$$

where $q(T)$ is the specific humidity of the air, $q_s(T_w)$ is the saturated specific humidity of the air, T is the dry-bulb temperature, T_w is the wet-bulb temperature, and λ is the psychrometric constant. This is equal to $C_{p/L}$, where C_p is the specific heat of air and L is the LATENT HEAT of vaporization of water. The value of the psychrometric constant varies according to the temperature and atmospheric pressure. At 68° F (20° C) and pressure of 14 lb in⁻² (100 kPa) it is 0.0009 oz ft⁻³ °F⁻¹ (0.489 g m⁻³ K⁻¹). The psychrometric equation can be converted to calculate the VAPOR PRESSURE or vapor density of the air.

psychrometry The measurement of DRY-BULB TEMPERATURE and WET-BULB TEMPERATURE by using a PSYCHROMETER and calculating the SPECIFIC HUMIDITY, VAPOR PRESSURE, or vapor density of the air from those values.

puelche (fog wind) A FÖHN WIND that occurs on the western side of the Andes. It sometimes produces violent SQUALLS.

puff of wind A BREEZE that is just strong enough to produce a patch of ripples on the surface of still water.

pulse radar *See* RADAR.

purga A strong type of BLIZZARD that occurs in winter in the tundra of northeastern Siberia. The wind blows with gale or even hurricane force and the air is filled with snow, some of it falling and some that has been swept up from the ground. Visibility is reduced almost to zero. The purga is very similar to the BURAN.

pyranometer (solarimeter) An instrument that measures solar radiation. Pyranometers are fairly robust and are more suitable for field use than the more accurate, but more delicate PYRHELIOMETERS. There are several pyranometer designs, but all are built around a sensor that detects the difference in temperature between two adjacent materials. This may be achieved with a sensor consisting of two concentric rings, the inner one painted black to give it a low

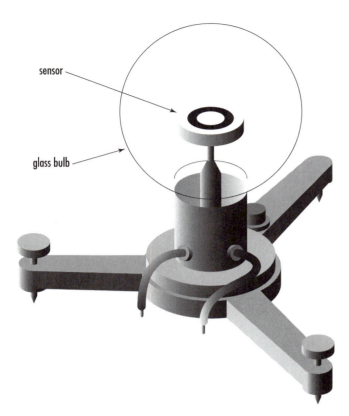

Pyranometer. An instrument for measuring solar radiation by comparing the temperatures of the adjacent black and white surfaces of a sensor. The sensor is contained in a glass sphere. This is the Eppley pyranometer.

pyrheliometer A very sensitive instrument that is used to measure the intensity of solar radiation. It contains a blackened surface positioned at right angles to the sunlight. In the Abbot silver disk pyrheliometer, which is the one most often used in the United States, the receiving surface is a disk of blackened silver supported on fine steel wires at the bottom of a copper

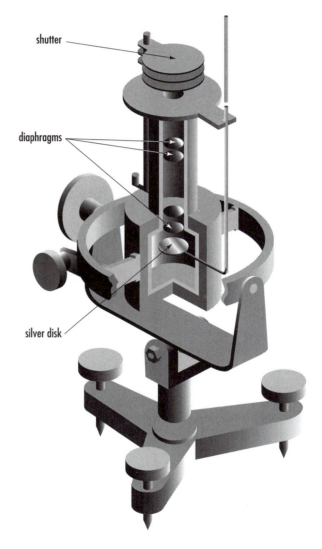

Pyrheliometer. The Abbot silver disk pyrheliometer measures the intensity of solar radiation. Sunlight enters past a system of shutters that open and close at two-minute intervals to allow the rate of change of solar intensity to be measured. Beneath the shutters light passes along a tube containing diaphragms that exclude light that does not come directly from the Sun and the immediately adjacent sky. At the bottom of the tube the light falls on a blackened silver disk attached to a thermometer.

ALBEDO and the outer one painted white to give it a high albedo. Up to 50 electrical sensors are in good contact with the underside of the rings, and these provide readings of the temperature difference between the high- and low-albedo surfaces. The sensor is enclosed in a glass bulb filled with dry air and designed either to allow radiation of all wavelengths to enter or to filter out particular wavelengths, such as those shorter than visible light. A pyranometer is usually set on the ground with its sensor surface horizontal and facing upward. A DIFFUSOGRAPH is a pyranometer modified to measure only diffuse light. Surface albedo is measured by comparing the readings from two pyranometers, one facing vertically upward and the other vertically downward.

pyrgeometer An instrument similar to a PYRANOMETER that measures INFRARED RADIATION.

tube. A very accurate thermometer is attached to the underside of the disk. Diaphragms arranged at intervals in the tube allow only direct sunlight and light from the sky immediately adjacent to the Sun to enter. The instrument can be used only when the sky is completely cloudless within a 20° radius of the Sun. As well as measuring solar intensity, the pyrheliometer measures the rate at which it changes. For this purpose there is a triple shutter above the tube. The shutter opens and closes to expose and shade the disk alter-

nately, at very precise two-minute intervals. The final readings must then be corrected for a standard air temperature of 68° F (20° C) and a disk temperature of 86° F (30° C). The Abbot pyrheliometer remains calibrated for many years and can be used to calibrate other instruments. Pyrheliometers are so sensitive and require such constant maintenance that they are used only at research laboratories and some of the principal weather stations.

QBO *See* QUASI-BIENNIAL OSCILLATION.

quasi-biennial oscillation (QBO) An alternation of easterly and westerly winds that occurs in the STRATO-SPHERE above the Tropics, between about 20° S and 20° N. The change happens on average every 27 months, but the period varies from less than two years to more than three years. Easterly winds become established at a height of about 12 miles (20 km), usually between about May and September. Then westerly winds in the upper stratosphere, above about 19 miles (30 km), begin to extend downward until by about January westerlies predominate. The QBO causes a MERIDIONAL CIRCULATION that affects the distribution of trace gases, including OZONE, in the stratosphere over the Tropics and subtropics, and some scientists suspect that in winter the QBO may be linked to the arctic POLAR VORTEX and to sudden periods of warm weather. This happens because sudden warming in high latitudes is associated with ROSSBY WAVES that start in the TROPOSPHERE and propagate upward into the stratosphere. Rossby waves are able to propagate in this way if the stratospheric winds are westerlies, but not if they are easterlies. KELVIN WAVES are also influenced by the QBO. The severity of winters in the Northern Hemisphere is linked to both the QBO and the SUNSPOT cycle. Winters are generally warmer when the stratospheric winds are westerlies than when they are easterlies. Scientists do not fully understand the QBO or the ways in which it affects the weather, however.

quasi-hydrostatic approximation The use of the HYDROSTATIC EQUATION as the vertical EQUATION OF MOTION. This assumes that there is some vertical motion of air, but that it is small.

quasi-stationary front A front that is moving at less than about 5 KNOTS (5.75 mph [9.25 km h[-1]).

Quaternary (Anthropogene, Pleistogene) The sub-era of the CENOZOIC era that covers approximately the last 1.64 million years and includes the PLEISTOCENE and HOLOCENE epochs (*see* GEOLOGICAL TIME SCALE).

QuikScat A NATIONAL AERONAUTICS AND SPACE ADMINISTRATION (NASA) satellite that was launched on June 19, 1999, from Vandenberg Air Force Base, California, on a *Titan II* rocket of the U.S. Air Force. It entered an elliptical orbit at a height of about 500 miles (800 km). QuickScat, short for *Quick Scatterometer,* replaced the NASA Scatterometer (NSCAT) that was lost in June 1997 when the satellite carrying it lost power. It carries the SEA WINDS RADAR instrument.

rabal A method that is used to measure the speed and direction of high-level winds above a weather station. The elevation and azimuth of a RADIOSONDE balloon are measured and recorded at regular intervals, using a theodolite.

radar An electromagnetic device that is used to detect and measure the motion of distant objects that are otherwise invisible, because it is dark or because they are obscured by a medium the radar can penetrate. The name is from *ra*dio *d*etection *a*nd *r*anging. Radar devices transmit an electromagnetic wave that is reflected from any object it strikes and receive the reflection. Information is obtained by comparing the signal and its reflection. Electromagnetic waves travel at the speed of light, so the time that elapses between the transmission and the reception of the reflection indicates the distance to the object (about 500 feet [150 m]) for every microsecond of delay). This type of measurement is made by pulse radar, in which short, intense bursts of radiation are transmitted, with a fairly long interval between bursts. Continuous-wave radar transmits a continuous pulse. In its basic form, this cannot be used to measure distance, but it can measure the speed at which the target object is moving, as the DOPPLER EFFECT. Frequency-modulated radar is a more advanced version of continuous-wave radar that can measure distance, because each part of the signal is tagged to make it and its reflection recognizable. *See also* SYNTHETIC APERTURE RADAR.

radar altimetry Measuring the topographical characteristics of a land surface by means of a radar ALTIMETER carried by an aircraft or space vehicle. The altimeter measures the distance between the vehicle carrying it and the surface vertically below it. Provided the vehicle remains at a constant height in relation to a DATUM LEVEL (not the ground surface) its distance to the surface varies according to changes in the ground elevation. With a series of passes the physical features of the landscape can be measured and plotted, and the plots used to compile a map.

radar climatology The use of recorded RADAR echoes showing clouds and precipitation in studies of the climate. If the records extend over several decades they can reveal any pattern in the distribution of cloud and precipitation over a particular region. This information is then incorporated into the overall picture, or model, of the climate of that region.

radar interferometry A technique that is used to measure very small changes in the shape of features on the solid surface of the Earth. These can give warning of an impending volcanic eruption. An instrument called an *interferometer* mounted on an observational satellite transmits RADAR waves that interact with each other to produce characteristic patterns known as *interference fringes*. The appearance of the fringes varies with very small changes in the distance traveled by the radar pulse and its echo.

radar meteorological observation A pattern of echoes that appears on the screen of a WEATHER RADAR. The pattern reveals such features as clouds and precipitation, with their distance, density, and direction of movement. It also shows severe STORMS, TROPICAL CYCLONES, and TORNADOES.

radar meteorology The use of RADAR in compiling a picture of the present state of the weather and in preparing weather forecasts. WEATHER RADAR reveals details of the size and type of clouds, location and intensity of precipitation, and direction and speed of movement of weather systems (*see* DOPPLER RADAR). THUNDERSTORMS, TORNADOES, and TROPICAL CYCLONES also produce clear and distinctive radar images.

radar wind A wind that is observed by RADAR. The radar tracks a RADIOSONDE balloon or a balloon carrying a radar reflector in order to determine the speed and direction of the wind at a known height.

radian (rad) The supplementary SYSTÈME INTERNATIONAL D'UNITÉS (SI) unit of plane angle, which is the angle subtended at the center of a circle by an arc equal to the radius of that circle. If the radius is r, an arc of length r subtends an angle of 1 rad and the circumference of the circle, $2\pi r$, subtends an angle of $2\pi r \div r$ rad $= 2\pi$ rad. The circumference subtends an angle of 360°; therefore, 360° $= 2\pi$ rad and 1 rad $= 57.296°$.

radiant flux density The amount of electromagnetic radiation that is emanating from a body, such as the Sun, or falling on a surface. It is measured in watts per square meter (W m^{-2}), LANGLEYS, or calories per square centimeter (cal cm^{-2}).

radiational index of dryness The value on which the CLIMATE CLASSIFICATION devised by MIKHAIL I. BUDYKO is based. The index is calculated from the ratio (R_o/Lr) of the net radiation that is available to evaporate moisture from a wet surface ($R\hat{o}$) to the amount of heat that is needed to evaporate the mean annual precipitation (Lr); L is the LATENT HEAT of vaporization. Areas with a moist climate have a value of less than 1 in the resulting index and dry areas have a value greater than 1.

radiation balance The "energy budget" of the Earth, in which the amount of energy the Earth receives from the Sun is set against the way that energy is distributed and eventually returned by radiating it back into space. It can be summarized as

$$R = (Q + q)(1 - a) - I$$

where R is the radiation balance, Q is the direct sunlight reaching the surface, q is the diffuse sunlight reaching the surface, a is the ALBEDO of the surface, and I is the outgoing long-wave radiation from the surface. On average, the surface of the Earth absorbs about 124 kilolangleys (kly) of radiation each year and radiates about 52 kly into space. This means the surface absorbs 72 kly more than it radiates.

The same equation can also be applied to the atmosphere. It absorbs about 45 kly of energy a year and radiates about 117 kly into space. The atmosphere therefore radiates 72 kly more than it absorbs. Consequently, the figures for the surface and atmosphere are in balance over the year. If this were not the case, the world would be growing either warmer or cooler (but *see* GLOBAL WARMING).

The balance is achieved by the transfer of energy from the surface to the atmosphere. Some of this transfer occurs directly. As the surface is warmed by the Sun the air in contact with it is also warmed by CONDUCTION and its warmth is transferred upward by CONVECTION. Most of the transfer is due to the EVAPORATION of water from the surface and its CONDENSATION in the atmosphere. Evaporation absorbs LATENT HEAT from the surface and condensation releases it into the air.

Solar energy is not received equally in all areas of the Earth. The Sun shines more intensely on the equator than it does on the Poles. This difference increases, because in latitudes higher than 40°, the atmosphere loses more energy in a year than the surplus absorbed by the surface and in latitudes lower than 40° the reverse is true and the surface absorbs more energy than the atmosphere loses. Despite this, the Poles do not grow colder year by year, nor the Tropics warmer. The balance is maintained by the horizontal transfer of heat. Without that transfer the radiation budget would balance at each latitude only if the mean temperature at the equator were 25° F (14° C) higher than it is and the mean temperature at the Poles 45° F (25° C) lower.

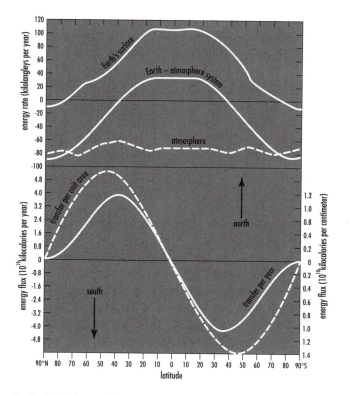

Radiation balance. **The upper graph shows the distribution of solar energy per year for the surface of the Earth, the atmosphere, and the Earth and atmosphere considered together. The lower graph shows the horizontal transfer of energy from low to high latitudes.**

Ocean currents (*see* GYRES) and the GENERAL CIRCULATION of the atmosphere transfer heat away from the equator. ADVECTION through the atmosphere accounts for a little more than half of the total transfer. Evaporation and condensation also increase the rate of this horizontal transfer of heat from low to high latitudes.

The total amount of heat transferred annually in this way reaches a maximum at about latitudes 40° N and S. The maximum amount transferred per unit area of the surface is at about 45° N and S. It is slightly greater in the Southern Hemisphere than in the Northern Hemisphere. Much of the transfer in the Tropics and subtropics occurs in the upper TROPOSPHERE through the HADLEY CELL circulation. The transfer in middle latitudes occurs mainly through the CYCLONES and FRONTAL SYSTEMS that move with the PREVAILING WESTERLIES. The rate of meridional transport varies with the temperature gradient. The gradient is steepest

in winter, and that is when the atmospheric circulation is most vigorous.

radiation cooling The loss of energy from the ground that takes place at night. During the day the surface absorbs solar radiation in the form of heat. As its temperature rises, the surface also emits INFRARED RADIATION (*see* BLACK BODY). The amount of energy radiated in this way is proportional to the temperature of the surface (*see* STEFAN–BOLTZMANN LAW), but during the day it is always less than the amount of solar radiation that is being absorbed. At night the blackbody radiation from the surface continues, but the absorption of solar radiation ceases. Consequently, there is a net loss of energy from the surface and therefore the surface cools. Its temperature continues to fall through the night, then starts rising again at dawn.

Most of the blackbody infrared radiation is absorbed by gases in the atmosphere (*see* GREENHOUSE EFFECT). These then reradiate it. On clear nights atmospheric absorption occurs throughout the TROPOSPHERE. It warms the air but allows the ground surface to cool. On cloudy nights the clouds form a barrier to the radiation. They reflect some of it and absorb some of it, reradiating a proportion of it downward. This greatly reduces the rate at which the surface cools and explains why in spring and fall frosts are more likely on clear nights than on cloudy nights.

radiation fog FOG that forms on clear nights when the RELATIVE HUMIDITY is fairly high. During the day, the ground absorbs heat from the Sun and warms. At night, it radiates its heat into the sky and because there is no cloud to absorb the radiated heat and warm the air below it, the ground cools sharply. Air adjacent to the ground is cooled by contact with it and its temperature falls to below the DEW POINT TEMPERATURE. Water vapor condenses and fog forms. If the air is still, the fog usually forms a very shallow layer—sometimes so shallow it reaches only to about the knees of people walking through it. If there is a slight breeze, turbulence carries the fog and the cold air to a greater height and the fog layer may extend to a height of up to about 100 feet (30 m).

The layer of cold, dense air containing the fog lies beneath warmer, less dense air. This prevents the fog from rising. If the ground is sloping, the fog tends to flow downhill and accumulate in valleys.

The following morning, the Sun warms the ground. This warms the air in contact with the ground and the water droplets constituting the fog begin to evaporate. As the ground continues to warm, air warmed by it is carried upward by CONVECTION, evaporating more of the fog. The fog clears in this way, from the bottom up, sometimes until all that remains is a layer of white cloud. This gives the impression that the fog is lifting, although that is not what is happening.

radiation inversion A low-level temperature INVERSION that forms at night as a consequence of RADIATION COOLING. As the ground cools, WATER VAPOR, CARBON DIOXIDE, and certain other atmospheric gases absorb some of the INFRARED RADIATION that it emits. This absorption raises the temperature of the air sufficiently to produce the inversion.

radiation night A night when the sky is clear. Cloud returns a proportion of surface INFRARED RADIATION, which reduces surface cooling. The absence of cloud allows the surface to cool rapidly. The drop in temperature close to the ground is especially marked when the air is still, because air movements would mix warmer air with the cold air at ground level.

radiation point *See* RADIATUS.

radiative convective model *See* CLIMATE MODEL.

radiative diffusivity The capacity of an atmospheric layer to distribute INFRARED RADIATION by DIFFUSION. This depends on the temperature and pressure within the layer and the amount of WATER VAPOR it contains.

radiative dissipation The loss of energy that is caused when fingers of warm air penetrate cooler air and cool by radiating heat. While the air is at a higher temperature than its surroundings, the temperature difference represents available POTENTIAL ENERGY that could be converted into KINETIC ENERGY. Radiation of its heat, however, causes the available potential energy to be dissipated before it can be converted.

radiatus A variety of clouds (*see* CLOUD CLASSIFICATION) that are arranged as broad, parallel bands that, as a result of perspective, appear to converge at a point on the horizon or at two points at opposite sides of the horizon. The points of convergence are called *radiation points*. Radiatus occurs with the cloud genera CIRRUS, ALTOCUMULUS, ALTOSTRATUS, STRATOCUMULUS, and CUMULUS. The name *radiatus* is derived from the Latin word *radiare*, which means "spoke" or "ray."

radioactive decay The process by which unstable atomic nuclei lose elementary particles and continue to lose them until the nuclei become stable. ISOTOPES of chemical elements that have nuclei subject to this type of decay are said to be radioactive.

For example, naturally occurring uranium (U) is a mixture of three isotopes: ^{238}U accounts for 99.28 percent, ^{235}U for 0.71 percent, and ^{234}U for 0.006 percent. All three are radioactive. ^{234}U and ^{238}U both decay in the same way and eventually become ^{206}Pb (lead-206), which is stable. ^{238}U has a half-life of 4.5 million years, and ^{234}U a half-life of 2.48 million years. They undergo ALPHA DECAY eight times and BETA DECAY six times. ^{235}U, with a half-life of 713 million years, undergoes seven alpha decay steps and four beta decay steps, and finally it also becomes a stable isotope of lead, ^{207}Pb.

It is impossible to predict when an individual nucleus will undergo decay, but the rate at which a particular isotope decays is very constant and can be used to determine the HALF-LIFE for that isotope. The constancy of radioactive decay means it can be used for RADIOMETRIC DATING, the best known version of which is RADIOCARBON DATING.

radiocarbon dating A method that is used to calculate the age of a sample taken from what was once a living organism by measuring the proportions of two CARBON ISOTOPES that the sample contains: carbon-12 (^{12}C) and carbon-14 (^{14}C).

^{12}C is the stable isotope; ^{14}C is formed in the atmosphere by the action of cosmic radiation on nitrogen (^{14}N). Cosmic radiation bombards the air with neutrons. Occasionally a neutron strikes the nucleus of a nitrogen atom and replaces one of its protons. This leaves the mass of the atom unchanged but alters it from $^{14}_{7}N$ to $^{14}_{6}C$. The ^{14}C is then oxidized to carbon dioxide (CO_2).

Plants absorb CO_2 and incorporate the carbon in sugars by the process of PHOTOSYNTHESIS. The plants do not discriminate between ^{12}C and ^{14}C, and so the two isotopes are present in plants in the same proportion as in the atmosphere. The carbon then passes to animals that eat the plants and their bodies also contain both isotopes.

^{14}C is unstable and undergoes radioactive decay with a HALF-LIFE of 5,730 years. While the plant or animal is alive the ^{14}C in its tissues is constantly being replenished, but after it dies it ceases to absorb carbon and so replenishment ceases. The ^{14}C continues to decay at a steady and known rate. This alters the $^{14}C:^{12}C$ ratio. Measuring the ratio in the sample and comparing it to that in the atmosphere therefore make it possible to calculate the time that has elapsed since the organism died. Obviously, the technique can be used only with material that once formed part of a living organism. It can date wood, linen, cotton, wool, hair, or bone, but not stone or metal.

When the technique was introduced in the late 1940s by the American chemist Willard Frank Libby (1908–80) the half-life of ^{14}C was thought to be 5,568 years. The year 1950 was set as the base (the "present") against which ages were to be reported, and before long material was being examined and then dated as having been formed so many years before the present (BP). Some years later the ^{14}C half-life was recalculated and all the previously announced dates had to be revised.

Then another difficulty was discovered. Radiocarbon dating was based on the assumption that the $^{14}C:^{12}C$ ratio in the atmosphere had remained constant for tens of thousands of years. This is now known to be untrue; the ratio varies a little. This means that ages measured by radiocarbon dating do not correspond to historical ages.

The difference can cause confusion. A human skeleton that was found in California in 2000 was reported as being 13,000 years old. At Monte Verde, a site in southern Chile, there are objects made by people and dated at 12,500 years. It sounds as though the California skeleton is older than the Monte Verde site, but this is not so. The California date was in calendar years, the Monte Verde date in radiocarbon years, and 12,500 radiocarbon years is equal to 14,700 calendar years.

It is important, therefore, to state what kind of years are being reported. Radiocarbon dates should always be described as *radiocarbon years BP*, or as *percent modern,* which means the proportion of ^{14}C in the

sample compared with that in samples that were formed in 1950, such as a piece of wood cut from a tree in that year.

Radiocarbon years are calibrated to convert them into calendar years. The calibration compares radiocarbon years and TREE RING years (*see* DENDROCHRONOLOGY). The $^{14}C:^{12}C$ ratio is measured in the sample. Then a tree ring with a similar ratio is found by searching through published data. The age of the tree ring is known and it must be the same as the age of the sample. The calibrated age is then reported as CalBC, CalAD, or CalBP, to indicate whether the age is measured as B.C. (B.C.E.), A.D. (C.E.), or BP.

The table compares some radiocarbon and calendar ages (in thousands of years).

Calendar	Radiocarbon
11	9.6
12	10.2
13	11.0
14	12.0
15	12.7
16	13.3
17	14.2
18	15.0
19	15.9
20	16.8

Radiocarbon dating is widely used in PALEOCLIMATOLOGY. It makes it possible for scientists to date plant material that they can link to particular climatic conditions.

(You can learn more about radiocarbon dating from the laboratory at the University of Oxford where this work is performed: www.rlaha.ox.ac.uk/orau/.)

radiometer   An instrument, carried on a satellite, that measures electromagnetic radiation. It may be passive or active. A passive instrument measures the radiation falling upon it. An active instrument emits a signal that is reflected back to it and compares the emission with its reflection. A radiometer may be designed to respond to any wavelength.

radiometric dating   A method used to determine the age of rocks and organic material that is based on the very regular rate of RADIOACTIVE DECAY. This regularity means that the proportions of "parent" (original, radioactive) and "daughter" (decay products) ISOTOPES present in the material can be used to calculate the time that has elapsed since the material was formed. At that time the material either contained the isotopes in different proportions or contained only the parent isotope if the daughters are produced only by the decay of the parent.

Several elements are used in radiometric dating. Organic material that is thousands of years old can be dated by RADIOCARBON DATING. Other decay series are used to date rocks, which are much older. The first to be introduced was the decay of uranium (half-life 4.51 billion years or 713 million years, depending on the isotope) to lead (Pb). Thorium-232 (^{232}Th), with a half-life of 13.9 billion years, decays to ^{207}Pb and this decay is also used for dating.

Potassium (^{40}K) decays to argon (^{40}Ar) with a half-life of 1.5 billion years. Rubidium (^{87}Rb) decays to strontium (^{87}Sr) by a single BETA DECAY, but there is uncertainty about the half-life, which may be 48.8 billion years or 50 billion years. Samarium (^{147}Sm), with a half-life of 250 billion years, undergoes ALPHA DECAY to become neodymium (^{143}Nd). All of these decays are used to date rocks.

radiosonde A package of instruments, carried aloft beneath a balloon, that measure atmospheric conditions and transmit the resulting data by radio to a surface receiving station. The first weather balloon to be equipped with a radio transmitter flew in 1927. Modern radiosondes came into service about 10 years later. Each of the weather stations that form part of the network monitoring the upper air releases one radiosonde at midnight and noon Greenwich time (Z) every day. Releasing all the balloons at the same time allows data from them to be compiled into a picture of conditions throughout the world at that time.

Midway between these launch times, at 0600 and 1800 Z, each station releases a balloon that carries only a radar reflector. This is tracked by radar to provide a profile of the wind speed and direction and the way this changes with height. A balloon of this type is called a *wind sonde*. A radiosonde that also carries a radar reflector, so it provides information on the wind at the same time as other meteorological data, is known as a *rawinsonde*. Wind direction is measured by noting the position of the sonde at intervals, usually of

one minute. Because of the wind, the balloon does not rise vertically, but the data it records are taken to represent a vertical profile.

There are about 90 upper air stations in the United States, 7 in Great Britain, and 2 in Ireland—approximately 1 upper air station for every 41,000 square miles (106,190 km²) in the United States and 1 for every 13,000 square miles (33,670 km²) in Great Britain and Ireland. In the world as a whole there are about 500 upper air stations.

The balloon is about 5 feet (1.5 m) in diameter and filled with helium. The instrument package is carried beneath the balloon at the end of a cable 98 feet (30 m) long. This length of cable prevents the contamination of instrument readings by effects from the balloon itself. Once released, the balloon rises at about 16 feet per second (5 m s^{-1}) to a height of 66,000–98,000 feet (20–30 km), where it bursts about 1–1.5 hours after launch. The instrument package returns to the surface by means of a parachute. About one-quarter of the packages released are recovered to be used again.

The package comprises a main body that contains an aneroid pressure capsule (ANEROID BAROMETER) to measure atmospheric pressure, a battery to supply power, electronic devices to convert data into a form suitable for radio transmission, and the radio transmitter. Above the main body there is an open structure containing a skin HYGROMETER and above that a plastic ring. This holds the fine wire of the electrical-resistance thermometer. Humidity readings from the hygrometer are ignored at heights above 33,000 feet (10 km) because the instrument is unreliable at the very low temperatures prevailing at this altitude.

Radiosonde data are augmented by more than 2,000 reports every day from aircraft.

Much bigger balloons are used for upper-atmosphere research. These are also filled with helium, but they are only partly filled before launch. As they ascend the helium expands to inflate the balloon fully. These research balloons are designed to return data from the middle and upper stratosphere.

radon (Rn) A radioactive element that is one of the NOBLE GASES. It is colorless, tasteless, and odorless. It has atomic number 86, relative atomic mass 222, and density 0.09 ounce per cubic inch (0.97 g cm^{-3}). It melts at −95.8° F (−71° C) and boils at −79.24° F (−61.8° C). Radon is formed by the RADIOACTIVE

DECAY of radium-226 and undergoes ALPHA DECAY. There are at least 20 ISOTOPES of radon, the most stable of which is radon-222, which has a HALF-LIFE of 3.8 days. Air contains very small but variable amounts of radon. Radium is a rare metal that occurs mainly in granites. Consequently, the concentration of radon is highest where the underlying rock is granite. As they decay, radon atoms ionize nearby atoms of atmospheric gases. This process increases the electrical conductivity of air, but the effect is extremely small. Most radon decay products, called *radon daughters,* also emit alpha radiation. In large doses, radon and its daughters are known to cause lung cancer, and so measures are taken to prevent the accumulation of radon inside buildings where the natural radon level is high.

raffiche   A mountain wind that blows in gusts in the Mediterranean region.

rain   Liquid precipitation consisting of droplets that are between 0.02 inch and 0.2 inch (0.5 and 5.0 mm) in diameter. The droplets vary considerably in size in heavy rain. Small raindrops are spherical, but those approaching 0.2 inch (5 mm) are variable in shape. Fine rain falls from NIMBOSTRATUS and STRATOCUMULUS when the cloud base is low. Droplets more than about 0.4 inch (1 mm) in size must fall through several thousand feet of cloud in order to grow to this size. They are usually associated with CUMULUS or CUMULONIMBUS clouds and fall as showers or in rainstorms.

rainbands   *See* BANDED PRECIPITATION.

rain before seven   The folk belief:

> *Rain before seven,*
> *Fine before eleven.*

This is often true, because bad weather that occurs very early in the morning has ample time to clear.

rainbow   An arch of colored bands that is the most familiar of optical weather phenomena. The colors are those of the spectrum—the colors that combine to make white light and into which white light can be broken by passing it through a prism. Not all the colors are visible in every rainbow, but when they are, they consist of violet, indigo, blue, green, yellow, orange, and red, with violet on the inside of the arch and red on the outside. Sometimes a secondary rainbow can be seen elevated about 8° above the primary, brighter, bow. The order of colors is reversed in the secondary bow.

A rainbow occurs when the Sun is behind the observer and is less than 42° above the horizon, and rain is falling in front at the same time. The observer sees sunlight that is reflected from the inside of the rear surface of raindrops, but that is also refracted twice, first as it enters each raindrop and again as it leaves. As light passes from air to the water of the raindrop its speed is slowed, but different wavelengths are slowed by different amounts. The violet wavelength is slowed most, red light is slowed least, and the colors between are slowed by intermediate amounts. The effect is to separate the colors. The angle between the sunlight striking the raindrops and the rainbow arch is 42° for red, on the outside of the arch, and 40° for violet, on the inside of the arch, with the other colors between.

Light from the rainbow approaches the observer at an angle of 40°–42° to the angle of the Sun's rays. If the Sun is more than 42° above the horizon, the red band at the top of the rainbow is below ground level and there is no rainbow. The lower the Sun, the higher the rainbow arc. From aircraft, a rainbow can sometimes be seen as a full circle, but it can never appear as more than a semicircle to an observer at ground level.

A secondary rainbow forms when light is reflected twice inside each raindrop. This increases to about 50° the angle by which red light is refracted and the second reflection reverses the order of colors. Consequently, the secondary rainbow is seen about 8° above the primary, with red on the inside.

rain cloud   Any cloud from which RAIN or DRIZZLE is likely to fall. The term has no precise meteorological meaning but usually refers to NIMBOSTRATUS. ALTOSTRATUS, STRATUS, and CUMULONIMBUS also produce rain, however, and so these may also be called rain clouds.

rain day   A day on which rain falls. This is usually taken to mean a period of 24 hours, beginning at 0900 Z (*see* GREENWICH MEAN TIME), during which at least 0.2 mm (0.08 inch) of rain falls. The numbers of rain days in one month and in one year indicate the distribution of rainfall through the year and provide a useful measure for comparing the climates of two places.

Seattle, Washington, for example, has an average of 51 rain days between April and September and 99 between October and March, indicating that its climate is rainier in winter than in summer. Las Vegas, Nevada, on the other hand, is dry throughout the year, with no more than 1 or 2 rain days in each month and a total of 18 rain days in the year. Taken together, the number of rain days and total annual rainfall give a clear picture of the seasonality of a climate. Seattle receives a total of 33.4 inches (848 mm) of rain a year on 150 rain days. Bombay, India, receives 71.3 inches (1,811 mm) a year but has only 72.7 rain days. The difference is that Bombay has a strongly seasonal MONSOON climate. Dividing the amount of rainfall by the number of rain days gives an idea of the intensity of the rainfall.

raindrops   RAIN is PRECIPITATION that reaches the ground as drops of water larger than those of DRIZZLE. The drops of water that fall from a cloud to reach the surface as rain vary in size, but there is a minimum size below which they are unable to survive their passage through the air below the CLOUD BASE. MIST and FOG consist of droplets about 100 μm (0.004 inch) in diameter, but these are really very large CLOUD DROPLETS. They form near the surface and are too small to fall, because air resistance and turbulence are sufficient to keep them aloft. Raindrops are at least about 1 mm (0.04 inch) across. The size limitation is due to the rate at which a drop of water falls in relation to the rate at which it evaporates.

The TERMINAL VELOCITY (V) of a drop of water falling through still air is equal to 8,000 times its radius ($V = 8 \times 10^3 r$), where r is the radius measured in SYSTÈME INTERNATIONAL D'UNITÉS (SI) units. A drop that is 1 mm (0.04 inch) in diameter therefore falls at about 4 m s^{-1} (157 inches per second). The base of a cloud marks the boundary between saturated and unsaturated air. If the base is at 60 m (200 feet), the drop spends 15 seconds falling through the dry air. Suppose, though, that the drop is half that size. It then spends 30 seconds in the dry air. If the air is turbulent, as it is inside CUMULIFORM clouds and around drops that are very much bigger, the terminal velocity is much slower ($V = 250 r^{1/2}$). A raindrop 5 mm (0.2 inch) across falls at about 0.4 m s^{-1} (16 inches per second) and takes 150 seconds to fall 60 m (200 ft).

The rate of evaporation depends on the temperature and RELATIVE HUMIDITY (RH) of the air beneath the cloud. This is variable, but even if it is about 95 percent, so the air is almost saturated, a drop that is less than about 30 μm (0.0012 inch) has evaporated completely by the time it has fallen a few inches. No drop smaller than about 100 μm (0.004 inch) in diameter is likely to reach the surface, because the base of most clouds is higher than 500 feet (150 m). That is the maximum distance such a drop can fall through air at 40° F (5° C) and RH 90 percent before it evaporates.

Bigger drops survive much better. A droplet 1 mm (0.04 inch) across can fall 42 km (26 miles) through air at this temperature and RH before it evaporates completely, and one 2.5 mm (0.1 inch) can fall 280 km (174 miles). It is not possible for raindrops to fall this distance, because above about 7 miles (11 km) the air is too dry for drops to form. Consequently they reach the surface as drops of water. Typical raindrops are between 0.08 inch (2 mm) and 0.2 inch (5 mm) in diameter, and they fall at 14–20 mph (23–33 km h^{-1}). Drops smaller than these that reach the ground are classed as drizzle.

Raindrops form from cloud droplets. These are typically about 20 μm (0.0008 inch) across. Depending on the temperature inside the cloud, they grow by collision (*see* COLLISION THEORY) or by the BERGERON–FINDEISEN MECHANISM. In order to grow to the size of a drizzle droplet, about 300 μm (0.1 inch) across, the cloud droplet must increase its volume more than 3,000 times, and to attain the size of a raindrop 2 mm (0.08 inch) across it must grow almost 1 million times bigger. (The volume of a sphere is equal to $4/3\pi r^3$, where r is the radius.)

rainfall frequency   A measure of how often rain falls in a particular place. This is of vital agricultural importance and determines the type of natural vegetation a region supports. Two places may both receive the same amount of rain in the course of a year, but the place where rainfall is distributed evenly throughout the year will have a quite different type of natural vegetation from the place where all the rain falls in one short season. The number and distribution of RAIN DAYS reveal the effectiveness of rainfall for agriculture and natural plant growth. Rainfall frequencies are also used to define RAINFALL REGIMES.

rainfall inversion   *See* PRECIPITATION INVERSION.

rainfall regime A CLIMATE CLASSIFICATION that is based on RAINFALL FREQUENCY. It was devised by the British climatologist W. G. Kendrew and published in his *The Climates of the Continents* (Oxford: Oxford University Press, first edition 1922, fifth edition 1961). Kendrew proposed six regimes.

(i) Equatorial regime. This comprises two seasons of heaviest rain, at or about the time the Sun is directly overhead, with drier periods between, but no dry season.

(ii) Tropical regime. There are two types of tropical regime. In the inner tropical, between the equator and latitude 10° N and S, maximum rainfall occurs twice in the year, separated by a long dry season. In the outer tropical regime, between 10° and the Tropics, the two wet seasons merge and the dry season is longer.

(iii) Monsoon regime. This has a marked summer maximum and a dry winter (*see* MONSOON). It occurs both inside and outside the Tropics.

(iv) Mediterranean regime. Most of the rain falls in winter, with either a single maximum in the middle of winter or two maxima, in spring and fall, and the summer is dry.

(v) Continental interior regime. This regime occurs in temperate latitudes. Most rain falls in summer. Winter rainfall is much less, but the winter is not completely dry.

(vi) West coastal regime. This regime occurs in temperate latitudes. Rain is abundant in all seasons, but heavier in the fall or winter than at other times.

rain forest A forest that grows where the rainfall is heavy and spread fairly evenly through the year. Rain forests grow in both the TROPICS and temperate latitudes.

Tropical rain forest was first defined in 1898 by the German botanist and ecologist Andreas Franz Wilhelm Schimper (1856–1901), who also coined the term (in German, *tropische Regenwald*). Schimper said the forest comprises evergreen trees that are at least 100 feet (30 m) tall, rich in thick-stemmed lianes (creepers), and with many woody and herbaceous (nonwoody) epiphytes growing on them. An epiphyte is a plant that grows on the surface of another plant but that is not a parasite. This definition still stands. The forest trees form a continuous canopy. There are also trees up to 100 feet (60 m) tall that stand high above the canopy.

Temperate rain forest develops outside the Tropics, wherever the annual rainfall exceeds 60–120 inches (1,500–3,000 mm). It consists of broad-leaved evergreen trees, often with coniferous species and with abundant climbers and epiphytes. This type of forest is found in coastal areas of the southeastern United States, northwestern North America, southern Chile, parts of Australia and New Zealand, and southern China and Japan.

rain forest climate In the THORNTHWAITE CLIMATE CLASSIFICATION, a climate in humidity province A, with a precipitation efficiency index greater than 127.

rain gauge The instrument that is used to measure the amount of rain that falls during a given period, usually one day. There are several ways to measure rainfall. The simplest is to leave an open-topped container exposed to the rain. This will collect rain, but

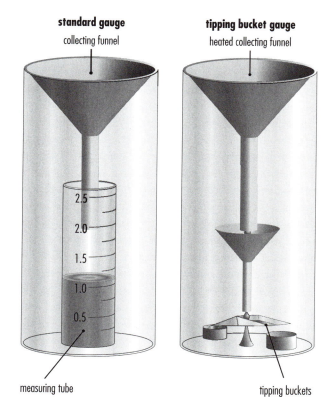

Rain gauge. **Two types of rain gauge that are widely used at weather stations. The standard gauge (left) has to be visited at regular intervals to be read and reset. The tipping-bucket gauge (right) allows the rainfall to be recorded automatically.**

water will also evaporate from it, especially after the rain has ceased falling, so it will not give an accurate reading. It is impossible to correct for this, because the rate of evaporation varies with the air temperature and wind speed.

The standard rain gauge is designed to minimize evaporation losses. It is called *standard* because it is the design that is approved internationally for use at WEATHER STATIONS. All standard rain gauges are made to the same specification and dimensions. When standard gauges are used, the scientists who use the data to prepare forecasts know that all the measurements reported to a meteorological center have been made in exactly the same way. If the staff at each station could decide for themselves how to make their measurements and what instruments to use, the data from one station would not be strictly comparable to those from another. What one station called 1.01 inches (25.65 mm) of rain, for example, another might call 1.00 inch (25.4 mm) and a third 1.02 inches (25.91 mm). The difference amounts to no more than a tiny one-hundredth of an inch (0.25 mm), but it would introduce an uncertainty into the data that is easily prevented by standardization.

The exterior of a standard gauge is a cylinder 20 cm (7.9 inches) in diameter that is mounted vertically with its top 1 meter (39.4 inches) above ground level. Rain falling into the cylinder enters a funnel that guides the water into a measuring tube inside the cylinder. The diameter of the measuring tube is 6.32 cm (2.49 inches); therefore, the cross-sectional area of the measuring tube is one-tenth that of the mouth of the funnel. Consequently, a column of water 10 inches (or millimeters) high in the measuring tube represents 1 inch (or millimeter) of rain entering the funnel. Multiplying by 10 makes it easier to read the rainfall accurately. The measuring tube may be calibrated, or the height of water in it may be measured with a measuring stick.

A standard gauge has to be visited at regular intervals, but it is also possible to measure and record rainfall automatically. The instrument most often used for this purpose is the tipping-bucket gauge.

Like the standard gauge, it is contained in a cylinder with a diameter of 20 cm (7.9 inches), but it has no measuring tube. Instead, rainwater entering the funnel is guided to a second, smaller funnel and from there into one of two buckets. Each bucket holds 0.25 mm (0.01 inch) of water. The two buckets are mounted on a rocker, like a seesaw. When 0.25 mm of water has flowed into a bucket its weight tips the bucket downward. The bucket then makes an electrical contact that is transmitted to a recording pen, which moves on a graph mounted on a rotating drum. At the same time, as the buckets tip the second bucket is positioned to collect water and the first bucket is emptied.

There are also gauges that automatically record the height of the water in the measuring tube. One design uses a float valve to do this and another measures the flow of an electric current through the water column.

Rainfall can also be measured automatically by a weighing gauge. In this device the collected water is fed into a cylinder that rests on a balance. Nowadays this is usually an electronic balance, but earlier instruments used a spring balance. The weight of the water is recorded on a graph.

All of these instruments are subject to errors. Although the standard gauge minimizes evaporation losses by enclosing the collecting funnel and measuring tube inside the outer cylinder, it cannot eliminate them entirely. Some water evaporates through the tube of the collecting funnel. The first rain to fall into the gauge wets the funnel and the film of water coating it does not reach the measuring tube, bucket, or weighing cylinder. This is called *wetting* and although the amount is no more than 0.04–0.08 inch (0.1–0.2 mm), where rainfall is extremely light this can be significant. Heavy rain can overflow tipping buckets, so they rock rapidly back and forth but underrecord the amount of rainfall.

Wind causes even greater inaccuracies. Eddies that form around the gauge carry a variable proportion of raindrops across the mouth of the funnel. Gauges are often placed inside a shield in order to minimize wind losses. The shield comprises a horizontal circular hoop, 20–40 inches (50–100 cm) in diameter, with baffles hanging vertically from it.

Trees, buildings, and other obstructions can shelter a rain gauge from rain that is falling obliquely. To prevent this, the distance between the gauge and the nearest obstruction should be at least equal to the height of the obstruction.

An optical rain gauge measures the intensity of rainfall. It transmits an infrared beam along a horizontal path 20–40 inches (50–100 cm) long to a detector. When the beam strikes a raindrop the radiation is scattered forward. The detector records the SCATTERING, from which the number of raindrops per second and in a unit volume of air can be counted. This provides a continuous record of rainfall intensity, but it does not measure the amount of rain directly.

rainmaking Any attempt to induce PRECIPITATION to fall from a cloud that otherwise might not release it. Several methods have been tried to achieve this. All of them aim to induce the formation of CLOUD DROPLETS and then to stimulate their growth. This can be done by injecting suitable material into the cloud. The process is called CLOUD SEEDING.

rainout The removal from the air of solid particles that are small enough to act as CLOUD CONDENSATION NUCLEI. Water vapor condenses onto the particles and the resulting CLOUD DROPLETS grow until they fall as precipitation, carrying the solid particles with them. Particles are also removed from the air by FALLOUT, IMPACTION, and WASHOUT.

rain shadow *See* OROGRAPHIC LIFTING.

rain squall A SQUALL that is accompanied by a short period of heavy rain. This falls from CUMULIFORM cloud, commonly CUMULONIMBUS, is carried by wind that blows outward from the center of the storm, and it often precedes the rain associated with the THUNDERSTORM.

rainy climate A climate in which the amount of rainfall is adequate to support the growth of plants that are not adapted to dry conditions. In the KÖPPEN CLIMATE CLASSIFICATION it is a climate classed as A (tropical forest) or C (warm temperate rainy). In the THORNTHWAITE CLIMATE CLASSIFICATION it is a climate in the HUMIDITY PROVINCES A and B.

rainy season A period each year when the amount of PRECIPITATION is much higher than it is at other times. The rainy season may occur in winter or in summer. Rainy winters occur around the Mediterranean Sea and on the western sides of continents in latitudes 30°–45° N and S. San Francisco, California, receives an average 22 inches (561 mm) of rain a year, of which 21 inches (534 mm) falls between October and April. Gibraltar, at the southern tip of Spain, has a very similar climate. Its average annual rainfall is 30 inches (770

mm), of which 28 inches (711 mm) falls between October and April.

In regions with a MONSOON climate it is the summer that is wet. Bombay, India, receives an average 71 inches (1,811 mm) of rain a year. Of this total, 67 inches (1,707 mm) falls between June and September.

rainy spell A period during which more rain falls than is usual for the place and time of year. In Britain the term is used more precisely to describe a period of 15 or more consecutive days during which the daily rainfall has been 0.008 inch (0.2 mm) or more.

raised beach A strip of land that contains rounded pebbles, sand, and seashells by which it can be recognized as having once been a beach, although it is now some distance above the seashore. The location of the shore has changed because either the land has risen as a result of movements in the Earth's crust, such as GLACIOISOSTASY, or the sea level has fallen. The accumulation of water in ICE SHEETS and GLACIERS is the most likely explanation for a fall in sea level (as opposed to a rise in land level).

random forecast A weather forecast in which the value for one feature is selected at random. This forecast is used to check the reliability of other forecasting methods. Many forecast methods are tested against a type of random forecast in which it is assumed that the weather will not change at all.

Rankine (R) A TEMPERATURE scale that is sometimes used as an alternative to the KELVIN SCALE, mainly in the United States. It is used, for example, in calculating DENSITY ALTITUDE, but its main use is in engineering. The scale was devised by the Scottish engineer and physicist William John Macquorn Rankine (1820–72). The Rankine degree is equal to 1° F, but the Rankine scale extends to ABSOLUTE ZERO: 0° R = −459.6° F. Water freezes at 492° R and boils at 672° R. The advantage of the Rankine scale is its compatibility with the Fahrenheit scale. Fahrenheit temperatures can be converted to Rankine temperatures without first converting the degrees themselves.

Raoult's law A law that was discovered by the French physical chemist François Marie Raoult

(1830–1901) in 1886. It states that when one substance (the solute) is dissolved in another (the solvent), the PARTIAL PRESSURE of the solvent vapor that is in equilibrium with the solution is directly proportional to the ratio of the number of solvent molecules to solute molecules. This means that the EQUILIBRIUM VAPOR density above the surface of a solution is lower than that above the surface of pure solvent and the more concentrated the solution, the greater is the difference. Consequently, more water enters the solution. CLOUD DROPLETS that form on HYGROSCOPIC NUCLEI are solutions that are often quite concentrated. Raoult's law shows that such droplets then grow rapidly by the CONDENSATION of more WATER VAPOR.

rasputitsa   A RAINY SEASON in Russia that occurs in spring and autumn nearly every year and lasts for several weeks. In autumn it turns the countryside into mud, makes unpaved roads impassable, and often causes serious flooding. The autumn season ends with the first frosts, which solidify the ground by freezing it. In spring, as the temperature rises, the frozen ground thaws and turns into mud once more. The spring rasputitsa ends as the continuing rise in temperature dries out the land.

ravine wind   A wind that blows along a narrow mountain valley or ravine. The wind is generated by a PRESSURE GRADIENT between the two ends of the valley. FUNNELING caused by the constricting effect of the valley sides accelerates the airflow, strengthening the wind.

raw   A colloquial description of weather that is cold, damp, and sometimes windy.

rawinsonde   *See* RADIOSONDE.

Rayleigh atmosphere   An idealized atmosphere that consists only of molecules and particles that are smaller than about one-tenth the wavelength of the solar radiation passing through it. The radiation would be subjected only to RAYLEIGH SCATTERING in such an atmosphere.

Rayleigh, Lord John William Strutt (1842–1919) English *Physicist* The scientist who explained, in 1871, why the sky is blue (*see* RAYLEIGH SCATTERING)

spent much of his life studying the properties of light and sound waves, waves in water, and earthquake waves, but his interests were much wider. His most famous discovery was in the field of chemistry, not physics. Rayleigh had become interested in the densities of different gases and found that when he measured the density of nitrogen in air it was always 0.5 percent greater than the density of that gas obtained from any other source. He eliminated all the reasons he could think of for this, but the disparity remained and in desperation in 1892 he published a short note in the scientific journal *Nature* asking for suggestions. He received a reply from the Scottish chemist William Ramsay (1852–1916), who solved the problem in 1894 by discovering the previously unknown gas argon. For their discovery, in 1904 Rayleigh was awarded the Nobel Prize in physics and Ramsay received the Nobel Prize in chemistry.

Rayleigh was born as John William Strutt at Terling Place, Langford Grove, Essex, on November 12, 1842. He was educated by a private tutor until he entered Trinity College, Cambridge, in 1861, and graduated in mathematics in 1865. After visiting the United States, he returned to England in 1868 and established a private laboratory at his home, Terling Place.

When his father died in 1873, when Rayleigh was 31, he acceded to the title, becoming the third baron Rayleigh. Lord Rayleigh, the name by which he is usually known, continued with his research, an occupation that was considered somewhat eccentric for an English aristocrat. Rayleigh was elected a Fellow of the Royal Society in the same year he acceded to his title, and in 1879 he succeeded James Clerk Maxwell (1831–79) as Cavendish Professor of Experimental Physics at Cambridge. After 1884 he spent most of his time working in his private laboratory. He held the post of professor of natural philosophy at the Royal Institution, London, from 1887 until 1905, but this position allowed him to remain at home for most of the time. Rayleigh was secretary to the Royal Society from 1885 until 1896 and its president from 1905 until 1908. From 1908 until his death he was chancellor of Cambridge University.

He died at Terling Place on June 30, 1919.

Rayleigh number   A DIMENSIONLESS NUMBER, calculated by LORD RAYLEIGH, that describes the amount of TURBULENT FLOW in air that is being heated from below by CONVECTION: $Ra = [(g \, \Delta \, \theta)/kv\theta](\Delta \, z)^3$, where

Ra is the Rayleigh number, *g* is the gravitational acceleration, $\Delta\theta$ is the POTENTIAL TEMPERATURE lapse in a layer of air with a depth Δz, *k* is the thermal conductivity of the air, and ν is the KINEMATIC VISCOSITY of the air. Convection always produces turbulence when the amount of heating is too great for energy to be transferred by conduction between molecules and when the resulting air motion is too vigorous for LAMINAR FLOW to be sustained by VISCOSITY. Turbulence occurs when the Rayleigh number exceeds approximately 50,000.

Rayleigh scattering   The way in which the direction of incoming solar radiation is altered by its interaction with air molecules and particles. The change in direction is due to the combined effects of DIFFRACTION, REFLECTION, and REFRACTION. This process was first observed experimentally by JOHN TYNDALL, but it was LORD RAYLEIGH who discovered the reason and showed that radiation is scattered most when the molecules and particles are smaller than the wavelengths of radiation being scattered. The angle through which radiation is scattered and its final direction are determined by the size of the scattering molecule or particle compared with the wavelength (λ) of the radiation, so that $\chi = (\pi d)/\lambda$, where χ is the size parameter and *d* is the diameter of the molecule or particle. The extent to which radiation is scattered is inversely proportional to the fourth power of the wavelength (λ^{-4}). This means the shorter wavelengths are scattered more than longer wavelengths and they are scattered in all directions, but with rather more scattering in the direction of the radiation than at right angles to it or in the opposite direction. The blue color of the sky is due to the fact that the shorter wavelengths are scattered most. Violet and indigo light, which have the shortest wavelengths of visible light, are absorbed by the atmosphere, so the sky is not violet or indigo. Blue scatters next and it is also close to the wavelength at which the Sun radiates most intensely. Blue light is scattered in all directions, coloring the entire sky (*see* SKY COLOR). The larger the particle, the more radiation is scattered in a forward direction. For spherical particles, such as CLOUD DROPLETS and some AEROSOLS, MIE SCATTERING occurs.

Re   *See* REYNOLDS NUMBER.

Réamur temperature scale   *See* TEMPERATURE.

Recent   *See* HOLOCENE EPOCH.

red shift   *See* DOPPLER EFFECT.

red sky   The observation that a red sky often indicates the weather that is likely to occur in the next few hours. This is summarized by the saying

Red sky at night, shepherd's delight;
Red sky in the morning, shepherd's warning.

The saying is not entirely reliable, because its accuracy depends on the rate at which weather systems are traveling, but it is correct more often than not. There are several versions of the rhyme (one substitutes *sailor* for *shepherd*) and the weather lore of many European cultures makes the link between a red sky and the approaching weather.

When the Sun is low in the sky, at dawn and sunset, its radiation travels obliquely through the atmosphere and so its path through the air is much longer than it is when the Sun is high in the sky. Blue light and green light are scattered repeatedly by collision with air molecules (*see* RAYLEIGH SCATTERING), allowing orange and red light to pass. If the air contains dust particles, these scatter light at the longer wavelengths and, because the particles are much larger than gas molecules, they scatter it predominantly in a forward direction. Then the sky in the direction of the Sun appears orange or red.

The red SKY COLOR indicates the presence of dust particles. Dust particles are soon washed to the surface by rain, and so their presence means the air is dry, and if the air is dry the weather must be fine. The Sun sets in the west. In midlatitudes most weather systems travel from west to east, so a red sky seen at sunset means fine weather is approaching and will probably arrive the following day. A red sky seen in the morning is in the east, where the Sun is rising. This means the fine weather has already passed and wet weather may be approaching from the west and, if so, it will arrive within a few hours.

reduced pressure   The AIR PRESSURE measured by a BAROMETER at a particular location after it has been corrected to bring it to a sea-level value. This reduction is necessary in order to make comparisons possible between the surface pressures at two places at different elevations.

At sea level the standard atmospheric pressure is 101.325 kPa (1013.25 mb [29.92 inches of mercury]). Pressure decreases with altitude and so when the sea-level pressure is 101.325 kPa the pressure at 5,000 feet (1,525 m) above sea level is 84.6 kPa and the pressure at 10,000 ft (3,000 m) is 70.1 kPa.

Barometric measurements taken at different elevations take no account of this effect. Therefore, comparisons are complicated. Suppose there are two stations, A and B. A is 10,000 feet above sea level and B is 5,000 feet above sea level. Station A reports its pressure as 82.7 kPa and station B reports 71.3 kPa. From these figures, and without considering the elevations of the two stations, it is not immediately obvious that the pressure at B is in fact higher than the pressure at A.

This is why all reported atmospheric pressures are corrected to their sea-level value. Sea level is used because, averaged over the year, its height is constant everywhere in the world (on shorter time scales it varies locally because of TIDES, STORM SURGES, ocean currents, and variations in atmospheric pressure). No such obvious and uncontroversial datum can be found for locations on land. The barometric reading is corrected by applying the known relationship between pressure and elevation. This reduces the measurement to its sea-level equivalent.

reflection   The "bouncing" of light when it strikes an opaque surface. It is always reflected at an angle equal to the angle at which it strikes the surface (*see* ANGLE OF INCIDENCE and ANGLE OF REFLECTION). Objects that are not themselves sources of light are made visible by the light reflected from them. If the surface is smooth, then all the light strikes it and is reflected from it at the same angle. If the surface is uneven, the angle at which light strikes it varies from place to place, and so does the angle at which it is reflected. Such uneven reflection scatters the light, producing multiple or distorted images.

refraction   The bending of light as it passes obliquely from one transparent medium to another, through which it travels at a different speed. The extent of the refraction is proportional to the difference in the speed of light in the two media and to the angle at which the light enters. The ratio of the speed of light in air to the speed of light in the medium is a constant for the medium, known as its *refractive index*. Air has a refractive

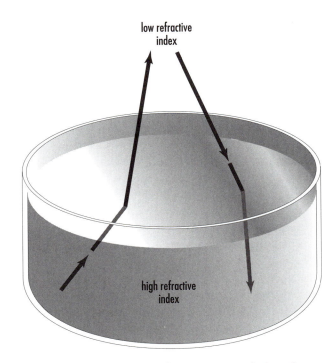

Refraction. **Light rays bend as they pass across the boundary between two transparent substances, such as air and water, through which light travels at different speeds.**

index of 1.0003, that of ice is 1.31, water is 1.33, window glass is 1.5. When light passes from a medium with a low refractive index into one with a high refractive index it is bent toward the vertical in relation to the surface between the two media. Light passing from a high to a low refractive index is bent in the opposite direction, away from a line vertical to the boundary.

refractive index   *See* REFRACTION.

regional climatology   *See* CLIMATOLOGY.

relative humidity (RH)   The ratio of the mass of water vapor that is present in a unit mass of dry air (the MIXING RATIO) to the amount that would be required to produce SATURATION (the SATURATION MIXING RATIO) in that air. It is written as a percentage and can be expressed as

mixing ratio ÷ saturation mixing ratio × 100

It is the measure of humidity that is most frequently used, and the one that is quoted in weather reports.

DRY-BULB TEMPERATURE (°C)	WET-BULB DEPRESSION (°C)														
	0.5	1.0	1.5	2.0	2.5	3.0	3.5	4.0	4.5	5.0	7.5	10.0	12.5	15.0	17.5
-10.0	85	69	54	39	24	10									
-7.5	87	73	60	48	35	22	10								
-5.0	88	77	66	54	43	32	21	11	0						
-2.5	90	80	70	60	50	41	31	22	12	3					
0.0	91	82	73	65	56	47	39	31	23	15					
2.5	92	84	76	68	61	53	46	18	31	24					
5.0	93	86	78	71	65	58	51	45	38	32	1				
7.5	93	87	80	74	68	62	56	50	44	38	11				
10.0	94	88	82	76	71	65	60	54	49	44	19				
12.5	94	89	84	78	73	68	63	58	53	48	25	4			
15.0	95	90	85	80	75	70	66	61	57	52	31	12			
17.5	95	90	86	81	77	72	68	64	60	55	36	18	2		
20.0	95	91	87	82	78	74	70	66	62	58	40	24	8		
22.5	96	92	87	83	80	76	72	68	64	61	44	28	14	1	
25.0	96	92	88	84	81	77	73	70	66	63	47	32	19	7	
27.5	96	92	89	85	82	78	75	71	68	65	50	36	23	12	1
30.0	96	93	89	86	82	79	76	73	70	67	52	39	27	16	6
32.5	97	93	90	86	83	80	77	74	71	68	54	42	30	20	11
35.0	97	93	90	87	84	81	78	75	72	69	56	44	33	23	14
37.5	97	94	91	87	85	82	79	76	73	70	58	46	36	26	18
40.0	97	94	91	88	85	82	79	77	74	72	59	48	38	29	21

To calculate the relative humidity:
1. determine the wet-bulb depression;
2. find the dry-bulb temperature in the left-hand column;
3. look along the top row to find the wet-bulb depression;
4. look down the column below the wet-bulb depression and along the row showing the dry-bulb temperature. The figure where the column and row intersect is the relative humidity.

For example, suppose the dry-bulb temperature is 7.5° C and the wet-bulb depression is 1.5° C. Look down the column that has 1.5 at the top and along the row that has 7.5 at the left. The column and row intersect at 80, therefore, the relative humidity is 80 %.

Note that relative humidity tables use temperatures in degrees Celsius (°C). To convert from Fahrenheit (°F), remember to distinguish between measurements of temperature and heat (the magnitude of degrees on the two scales). For the dry-bulb temperature, $°C = (°F - 32) \times 5 \div 9$.
For the wet-bulb depression, $°C = °F \times 5 \div 9$.

Relative humidity. **To calculate the relative humidity: (1) Calculate the wet-bulb depression (by subtracting the wet-bulb temperature from the dry-bulb temperature); (2) find the dry-bulb temperature in the column on the left; (3) find the wet-bulb depression in the row along the top; (4) look down the column below the wet-bulb depression and along the row from the dry-bulb temperature; the two intersect at the relative humidity.**

This is because it tells us what is important about the humidity from a meteorological point of view—how close the air is to saturation and, therefore, the likelihood that clouds will form or that precipitation will fall.

It is also easily misunderstood, because it gives no direct indication of the actual quantity of water vapor present. This is due to the fact that the moisture-holding capacity of the air varies with the temperature. Warm air can hold much more water vapor than cool air and yet have a much lower relative humidity. On a January day in Boston, Massachusetts, for example, the temperature might be 36° F (2° C) and the RH 63 percent. On the same day at Phoenix, Arizona, the temperature might be 65° F (18° C) and the RH 40 percent. It seems that the air over Boston contains more moisture (RH 63 percent) than the air over Phoenix (RH 40 percent). In fact, though, the reverse is true. Each cubic yard of the Boston air contains 0.16 ounce of water vapor (3.5 g m^{-3}) and each cubic yard of Phoenix air contains 0.28 ounce (6.1 g m^{-3}).

Its dependence on temperature is also its major disadvantage to atmospheric scientists. It does not yield values from which the moisture content of the air can be compared easily over different places or over the same place at different times.

RH is measured directly, using a HAIR HYGROMETER, or more accurately by using a PSYCHROMETER to determine the WET-BULB DEPRESSION. The RH is then read from a table containing RH values calculated from the dry-bulb temperature and wet-bulb depression.

relative hypsography *See* THICKNESS PATTERN.

relative isohypse *See* THICKNESS LINE.

relative pollen frequency (RPF) The number of pollen grains of a particular species, genus, or family of plants that can be counted in a given volume of sediment and, if the rate of deposition is known in a unit of time, expressed as a percentage of either the total amount of pollen present or the total amount of tree pollen present. This is the most widely used method for comparing pollen diagrams, but for some purposes measuring the absolute pollen frequency is preferred. The amount and type of pollen present in a sediment that can be dated, for example by radiometric dating, allow the vegetation at that time to be identified. This gives a reliable indication of the type of climate at the time.

relative vorticity (ζ) The VORTICITY about a vertical axis that a mass of fluid possesses by virtue of its own motion relative to the surface of the Earth. In the Northern Hemisphere a CYCLONE possesses positive (counterclockwise) vorticity and an ANTICYCLONE possesses negative (clockwise) vorticity. In the Southern

Substance	Symbol	Atmospheric residence time
nitrogen	N_2	42 million years
oxygen	O_2	1,000 years
CFC-114	$C_2F_4Cl_2$	300 years
HCFC-23	CHF_3	250 years
CFC-12	CF_2Cl_3	100 years
CFC-113	$C_2F_3Cl_3$	85 years
H-1301	CF_3Br	65 years
carbon dioxide	CO_2	55 years
CFC-11	$CFCl_3$	45 years
carbon tetrachloride	CCl_4	35 years
H-1211	CF_2ClBr	16 years
HCFC-22	CHF_2Cl	12 years
methane	CH_4	11 years
HCFC-124	C_2HF_4Cl	5.9 years
methyl chloride	CH_3Cl	1.5 years
HCFC-123	$C_2HF_3Cl_2$	1.4 years
water	H_2O	10 days
ammonia	NH_3	7 days
smoke		hours
large particles		minutes

Hemisphere a cyclone possesses negative vorticity and an anticyclone positive vorticity.

remote sensing Obtaining information about a subject without direct physical contact with it. Instruments carried by a RADIOSONDE are in contact with the air they are monitoring, so although the meteorologists who receive data from them are not in contact with that air, the sensing itself is direct, not remote. An orbiting satellite, on the other hand, transmits atmospheric data from outside the atmosphere. This sensing is remote and works by acquiring images in various parts of the electromagnetic spectrum, such as the microwave, infrared, visible light, and ultraviolet wave bands. Scientists on the ground are able to interpret these images to obtain information about various aspects of climate, such as temperatures, clouds, and humidity.

Renaldini, Carlo (1615–1698) Italian *Physicist* An aristocrat from Ancona, on the Adriatic coast of northern Italy, Renaldini became professor of philosophy at the University of Pisa. He was also a member of the Accademia del Cimento (Academy of Experiments), in Florence. The accademia was the principal institution studying the atmosphere. In 1694 Renaldini proposed a method for calibrating thermometers. This process was causing difficulty, because when heated from freezing to boiling temperature water does not expand at a constant rate and there seemed no reason to suppose that any other liquid would do so. It was agreed that the freezing and boiling temperatures of water should mark two ends of the temperature scale, and Renaldini suggested the intermediate points might be identified by mixing boiling and ice-cold water in varying proportions. Equal weights of water at 32° F (0° C) and 212° F (100° C) would reveal the point at which to mark the halfway point, 122° F (50° C): 20 parts of boiling water mixed with 80 parts of freezing water would indicate 68° F (20° C), and so on, with each degree rise in temperature corresponding to the addition of the same amount of heat. The method proved difficult to put into practice and did not overcome the difficulty of depending on the behavior of a particular liquid.

réseau The name that is used by the WORLD METEOROLOGICAL ORGANIZATION to describe the global network of weather stations that have been chosen to represent the world climate. The full name is *réseau mondiale*, which is French for "global network."

reshabar A strong, dry, northeasterly KATABATIC WIND that blows down the sides of the mountain ranges in Kurdistan, in southern Iraq and Iran. The name means "black wind." It is hot in summer and cold in winter.

residence time The atmosphere is a dynamic system. DUST particles, mineral grains, POLLEN, SPORES, as well as molecules of WATER VAPOR and the other gases that constitute the air are constantly entering the atmosphere and leaving it. The residence time of any individual atom, molecule, or particle is the length of time that it remains in the air.

Residence time (R) is calculated as the total mass (M) of the substance divided by the FLUX (F), which is the rate at which it is entering and leaving the air ($R = M/F$). The larger the total mass the longer the residence time. NITROGEN is the principal atmospheric gas. Although a very large amount of nitrogen leaves the air every day by NITROGEN FIXATION and enters it by denitrification (see NITROGEN CYCLE), the atmosphere contains approximately 4,466 million million tons (4.466×10^{15}, or 4.06×10^{18} kg). The flux therefore represents only a minute fraction of the total mass of nitrogen in the air. Consequently, the average time a nitrogen molecule spends in the air between being released by denitrification and removed again by nitrogen fixation is about 42 million years.

The atmosphere contains about 1,200 million million tons (1.2×10^{15}, or 1.09×10^{18} kg) of OXYGEN. Oxygen is removed from the air mainly by respiration and is returned to it by the process of PHOTOSYNTHESIS. An oxygen molecule remains in the air for an average of about 1,000 years.

CARBON DIOXIDE is present in trace amounts and is cycled rapidly by photosynthesis and respiration. A CO_2 molecule remains airborne for an average of about 55 years. METHANE has a residence time of about 11 years. Methyl chloride (CH_3Cl), which is produced by chemical reactions in seawater and is the largest natural source of atmospheric chlorine, has a residence time of 1.5 years. HALONS have residence times of 16 years for H-1211 (CF_2ClBr) and 65 years for H-1301 (CF_3Br). CFC-11 and CFC-12, which are the most common chlorofluorocarbons (CFCs), have residence times of 45 years and 100 years, respectively.

There is even more water on the Earth than there is nitrogen. The total amount of water is about 1.54×10^{18} tons ($1,400 \times 10^{18}$ kg). The residence time of a water molecule is shorter than that of a nitrogen molecule, however, because the flux is much greater. Much more water is moving through the HYDROLOGICAL CYCLE than through the nitrogen cycle. A water molecule spends an average 4,000 years in the ocean; about 400 years in lakes, rivers, and GROUND WATER and as ice; and only about 10 days in the atmosphere.

Solid particles remain in the air for quite short periods. They obey the same law as gas molecules, but their total mass is much smaller than that of the atmospheric gases and their flux rate is greater. Some particles act as CLOUD CONDENSATION NUCLEI and are removed by RAINOUT or SNOWOUT. Others are swept from clouds and from the air beneath clouds by PRECIPITATION and so are removed from the air by WASHOUT. Washout is much more efficient than rainout at removing particles. In addition, solid particles settle by gravity and the rate at which they do so must also be taken into account. Most particles bigger than 1μm (0.00004 inch) in diameter are removed from the air by settling and those more than 10 μm (0.0004 inch) usually remain in the air for only a few minutes. Smoke particles usually remain airborne for a few hours.

resistance hygrometer   A HYGROMETER that measures RELATIVE HUMIDITY directly, with an accuracy of ±10 percent. It exploits the property that the electrical resistance of certain materials, such as carbon black, changes with variations in the relative humidity. An electrical conductor is coated with the material and changes in the current flow are measured. Because radiosondes and satellites measure temperature at the same time as they measure relative humidity, other types of HUMIDITY values can be calculated from resistance-hygrometer data. As a result of the widespread reliance on radiosonde and satellite measurements, this is now the most widely used type of hygrometer.

resonance trough   A TROUGH that forms at some distance from a major trough and that is related to it. The distance between a resonance and dominant trough is measured in WAVELENGTHS of ROSSBY WAVES. The trough that forms in winter over the Mediter-

ranean Sea may be a resonance trough between the major troughs that lie above the eastern coasts of North America and Asia.

respiration   The process by which living organisms release energy by the oxidation of carbon. Breathing is the pumping action by which vertebrate land animals draw air into their lungs to obtain oxygen and expel carbon dioxide.

The carbon is in the form of glucose ($C_6H_{12}O_6$), a sugar that is produced in green plants and some bacteria and cyanobacteria by the process of PHOTOSYNTHESIS. In Archea, some BACTERIA, and some fungi respiration is anaerobic (takes place in the absence of oxygen). The anaerobic reaction in the case of yeast, which is a fungus, is known as *fermentation* and it can be summarized as

$$C_6H_{12}O_6 \rightarrow 2C_2H_5OH + 2CO_2$$

C_2H_5OH is ethanol, which is also called *ethyl alcohol* or just *alcohol*.

Aerobic respiration, which requires the presence of oxygen, can be summarized as

$$C_6H_{12}O_6 + 6O_2 \rightarrow 6CO_2 + 6H_2O$$

Plants and all animals practice aerobic respiration. Species that dwell on land obtain their oxygen directly from the atmosphere. Aquatic organisms rely on oxygen that is dissolved in the water (they do not obtain oxygen by splitting water molecules).

In fact, the reaction in both cases takes place as a sequence of steps and is a great deal more complicated than this summary makes it appear. The energy that is released is used to attach a phosphate group to adenosine diphosphate (ADP), making it into adenosine triphosphate (ATP). ATP is transported through the organism and wherever energy is required a phosphate group is discarded (ATP becomes ADP) with a release of energy. All the energy used by living organisms from bacteria to trees to people is transported and released by the ADP ↔ ATP reaction.

So far as the atmosphere is concerned, the process of respiration returns to the air the CO_2 that is removed by photosynthesis, and aerobic respiration removes from the air the oxygen that is released by photosynthesis. Together photosynthesis and respiration maintain a constant atmospheric concentration of these two gases.

response time The time that elapses between a change in the amount of energy that is available in one part of the climate system and the effect that energy produces. This varies greatly according to the type of surface.

For example, in early spring there is an increase in the amount of solar heat that reaches the ground. If snow and ice cover the ground, however, most of that heat is reflected. This reflection greatly reduces its warming effect and the extent to which it warms the air in contact with the surface. The climate responds slowly to the increase in energy. Where the ground is free of ice and snow it warms rapidly and the response time of the climate is short. The oceans absorb a large amount of heat before the water temperature increases. This is because of the high HEAT CAPACITY of water. Air in contact with the sea surface warms eventually, but the absorption of heat delays that response. Ocean currents also transport the warmed water over long distances and it may also be carried deep below the surface and held there for years before returning to the surface and warming the air. In this case the response time is very long.

resultant wind The average speed and direction of the wind at a particular place over a specified period. It is calculated by recording the wind speed and direction at intervals throughout the period, then calculating their mean values.

return period The frequency with which a rare natural phenomenon may be expected to occur. It is based on recorded occurrences in the past and is then expressed as a range of values. These values also represent the statistical probability that the phenomenon will occur in any particular year. For example, records and calculations may indicate that a certain area is likely to be flooded once every 10 years. There is therefore a 10 percent chance that it will be flooded in any particular year and the flood that affects it is known as *a 10-year flood*. Ten years is then said to be the return period for that event. With this method, there can be 50-year, 100-year, or 1,000-year floods. The less frequent the event, the more severe it is when it happens. In 1952 the English village of Lynmouth was severely damaged by flooding. This was an event so unlikely that it was classed as a *50,000-year flood*. Windstorms, blizzards, droughts, and other types of hazardous weather can be assigned probabilities in the same way.

(You can learn more about flood risk and the Lynmouth flood in Michael Allaby, *Dangerous Weather: Floods* [New York: Facts On File, 1998].)

return stroke The most visible part of a LIGHTNING flash, which carries positive electrical charge toward the cloud from which the STEPPED LEADER originated. In the case of forked lightning, the visible flash travels not from the cloud to the ground, but from the ground to the cloud. The stepped leader, carrying negative charge, ionizes the air along a LIGHTNING CHANNEL. As it approaches the region of positive charge the return stroke sets out in the opposite direction and meets it. This neutralizes the stepped leader, but it may not neutralize the entire cloud, so the flash continues, its next stage initiated by a DART LEADER.

reverse osmosis *See* OSMOSIS.

revolving storm A STORM in which the air moves in a CYCLONIC direction, so it rotates about a low-pressure center. TROPICAL CYCLONES are revolving storms. Revolving storms are contrasted to convectional storms that are produced by CUMULONIMBUS clouds that are isolated or that form a SQUALL LINE.

Rex A TYPHOON that struck northern Japan in August 1998. It caused floods and landslides in which at least 11 people died, and 40,000 were forced to evacuate their homes.

Reynolds effect A process by which CLOUD DROPLETS grow in WARM CLOUDS that was discovered by OSBORNE REYNOLDS. In a cloud where some droplets are warmer than others, water evaporates from the warmer droplets and condenses onto the cooler droplets.

Reynolds number (*Re*) A DIMENSIONLESS NUMBER that is used to measure the extent to which a fluid flows smoothly (*see* LAMINAR FLOW) or turbulently (*see* TURBULENT FLOW). For a fluid flowing through a pipe it is calculated by

$$Re = v\rho l/\eta$$

where v is the flow VELOCITY, ρ is the DENSITY of the fluid, l is the radius of the pipe, and η is the VISCOSITY of the fluid. For air, Re can be calculated by

$$Re = LV/v$$

where L is the distance over which the air is moving, V is its velocity, and v is its KINEMATIC VISCOSITY, which is equal to approximately 16×10^{-5} ft^2 s^{-1} (1.5×10^{-5} m^2 s^{-1}). If Re is less than about 1,000, the flow is dominated by VISCOSITY. If Re is greater than about l,000, the flow is dominated by turbulence. Except on a very small scale, such as the flow around a spherical ball, Re is usually much larger than 10^3. Typically it is 10^6 or 10^7. The number was discovered by the physicist OSBORNE REYNOLDS.

Reynolds, Osborne (1842–1912) English *Physicist and engineer* Osborne Reynolds was born on August 23, 1842, in Belfast, Northern Ireland. His father was a mathematician who became a schoolteacher at Dedham Grammar School, in Suffolk, England. This necessitated the family's moving to England, but in a sense the family was returning home. Osborne's great-grandfather and great-great grandfather had both held the position of rector in the parish of Debach-with-Boulge, to the northeast of Ipswich. Later, Osborne's father also became rector there.

Reynolds began his education at Dedham Grammar School. A proficient mathematician, he left school at 19 and went to work for an engineering company, where he was able to apply his knowledge of mathematics. Having gained some practical experience, Reynolds enrolled at Queens' College, Cambridge, to study mathematics. He graduated in 1867 and was immediately awarded a fellowship at Queens'. He then moved to London to work for John Lawson, a civil engineer, but left in 1868 because he had been elected the first professor of engineering at Owens College, Manchester (now the University of Manchester). Reynolds was elected a Fellow of the Royal Society in 1877 and received the society's Royal Medal in 1888. This award was followed by many more medals and honorary degrees.

Reynolds conducted research into the movements of comets and atmospheric phenomena caused by electricity, but his most important work concerned the flow of water through channels, including wave and tidal movements in rivers and estuaries. His discoveries led to his formulation of what is now known as the *REYNOLDS NUMBER*. The Reynolds number is widely used by meteorologists and atmospheric physicists in calculations of the TURBULENT FLOW of air. Account must be taken of turbulence when designing buildings, aircraft, or any other objects that move through the air or that the air moves around. Turbulence also affects the rate at which atmospheric pollutants disperse.

He retired in 1905 and died at Watchet, Somerset, on February 21, 1912.

RH *See* RELATIVE HUMIDITY.

RI *See* NATIONAL RAINFALL INDEX.

Richardson, Lewis Fry (1881–1953) English *Mathematician and meteorologist* In 1922, the book *Weather Prediction by Numerical Process*, which described a system for NUMERICAL FORECASTING, which is the preparation of weather forecasts by the use of mathematical techniques for solving equations, was published. Lewis Fry Richardson, its author, had been developing the system since 1913, and his book contained a worked example to show how it might be done.

Unfortunately, Richardson was half a century ahead of his time. Accurate weather forecasting requires detailed information about conditions in the upper atmosphere. Modern meteorologists have access to these data, but they were not available in the 1920s. Nor did Richardson and his contemporaries have access to modern computers or even pocket calculators. His scheme involved so many calculations, all of which had to be performed with paper and pencil, that by the time they were completed the conditions they were forecasting would have come and gone. Had it been possible to produce them fast enough, the forecasts would not have been very reliable. The equations Richardson used work only under certain atmospheric conditions, a fact of which he was unaware. These deficiencies made the method impractical and little attention was paid to it. Today, upper-air data and adequate computing power are available. Richardson's book was republished in 1965 and a modified version of his method is now used for making large-scale, long-range weather forecasts.

Lewis Fry Richardson was born into a family of tanners on October 11, 1881, at Newcastle-upon-Tyne, an industrial city in the northeast of England. It was a Quaker family and Richardson remained a committed Quaker throughout his life. He was educated at a school in York and then studied natural science (physics, mathematics, chemistry, biology, and zoology) at King's College, at the University of Cambridge. In 1927, at the age of 47, he was awarded a doctorate in mathematical psychology by the University of London. (Mathematical psychology is the name given to any theoretical work that uses mathematical methods, formal logic, or computer simulation.)

From 1903 to 1904 he worked for the National Physical Laboratory, a government research institution, and in 1912, after the sinking of the *Titanic*, he conducted experiments with echo sounding. In 1913 he went to work at the METEOROLOGICAL OFFICE.

His mathematical work, concerned mainly with the calculus of finite differences, showed great originality. Later in life he even sought to use mathematics to identify and clarify the causes of war, an interest that arose from his religious beliefs.

During the First World War, Richardson served among fellow Quakers in the Friends Ambulance Unit. He was sent to France, where he tried to practice Esperanto, another interest of his, on German prisoners of war. After the war he returned to the Meteorological Office.

As a member of the Society of Friends, Richardson was a pacifist, and in 1920 he resigned from the Meteorological Office when it was absorbed into the Air Ministry. He became head of the Physics Department at Westminster Training College, and in 1929 he was appointed principal of Paisley Technical College (now the University of Paisley) in Scotland. He remained at Paisley until his retirement in 1940.

Retirement gave him time to develop his ideas on eradicating sources of conflict. His two books on the subject, *Arms and Insecurity* and *Statistics of Deadly Quarrels,* were published posthumously in 1960, attracting interest among religious and pacifist groups.

During his lifetime he was best known for his studies of atmospheric turbulence. This led to his proposal of what came to be called the *RICHARDSON NUMBER* for predicting whether turbulence would increase or decrease.

In 1926 Richardson was elected a Fellow of the Royal Society of London. He died at Kilmun, Argyll, Scotland, on September 30, 1953.

(You can learn more about Lewis Fry Richardson from his home page at the University of Paisley: maths.paisley.ac.uk/lfr.htm, and from www.mpae.gwdg.de/EGS/egs_info/richardson.htm.)

Richardson number   A mathematical value, devised by the English mathematician and meteorologist LEWIS FRY RICHARDSON (1881–1953), that makes it possible to predict whether atmospheric turbulence (*see* TURBULENT FLOW) is likely to increase or decrease. The Richardson number (Ri) is calculated from the strength of the WIND SHEAR, BUOYANCY of the air, and POTENTIAL TEMPERATURE. It represents the ratio of the rate at which the KINETIC ENERGY of the turbulent motion is being dissipated by buoyancy due to natural or FREE CONVECTION to the rate at which kinetic energy is being produced by mechanical or FORCED CONVECTION. If Ri is greater than 0.25, turbulence decreases and disappears. If Ri is less than 0.25, turbulence increases.

Richardson's jingle   Two lines of verse, written by LEWIS FRY RICHARDSON, that summarize the way large atmospheric EDDIES produce smaller eddies:

*Big whirls have little whirls that feed on their velocity
And little whirls have smaller whirls, and so on to viscosity.*

ridge   A long, tonguelike protrusion of high pressure into an area of lower pressure. The waves in the POLAR FRONT JET STREAM associated with the INDEX CYCLE that extend toward the North Pole are also called *ridges.*

rime frost   A layer of ice that is white and has an irregular surface. It forms when water droplets in SUPERCOOLED FOG or DRIZZLE freeze on contact with a surface that is at or below freezing temperature. Rime can also form by the DEPOSITION of water vapor. Once the process has commenced, water freezes or is deposited onto ice crystals that are already present. If there is a wind, rime ice forms only on the sides of structures that face into the wind. Rime ice grows into elaborate, delicate feathery shapes.

ring around the Moon   *See* HALO.

ring vortex   *See* MICROBURST.

Riss glacial A GLACIAL PERIOD in the European Alps, named after an alpine river, that is equivalent to the WOLSTONIAN GLACIAL of Britain and the Saalian glacial of northern Europe and partly coincides with the ILLINOIAN GLACIAL of North America. The Riss began about 200,000 years ago and ended about 130,000 years ago. It followed the MINDEL–RISS INTER-GLACIAL. The INTERGLACIAL that followed it is known as the *RISS/WÜRM INTERGLACIAL* in the European Alps and as the *EEMIAN INTERGLACIAL* in northern Europe.

Riss/Würm interglacial An INTERGLACIAL that occurred in the European Alps from about 130,000 years ago until 70,000 years ago. It was equivalent to the EEMIAN INTERGLACIAL of northern Europe and IPSWICHIAN INTERGLACIAL of Britain. It was the last interglacial before the onset of the Würm glacial, which is equivalent to the DEVENSIAN GLACIAL in Britain and the WISCONSINIAN GLACIAL in North America.

Rita Two TYPHOONS, the first of which struck the Japanese island of Shikoku in August 1975, one week after Typhoon PHYLLIS. Rita killed 26 people and injured 52. The second Typhoon Rita, rated category 4 on the SAFFIR/SIMPSON SCALE, struck the Philippines on October 26, 1978. Nearly 200 people were killed and about 10,000 homes were destroyed.

RMS *See* ROOT-MEAN-SQUARE.

Rn *See* RADON.

Ro *See* ROSSBY NUMBER.

roaring forties *See* POLAR WET CLIMATE.

rock mushroom *See* MUSHROOM ROCK.

ROFOR An internationally recognized abbreviation for a ROUTE FORECAST that is prepared for pilots who are about to fly along a particular route.

ROFOT An internationally recognized abbreviation for a ROUTE FORECAST that uses English units of measurement (miles, knots, degrees Fahrenheit, etc.)

roll vortex *See* BOLSTER EDDY.

Römer temperature scale A temperature scale that was devised in about 1701 by the Danish astronomer, physicist, and instrument maker Ole Christensen Römer (1644–1710). In 1701, Isaac Newton (1642–1727) had suggested that average body temperature and the temperature of freezing water should be used as the two FIDUCIAL POINTS against which thermometers should be calibrated. Römer never published a description of the method he used, but in 1708 DANIEL FAHRENHEIT visited him, watched him at work, and wrote his own account.

Römer inserted the thermometer into freezing water and marked the point reached by the alcohol in the thermometer. He then placed the thermometer into tepid water, which Fahrenheit wrote was at blood heat (*blutwarm*). He then added half of the distance between these points below the lower fiducial point and marked this lowest point as 0. There is some confusion about the lower fiducial point, however. Some historians hold that Römer used a mixture of water, ice, and ammonium chloride to determine the lower fiducial point and called that 0, others that he used melting snow only and called that point $7^1/_2$. In either case, he divided the distance between 0 and the upper fiducial point into $22^1/_2$ parts.

On the Römer scale water freezes at $7^1/_2°$, boils at $60°$, and average body temperature is $22^1/_2°$. The scale is no longer used but is important historically, because it is the one on which Fahrenheit based his scale (*see* FAHRENHEIT TEMPERATURE SCALE).

ROMET An internationally recognized abbreviation for a ROUTE FORECAST that uses metric units (kilometers, kilometers per hour, degrees Celsius, etc.).

root-mean-square (RMS) A method for determining the value of a quantity that is fluctuating. It is calculated by sampling the values, squaring them, averaging them, and then finding the square root of the average.

Rossby, Carl-Gustav Arvid (1898–1957) Swedish-American *Meteorologist* Carl-Gustav Rossby was born in Stockholm, Sweden, on December 28, 1898. He was educated at the University of Stockholm; in 1918, after graduating with a degree in theoretical mechanics, he moved to the BERGEN GEOPHYSICAL INSTITUTE. There he worked with VILHELM BJERKNES on oceanographic as well as meteorological problems.

When Bjerknes moved to Germany in 1921 to take up a position at the University of Leipzig, Rossby followed him and spent a year there. In 1922 Rossby returned to Stockholm to join the Swedish Meteorological Hydrologic Service.

During the next three years Rossby traveled as the meteorologist on several oceanographic expeditions. He also studied mathematics at the University of Stockholm and graduated in 1925 with a licentiate (a European degree one rank below a doctorate).

In 1926 Rossby visited the United States with a scholarship from the Scandinavian–American Foundation. He joined the staff of the Weather Bureau in Washington, D.C. At the time this was the only meteorological center in the United States. While working there Rossby wrote several papers on turbulence in the atmosphere and on the dynamics of the STRATOSPHERE.

In 1927 he moved to California. Sponsored by the Daniel Guggenheim Fund for the Promotion of Aeronautics, Rossby established experimentally the first weather service that was designed expressly for the benefit of aviators. It became the model on which later aviation weather services were based.

The following year he received his first important academic appointment when he was made the country's first professor of meteorology, at the Massachusetts Institute of Technology (MIT). He devised the first university meteorological program, and during the 11 years he spent at MIT he continued to pursue his research interests in the thermodynamics of AIR MASSES, turbulence in the atmosphere and oceans, and BOUNDARY LAYERS. Later he became increasingly interested in large-scale atmospheric movements.

In 1939 he was appointed assistant chief of research and education at the UNITED STATES WEATHER BUREAU, but he left in 1940 to become chairman of the Institute of Meteorology at the University of Chicago. It was soon after his arrival in Chicago that Rossby developed his theory describing the long atmospheric waves that now bear his name.

During World War II Rossby organized training for military meteorologists, while continuing his research on long waves. His wartime work took him to many parts of the world and into personal contact with many British as well as American meteorologists. After the war he was able to recruit many of these scientists to the University of Chicago, where together they played an important part in developing the mathematics needed to introduce NUMERICAL FORECASTING using electronic computers.

In 1947 Rossby was invited to establish a department of meteorology at the University of Stockholm. This was funded by American and Swedish foundations as well as by international bodies including UNESCO and it attracted students from many countries. Appropriately, the institute was called the *International Institute of Meteorology*. Rossby divided his time between working at the institute in Stockholm and at the outpost of the University of Chicago that was opened at the Woods Hole Oceanographic Institute. His work at Stockholm was concerned mainly with the development of numerical forecasting methods for European weather services.

Rossby died in Stockholm on August 19, 1957.

Rossby number (*Ro*)   A DIMENSIONLESS NUMBER that was discovered by the meteorologist CARL-GUSTAV ARVID ROSSBY. The Rossby number is the ratio of the ACCELERATION of moving air due to the PRESSURE GRADIENT and the CORIOLIS EFFECT: *Ro* = (relative acceleration)/(Coriolis effect). It is given by

$$Ro = U/\Omega L$$

where *U* is the horizontal wind VELOCITY, Ω is the ANGULAR VELOCITY of the Earth, and *L* is the horizontal distance over which the wind travels.

Rossby waves (long waves, planetary waves)   Waves that develop in moving air in the middle and upper troposphere. They have wavelengths of 4,000–6,000 km (2,485–3,728 miles) and are named after CARL-GUSTAV ARVID ROSSBY, who discovered them.

rotating cups anemometer   *See* ANEMOMETER.

rotor   *See* ROTOR CLOUD.

rotor cloud   A cloud that forms at the crest of the first of a series of LEE WAVES if the downwind side of the mountain is very steep. It is the forced movement of stable air across a mountain that triggers the development of lee waves. A CAP CLOUD often forms at the mountain peak, where water vapor condenses in the air that has been cooled in an ADIABATIC process. LENTICULAR CLOUDS may form farther downwind. On the lee side of a steep slope, the smooth flow of air associated with the lee waves breaks down and beneath the crest

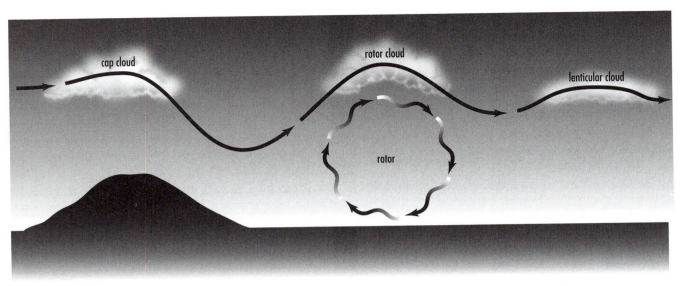

Rotor cloud. **A small cumulus cloud that forms downwind of a steep-sided hill at the crest of the wave induced in the flow of air by its passage over the hill**

of the first wave the air may be rotating. This is called a *rotor*. Air within it is extremely turbulent and occasionally a rotor may cause the wind direction at ground level to be the reverse of that in the waves themselves. A rotor can be very dangerous for aircraft. A rotor cloud develops at the top of the rotor if air is sufficiently humid for its water vapor to condense. The air within the cloud retains the turbulent motion of the rotor, producing a cloud that is much more CUMULIFORM in shape than either the cap or lenticular clouds.

roughness length *See* AERODYNAMIC ROUGHNESS.

route component The component of the wind direction that is aligned parallel to the track over the land or sea surface that an aircraft pilot plans to follow. The component is calculated from the average wind direction along the entire track at the altitude the aircraft will fly. If the component is in the direction of the flight (a tail wind) it is positive, and it is negative if it is in the opposite direction (a headwind).

route forecast A weather forecast that is prepared for pilots. It provides them with details of weather conditions at various altitudes along the route they are planning to fly. Data from the route forecast are fed into a computer. This calculates the heading on which

the aircraft should fly in order to follow the desired track over the ground or sea surface and the speed at which the aircraft will travel in relation to the surface (the ground speed) when it flies at its designated cruising airspeed.

Rowland, Frank Sherwood (1927–) American *Chemist* F. Sherwood Rowland shared the 1995 Nobel Prize in Chemistry with MARIO MOLINA and PAUL CRUTZEN. The prize was in recognition of the fact that these three scientists had shown first that the amount of OZONE in the OZONE LAYER might decrease as a result of reactions with chemicals released by human activity, and later that chlorofluorocarbons (CFCs) were the chemical compounds primarily involved.

F. Sherwood Rowland was born on June 28, 1927, in Delaware, Ohio, where his parents had moved the previous year when his father took up the post of professor of mathematics at Ohio Wesleyan University in the city. As a small boy, Sherwood developed a keen interest in naval history.

He was educated at schools in Delaware, and in 1943 he enrolled at Ohio Wesleyan University, studying chemistry, physics, and mathematics, but in June 1945, before completing his course, Rowland enlisted in the navy. He was discharged 14 months later in California, hitchhiked the 2,000 miles (3,200 km) back to Delaware, and resumed his studies. These were combined with sport, an interest that began when he was in high school. He played tennis, basketball, and baseball for university teams.

He graduated in 1948 and the same year entered graduate school in the Department of Chemistry at the University of Chicago. He obtained his Ph.D. in 1952 and in June of that year married Joan Lundberg, a fellow graduate student.

In September 1952 Rowland became an instructor in the Chemistry Department at Princeton University, where he remained until 1956. He spent three summers, from 1953 to 1955, working on the use of tracer chemicals in the Chemistry Department of the Brookhaven National Laboratory. From 1956 until 1964 he was an assistant professor at the University of Kansas and in 1964 he was appointed professor of chemistry at the University of California, Irvine, where he is now Donald Bren Research Professor of Chemistry and Earth System Science. His link with Brookhaven continued until 1994.

It was in January 1972 that he attended a workshop on chemistry and meteorology in Fort Lauderdale, Florida, where he heard the English chemist JAMES LOVELOCK describe how he had detected minute concentrations of a CFC in the air. Lovelock thought their great chemical stability would make CFCs useful for tracing the movement of AIR MASSES, but Rowland realized that no molecule can remain inert for ever. At high altitudes it will be broken apart by sunlight. He began to wonder what would happen to CFCs when they decayed. In 1973 MARIO MOLINA joined his group as a research associate and the two men set about studying the fate of airborne CFC molecules.

Paul Crutzen had already drawn attention to the possibility of depleting stratospheric ozone. Rowland and Molina published the results of their research as a paper in *Nature* in 1974. They had calculated that the breakdown of CFC molecules would release chlorine in a form that would destroy ozone molecules. Their paper stimulated a federal investigation of the situation and in 1978 the use of CFCs as propellants in spray cans was banned in the United States. The phasing out of the use of CFCs was agreed on internationally in 1987 under the terms of the MONTREAL PROTOCOL ON SUBSTANCES THAT DEPLETE THE OZONE LAYER. It was

for this work that Rowland, Molina, and Crutzen were awarded their Nobel Prize.

(You can learn more about Professor Rowland and his research interests at fsr10.ps.uci.edu/GROUP/people/drowland.html and about his life at www.nobel.se/chemistry/laureates/1995/rowland-autobio.html.)

Roxanne A HURRICANE, rated at category 3 on the SAFFIR/SIMPSON SCALE, that struck the island of Cozumel, off the Mexican coast, in October 1995, generating winds of 115 mph (185 km h⁻¹). It killed 14 people and tens of thousands were forced to flee their homes.

RPF *See* RELATIVE POLLEN FREQUENCY.

Ruby A TYPHOON that struck the Philippines on October 24 and 25, 1988. It caused floods and landslides in which about 500 people died. The damage was estimated at $52 million.

Rumbia A TROPICAL STORM that struck southern Mindanao, Philippines, on November 30, 2000. It generated heavy rain and high waves that caused flooding. More than 1,600 people were forced to leave their homes.

runoff Water that falls to the ground as PRECIPITATION, including melting snow, frost, dew, and fog droplets, and that flows across the ground surface directly into rivers or lakes. As it crosses the ground, some of the water filters into the soil; eventually it joins the GROUNDWATER. The remainder, or net runoff (symbolized by Δr), is not available to plants. Measurement of the amount of runoff is used in calculating the WATER BALANCE for an area.

run of wind The "length" of a wind, from which its speed can be calculated. Cup ANEMOMETERS spin around a vertical axis. The number of revolutions the instrument makes in a measured period can be converted into the horizontal distance (D) one of the cups has traveled by:

$$D = \pi d R$$

where d is the diameter of the circle traveled by the cup and R is the number of revolutions. This can then be converted into the speed (S) by

$$S = D/T$$

where T is the time that elapses. T can be of any length, from a few minutes to a day, and the result of the calculation is the average wind speed over that period. This is a more useful value than the wind speed measured at any particular moment, because the speed of the wind changes constantly and all instruments experience a lag in registering the changes. A typical cup anemometer experiences a lag of about 8 seconds in registering a change of speed at about 4 knots (4.6 mph [7.4km h^{-1}]) but a smaller lag at higher speeds. About 50 feet (15 m) of air must pass the anemometer before it will give an accurate reading.

runway observation A measurement or assessment of a meteorological condition that is made at a specified position on or close to an airfield runway. Runway observations are usually made of WIND SPEED and direction, PRESSURE (which affects pressure altimeters), TEMPERATURE, PRECIPITATION, CLOUD BASE, and horizontal VISIBILITY, which are the factors of most importance to aircraft landing or taking off.

runway temperature The air TEMPERATURE measured about 4 ft (1.2 m) above the surface of an airfield runway. This is used in calculating the DENSITY ALTITUDE. Runway temperature is reported to pilots if the density altitude may differ from the value they expect.

runway visibility The horizontal VISIBILITY along an airfield runway, measured by an observer in a specified position looking along the runway in the direction of take-off and landing.

runway visual range The greatest distance along an airfield runway at which the runway lights are visible to the pilot of an airplane that has just touched down and is decelerating in its landing run.

rural boundary layer The layer of air in a rural area adjacent to a large city that lies between the top of tall vegetation and the uppermost limit of the region within which the climatic properties of the air are modified by the surface below. The rural boundary layer is markedly thinner than the nearby URBAN BOUNDARY LAYER. Air in it is stable (*see* STABILITY OF AIR) and capped by an INVERSION. Air moving outward from the city is carried above it, so the urban and rural bodies of air do not mix.

Ruth Two TYPHOONS, the first of which struck Vietnam on September 15, 1980. It killed at least 164 people. The second Typhoon Ruth, rated as category 5 on the SAFFIR/SIMPSON SCALE, struck Luzon, Philippines, on October 27, 1991. It generated winds of up to 143 mph (230 km h^{-1}) and killed 43 people.

Saalian glacial *See* RISS GLACIAL; WOLSTONIAN GLACIAL.

Saffir/Simpson hurricane scale For more than a century, wind force was reported by using the scale that had been devised by Admiral FRANCIS BEAUFORT. The BEAUFORT WIND SCALE is still in use, but it has one major disadvantage: it is designed for temperate regions, where winds stronger than 75 mph (120.6 km h⁻¹) are very uncommon. All such winds are classed on the Beaufort scale as being of hurricane force.

This is inadequate for those parts of the world that experience real TROPICAL CYCLONES—called *HURRICANES, BAGYO, CYCLONES,* or *TYPHOONS,* depending on where they occur. All of these cyclones produce winds of greater force than the 75 mph, force 12, of the Beaufort scale but vary considerably in the winds they generate and, therefore, in the damage they are capable of inflicting.

To address this difficulty, meteorologists at the U.S. WEATHER BUREAU (now part of the NATIONAL WEATHER SERVICE) introduced in 1955 an extension to the Beaufort scale: the Saffir/Simpson scale, named after the scientists who devised it. It adds five more points to the wind scale, but it also conveys more information than the Beaufort scale. As well as wind speed and a general description of the type of possible wind damage, it includes the atmospheric pressure at the center of the storm and the height of the STORM SURGE. The pressure indicates the intensity of the storm—the lower the pressure the more violent the storm—and information about the anticipated storm surge is vital, because tropical cyclones begin at sea and mainly affect coastal areas. Tropical cyclones everywhere are now classified according to the Saffir/Simpson scale.

SAGE *See* STRATOSPHERIC AEROSOL AND GAS EXPERIMENT.

Saharan depression A type of DEPRESSION that forms in winter over the western Mediterranean. About 17 percent of all the winter depressions in this region are of the Saharan type. About 9 percent form over the Atlantic and about 74 percent are GENOA-TYPE DEPRESSIONS. Saharan depressions form in the LEE of the Atlas Mountains when cold MARITIME POLAR AIR crosses the warmer sea surface and becomes unstable. Saharan depressions are the most important source of rainfall in late winter and early spring.

Saharan high The SUBTROPICAL HIGH that lies permanently over the Sahara Desert. It is produced by the subsidence of air on the high-latitude side of the HADLEY CELL. The subsiding air is warm and dry, and because of the ANTICYCLONE the PRESSURE GRADIENT drives air out of the desert. This prevents moist air from flowing into the region and therefore maintains the arid conditions. The highest temperature ever recorded on Earth was 136° F (57.8° C) at Azizia, Libya, on September 13, 1922.

Saffir/Simpson Hurricane Scale

Category	Pressure at center mb in of mercury cm of mercury	Wind speed mph km h^{-1}	Storm surge feet meters	Damage
1	980 28.94 73.5	74–95 119–153	4–5 1.2–1.5	Trees and shrubs lose leaves and twigs; mobile homes destroyed
2	965–979 28.5–28.91 72.39–73.43	96–110 154.4–177	6–8 1.8–2.4	Small trees blown down; exposed mobile homes severely damaged; chimneys and tiles blown from roofs
3	945–964 27.91–28.47	111–130 178.5–209	9–12 2.7–3.6	Leaves stripped from trees; large trees blown down; mobile homes demolished; small buildings damaged structurally
4	920–944 27.17–27.88	131–155 210.8–249.4	13–18 3.9–5.4	Extensive damage to windows, roofs, and doors; mobile homes destroyed completely; flooding to 6 miles (10 km) inland; severe damage to lower parts of buildings near exposed coasts
5	920 or lower below 17.17 below 69	more than 155 more than 250	more than 18 more than 5.4	Catastrophic; all buildings severely damaged, small buildings destroyed; major damage to lower parts of buildings less than 15 ft (4.6 m) above sea level to 0.3 mile (0.5 km) inland

Sahel   The region that lies along the southern margin of the Sahara Desert, in northern Africa. The region extends from Senegal in the west to Sudan and Ethiopia in the east and includes parts of Senegal, Guinea-Bissau, Mauritania, Mali, Burkina Faso, Niger, Nigeria, Chad, Central African Republic, Sudan, Eritrea, and Ethiopia.

The Sahel forms a transitional zone between the desert to the north and the humid tropical grasslands to the south. The vegetation comprises short grasses, tall herbs, and thorn scrub with species such as acacias (*Acacia* species) and baobab trees (*Adansonia digitata*), but the plants are scattered and there are few places where the vegetation cover is continuous.

The climate is semiarid and strongly seasonal, with a short rainy season in summer. Niamey, Niger, receives an average 22 inches (554 mm) of rain a year, but no rain at all falls between the end of October and the beginning of March. Most of the rain, about 20 inches (495 mm), falls in June, July, August, and September. N'Djamena, the capital of Chad, receives no rain from the end of October until the beginning of April and of the 29 inches (744 mm) it receives in an average year, 24 inches (610 mm) falls in July, August, and September. August is the wettest month.

Temperatures change little through the year and the climate is hot. April is the warmest month at N'Djamena, when the average daytime temperature is 107° F (42° C) and has been known to reach 114° F (46° C). In December, the coldest month, the average daytime temperature is 92° F (33° C) and 101° F (38° C) has been recorded. It is much cooler at night, but the lowest nighttime temperature recorded is 47° F (8° C) and the average temperature at night in December and January is 57° F (14° C). Niamey experiences almost identical temperatures.

Averages can be misleading, however. It is the northward movement of the INTERTROPICAL CONVERGENCE ZONE (ITCZ) that carries the tropical rain belt to the Sahel and causes the summer rains, but occasionally the ITCZ remains to the south. When this happens the rains are lighter than usual or, if the ITCZ remains a long way to the south, they fail altogether. During the late 1960s the Sahel experienced a sequence of years when the summer rains were light. This produced DROUGHT. Then, in 1972 and 1973, the rains failed completely; the rainfall did not return to normal until the 1980s.

It was not the first time the Sahel had been afflicted with severe drought. Several occurred in the 17th century and caused serious famines. Those droughts were associated with the coldest part of the LITTLE ICE AGE. No one knows what caused the drought in the 20th century, although it did coincide with the latter part of a period when temperatures were falling sharply throughout the Northern Hemisphere.

Some of the people of the Sahel grow crops around the oases. Others live a seminomadic life, taking their herds and flocks of cattle, sheep, goats, and camels to traditional seasonal pastures. The drought that peaked in the 1970s proved devastating. It is estimated that between 100,000 and 200,000 people and up to 4 million cattle died. Countless more people were forced to migrate south across the national frontiers that are a

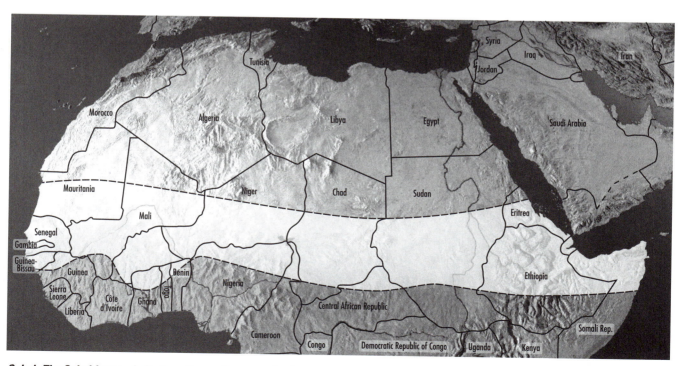

Sahel. The Sahel forms a belt along the southern margin of the Sahara Desert, extending from Senegal to Sudan, Eritrea, and Ethiopia.

legacy of European colonialism. It was the Sahel drought that alerted the international community to the difficulties facing people who live along the borders of deserts.

It is not true that the drought was caused by overgrazing. It was an entirely natural event and was not the fault of the people living in the region. Overgrazing did exacerbate its effects, however. As the pastures failed, livestock was crowded into increasingly smaller areas, where they did destroy the sparse vegetation. Governments also encouraged nomadic people to settle in permanent villages. This gave them access to medical care and schools, but it also placed excessive pressure on the grazing around the villages.

(You can read more about the Sahel and the drought of the 1970s in Michael Allaby, *Ecosystem: Deserts* [New York: Facts On File, 2001].)

Saint Bartholomew's Day A CONTROL DAY that falls on August 24. According to a folk belief,

> *If St. Bartholomew be fair and clear,*
> *Then a prosperous autumn comes that year.*

Saint Elmo's fire *See* POINT DISCHARGE.

Saint Hilary's Day A CONTROL DAY that falls on January 13. According to a folk belief, this is the coldest day of the year and so the rest of the winter will be warmer.

Saint Luke's summer According to a British folk belief, a period of mild weather that occurs around Saint Luke's Day (October 18) and often ends at SAINT SIMON AND SAINT JUDE.

Saint Martin's summer According to a British folk belief, a period of mild weather that occurs around Saint Martin's Day (November 11).

Saint Mary's Day A CONTROL DAY that falls on July 2. According to a folk belief, if it rains on this day the rain will continue for a further month.

Saint Michael and Saint Gallus A pair of CONTROL DAYS that fall on September 29 and October 16, respectively. According to a folk belief,

> *If it does not rain on St. Michael and Gallus,*
> *The following spring will be dry and propitious.*

Saint Paul's Day A CONTROL DAY that falls on January 25. According to a folk belief,

> *If St. Paul's Day be fair and clear,*
> *Then it betides a happy year.*

Saint Simon and Saint Jude A CONTROL DAY that falls on October 29, the day both saints are celebrated. According to a folk belief, it is the day when the weather is bad and possibly dangerous, with gales and storms at sea.

Saint Swithin's Day Possibly the most famous of all CONTROL DAYS, Saint Swithin's Day is July 15. The weather on this day will continue for a further 40 days.

> *Oh St. Swithin if thou'll be fair,*
> *For forty days shall rain nae mair,*
> *But if St. Swithin's thou be wet,*
> *For forty days it raineth yet.*

The belief arose because before his death in 862, Swithin, bishop of Winchester, was supposed to have expressed a wish to be buried in the churchyard (ordinarily a bishop would be buried inside the cathedral) so that the rain might fall upon his grave. Later Swithin was canonized and the canonization ceremony was arranged for July 15, 964. When the monks tried to exhume his body in order to take it into the cathedral for the ceremony the legend asserts that it rained and continued to rain for 40 days, so the canonization had to be postponed. It was eventually carried out, and Swithin was interred inside the cathedral. Checks on weather records have found no correlation between rain on July 15 and the weather over the following 40 days and very little is known reliably about Swithin.

Saint Vitus's Day A CONTROL DAY that falls on June 15. According to a folk belief,

> *If St. Vitus Day be rainy weather,*
> *It will rain for thirty days together.*

Sally A TYPHOON that passed Hong Kong on September 10, 1996, then crossed the coast of Guangdong, China, with winds of up to 108 mph (174 km h⁻¹). It killed more than 130 people and destroyed nearly 400,000 homes.

SALR   *See* SATURATED ADIABATIC LAPSE RATE.

salt crystal   The solid form that common salt (sodium chloride, NaCl) takes when it goes out of solution. The basic crystal is cubic in shape, about 10 μm (0.0004 in) along each side, and it grows by the addition of more cubes.

Salt enters the air when drops of sea spray evaporate. The tiny crystals are then carried by air currents. Salt is hygroscopic: that is, it dissolves in water that its crystals absorb from the atmosphere and airborne salt crystals act as HYGROSCOPIC NUCLEI. These are the most efficient of all CLOUD CONDENSATION NUCLEI. Water condenses onto a salt crystal at a RELATIVE HUMIDITY as low as 75 percent. Salt crystals are sometimes used in CLOUD SEEDING, in which their effect is to increase the range of size of CLOUD DROPLETS. This increases the likelihood of PRECIPITATION, because some of the droplets grow large enough to fall and continue growing by collision and coalescence (*see* COLLISION THEORY) as they do so.

The efficiency of using salt crystals for cloud seeding is initially due to the readiness with which they dissolve in water. When a crystal dissolves the resulting droplet is a fairly strong saline solution. The VAPOR PRESSURE is always lower over a solution than it is over pure water, so water evaporates more slowly from the solution. Once they form, therefore, saline cloud droplets resist EVAPORATION. More water condenses onto them and as the droplets grow the salt solution is diluted. This increases the vapor pressure.

It would also increase the rate of evaporation if it were not for a counteracting effect that is due to the size of a droplet. Water molecules are linked by weak HYDROGEN BONDS. Molecules at the surface are held less firmly than molecules elsewhere, because there are no molecules above them to which they can be linked. The attraction between molecules at the surface is strongest if the surface is flat. It is weaker over a curved surface by an amount that is proportional to the degree of curvature. Consequently, small droplets evaporate faster than large droplets.

As the saline droplet begins to grow and the vapor pressure over it increases, it also grows larger and its

Salt crystal. Salt crystallizes into very small cubes that dissolve to form tiny droplets of salt water. The vapor pressure over the solution is weak, reducing the rate of evaporation. As the droplets grow the solution becomes more dilute and the vapor pressure increases, tending to increase the rate of evaporation. As the droplets grow, their surface curvature decreases. This increases the intermolecular forces at the surface and reduces the rate of evaporation.

surface becomes less curved. This reduces the rate of evaporation from it.

The overall result is that salt crystals readily cause water vapor to condense and the resulting droplets tend to survive long enough to grow. They continue to grow until they attain a size that is in equilibrium with the amount of moisture that is present in the cloud.

salt haze   A thin HAZE that is caused by the CONDENSATION of WATER VAPOR onto SALT CRYSTALS when the RELATIVE HUMIDITY of the air is below about 90 percent. Salt hazes form in humid climates. If the humidity is higher, the HAZE DROPLETS quickly grow into CLOUD DROPLETS.

samoon   A hot, dry, northwesterly FÖHN WIND that blows from Kurdistan across Iran.

sand auger   *See* DUST WHIRL.

sand devil   *See* DUST WHIRL.

sand storm   A wind storm that lifts sand grains into the air and transports them, often for long distances. The THRESHOLD VELOCITY for dry, medium-sized sand grains about 0.01 inch (0.25 mm) in diameter is 12 mph (19 km h⁻¹), and winds of more than 15 mph (24 km h⁻¹) raise enough sand to cause a sand storm provided the air is unstable (*see* STABILITY OF AIR). Wherever there is loose, dry sand a wind of this speed raises sand high enough to reduce visibility greatly and drive it with enough force to make exposure to it extremely uncomfortable. The wind blows horizontally, however, and although collisions between sand grains and the TURBULENCE of the air can raise the sand a short distance above the ground, they cannot lift it high enough for it to travel far. In unstable air, however, upcurrents can lift the sand to a considerable height. A sand storm is produced in the same way as a DUST STORM, from which it differs only in the size of particles involved.

Sangamonian interglacial   The INTERGLACIAL that followed the ILLINOIAN GLACIAL and preceded the most recent (WISCONSINIAN) GLACIAL PERIOD in North America. The Sangamonian began about 120,000 years ago and ended about 75,000 years ago.

Santa Ana   A wind of the FÖHN type that occurs in southern California, most commonly in autumn and winter. It forms part of the clockwise movement of air around a strong ANTICYCLONE centered over the Great Basin. This carries air from the desert of Arizona and Nevada toward the Pacific. As it crosses the Coastal Range FUNNELING intensifies it. Funneling is especially marked through the Santa Ana Canyon, which gives the wind its name. Then, as it descends on the western side of the mountains, the air warms in an ADIABATIC process. The wind is dry, carries a large amount of dust, and often blows at 40 mph (64 km h⁻¹) with gusts of twice that speed. Its temperature is often close to 90° F (32° C) and sometimes higher. It has a strongly desiccating effect on vegetation and fans and drives fires.

SAR   *See* SYNTHETIC APERTURE RADAR.

Sarah   A TYPHOON that struck Taiwan on September 11, 1989. It broke a Panamanian-registered freighter in half and killed 13 people.

Sargasso Sea   An area in the western North Atlantic Ocean that lies approximately between latitudes 20° N and 35° N and longitudes 30° W and 70° W. The sea is roughly elliptical in shape and occurs inside a system of ocean currents that rotate clockwise. The GULF STREAM, CANARY CURRENT, and NORTH ATLANTIC DRIFT flow around its edges. Waters of the sea are relatively calm. Winds are light, the EVAPORATION rate is high, and RAINFALL is low. The water is very clear and warm, with an average temperature of 64° F (18° C). The high rate of evaporation combined with low PRECIPITATION produce water with a salinity of 36.5–37.0 parts per thousand (‰). The average salinity of seawater is about 35.0 parts per thousand. The sea is famous for its abundance of gulf weed, a floating brown seaweed (several *Sargassum* species). It is not true that ships have ever been caught and trapped by the weed. The Sargasso Sea is the breeding ground of the American and European eels (*Anguilla rostrata* and *A. anguilla,* respectively). The larvae of European eels drift with the Gulf Stream, taking about three years to reach the cool, shallow waters off the coast of Europe, where they turn into elvers and migrate into rivers. American eels breed in the western part of the Sargasso Sea, and their larvae take only one or two years to reach the mouths of rivers.

sastruga   A wave in the snow and ice of Antarctica that forms where the KATABATIC WINDS blow constantly. Sastruga are aligned parallel to the wind. They are

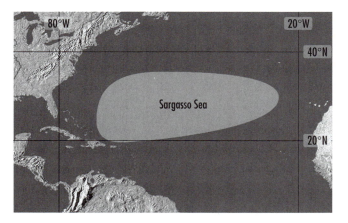

Sargasso Sea. The roughly elliptical Sargasso Sea lies in the western North Atlantic and is enclosed by ocean currents that flow clockwise around it.

usually about 2 inches (5 cm) high but in places they can be more than 6 feet (1.8 m) high.

satin ice *See* ACICULAR ICE.

saturated adiabat (moist adiabat, wet adiabat) A line on a TEPHIGRAM that marks the constant WET-BULB POTENTIAL TEMPERATURE. It makes an angle of about 45° to the DRY ADIABAT in the lower TROPOSPHERE but approaches the dry adiabat at lower temperatures and pressures until the two are almost, but never quite, parallel.

saturated adiabatic lapse rate (SALR) The rate at which rising, saturated air cools in an ADIABATIC manner and descending air warms. As it rises, unsaturated air cools at the DRY ADIABATIC LAPSE RATE (DALR). When it reaches the CONDENSATION LEVEL, however, the water vapor it carries begins to condense. This releases LATENT HEAT of condensation, which warms the air and reduces the rate at which its temperature

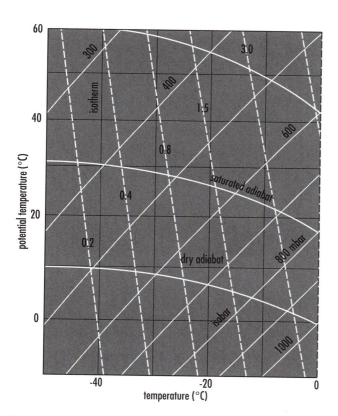

Saturated adiabat. The saturated adiabat marks the line of constant wet-bulb potential temperature.

decreases with height. As it rises further, the air cools at the SALR, which is lower than the DALR. Unlike the DALR, the SALR is not the same under all conditions. It varies according to the temperature of the air. This is because the temperature determines the amount of water vapor the air can hold and, therefore, the amount of condensation that occurs. In very warm air, the SALR may be as low as about 2.7° F per 1,000 feet (5° C km^{-1}), but at an air temperature of $-40°$ F ($-40°$ C) it is about 5° F per 1,000 feet (9° C km^{-1}). An average value is about 3° F per 1,000 feet (6° C km^{-1}).

saturated adiabatic reference process An idealized representation of the way moist air behaves that is used as a standard, or reference, with which events in the real atmosphere can be compared. In fact it is a fairly accurate description of what usually happens.

The reference process assumes that rising air that is cooling in an ADIABATIC manner and is saturated remains very close to SATURATION. Water vapor begins to condense when the temperature of the rising air falls to its DEW POINT TEMPERATURE. CONDENSATION releases LATENT HEAT, which sustains the BUOYANCY of the air. As the air continues to rise its temperature also continues to fall. This chills the air between CLOUD DROPLETS sufficiently to cause the condensation of excess water vapor, maintaining the air at saturation. It is assumed that the moisture condenses as liquid, not ice. This is realistic. In real clouds, water droplets often remain liquid until the temperature falls below about $-13°$ F ($-25°$ C) (*see* SUPERCOOLING).

The process is reversible. As the temperature of subsiding air rises, the resulting EVAPORATION of droplets adds enough water vapor to the air to maintain it at saturation. This is close to the process that has been observed in the downdrafts of CUMULONIMBUS clouds.

saturated air Air in which the RELATIVE HUMIDITY is 100 percent. This does not refer to the actual amount of moisture present in the air, but only to the amount present as a percentage of the greatest amount air at a particular temperature can hold at that temperature.

saturation The condition in which the moisture content of the air is at a maximum. If additional water molecules enter saturated air, then an equal number must leave it, by condensing into liquid water or being deposited as solid ice (but *see* SUPERSATURATION).

Over the surface of water, water molecules are constantly escaping into the air by evaporation. Once in the air, they add to the VAPOR PRESSURE. As they move through the air, a proportion of them strike the surface of the liquid water. Their energy of motion (kinetic energy) is absorbed by the relatively denser mass of water molecules constituting the liquid and the impacting molecules no longer possess the energy needed to escape into the air. They enter the water mass. Molecules also escape from and merge with an ice surface in the same way (*see also* SUBLIMATION).

As evaporation continues, the vapor pressure increases in the air above the water surface. Eventually, however, a point is reached at which the number of water molecules evaporating from the liquid surface every second precisely balances the number recondensing into it over the same period. The amount of water vapor present in the air cannot increase beyond this point. The vapor pressure at which this occurs is known as the *SATURATION VAPOR PRESSURE* and at the saturation vapor pressure the air is said to be *saturated*.

Strictly speaking, it is not the air that is saturated, but the water vapor. This is easier to understand if the air is likened to a dry sponge onto which water is sprinkled. As they fall, the water drops disappear into the sponge, which grows steadily moister. Continue with this for long enough and all the tiny air spaces in the sponge are filled with water. The sponge can then hold no more and water starts to drip out of it. This has no effect whatever on the material from which the sponge is made—it is the millions of air spaces that are filled with water. These lie between the cells or inside the bubbles that constitute the foam but are not the solid matter of the sponge itself. Consequently, it is the water that is saturated, and not the sponge. Similarly, atmospheric water vapor comprises molecules that move among the other molecules of the air and the dry air is entirely unaffected by their presence. It is usual, however, to think of the air as being saturated.

Raising the TEMPERATURE of the water molecules increases their kinetic energy. In practice, collisions between air molecules and molecules of water vapor ensure they all have much the same kinetic energy. Water molecules therefore acquire more energy as the air temperature increases. This increases the ease with

Temperature °F (°C)	Saturation mixing ratio g kg^{-1}
104 (40)	47
95 (35)	35
86 (30)	26.5
77 (25)	20
68 (20)	14
59 (15)	10
50 (10)	7
41 (5)	5
32 (0)	3.5
14 (−10)	2
−4 (−20)	0.75
−22 (−30)	0.3
−40 (−40)	0.1

which they are able to escape from the liquid surface, so the quantity of water vapor increases until a new balance is reached at the higher temperature, with a raised saturation vapor pressure. The result is that the amount of water vapor air can hold increases with temperature. In fact, it does so extremely rapidly. Air at 80° F (27° C) holds more than four times the amount of water vapor air at 40° F (4° C) can hold.

saturation deficit (vapor-pressure deficit) The difference between the actual VAPOR PRESSURE and the SATURATION VAPOR PRESSURE at the same temperature. This is also the amount of water vapor, usually measured in grams per cubic meter (g m^{-3}), that must be added to the air to raise it to SATURATION at the existing temperature and pressure.

saturation mixing ratio The value of the MIXING RATIO of saturated air at a particular temperature and pressure. The table shows variations in the saturation mixing ratio with temperature at mean sea-level pressure.

saturation ratio The ratio of the actual SPECIFIC HUMIDITY of a body of air to the specific humidity of saturated air at the same temperature and pressure.

saturation vapor pressure The VAPOR PRESSURE at which the water vapor in the layer of air immediately above the surface of liquid water is saturated at a given

Saturation vapor pressure

Temperature °F (°C)	Pressure mb (Pa)
−58 (−50)	0.039 (3.94)
−40 (−40)	0.128 (12.83)
−22 (−30)	0.380 (37.98)
−4 (−20)	1.032 (103.2)
14 (−10)	2.597 (259.7)
32 (0)	6.108 (610.78)
50 (10)	12.272 (1227.2)
68 (20)	23.373 (2337.3)
86 (30)	42.430 (4243.0)
104 (40)	73.777 (7377.7)

temperature. The table gives a number of representative values. This shows that as the air temperature increases the amount of water vapor needed to saturate it also increases, demonstrating the relationship between the temperature of air and its capacity to hold water vapor. It reaches sea-level atmospheric pressure (1013 mb, 101.3 kPa) at 212° F (100° C). The saturation vapor pressure over an ice surface is lower than that over the surface of water supercooled to the same temperature, because the stronger bonds between water molecules in ice reduce the rate at which they enter the air by SUBLIMATION. As a result, air over an ice surface holds less water vapor than air over a liquid surface at the same temperature.

Saussure, Horace Bénédict de (1740–1799) Swiss *Physicist* Horace Bénédict de Saussure was born at Conches, near Geneva, Switzerland, on February 17, 1740. His education began in 1746, when he enrolled at the public school in Geneva. He entered the Geneva Academy in 1754 and graduated in 1759, having presented a dissertation on the physics of fire. He was then 19 years old.

It was in the following year that he paid his first visit to Chamonix, a small resort in southeastern France standing at the foot of Mont Blanc. This is the tallest mountain in Europe, its peak 15,771 feet (4,810 m) above sea level, and when de Saussure visited it in 1760 no one had ever managed to climb it. De Saussure was fascinated by the mountain and toured the neighboring parishes offering "a considerable reward"

to anyone who discovered a practicable route to the summit. In fact, Mont Blanc was first climbed 26 years later, on August 8, 1786, by Jacques Balmat and Gabriel Paccard. De Saussure climbed it himself in the summer of 1787 and reached the summit at 11 A.M. on August 3, accompanied by a number of guides and his personal valet.

De Saussure returned to the mountain many times over the succeeding years. His interest was not primarily in mountaineering as a sport, but in alpine plants, geology, and meteorology. From 1773 he began to spend increasing amounts of time in the area and climbed many of the mountains. The first volume of his most famous book, *Voyages dans les Alpes* (Journeys in the Alps), was published in 1779. The remaining three volumes were published between then and 1796. In them he described seven of his alpine journeys.

In 1761 de Saussure applied unsuccessfully for the vacant professorship of mathematics at the Geneva Academy. The following year he applied again, this time for the professorship of philosophy, and was successful. The new professor delivered his inaugural lecture in October. In 1772 he was elected a Fellow of the Royal Society of London and in the same year he founded the Society for the Advancement of the Arts in Geneva.

By this time his reputation as a physicist was well established. In 1767 he is credited with having constructed the first solar collector. This was a box with a glass top and heavily insulated sides, which he used to discover why it is always cooler in the mountains than it is at lower levels. He took his box to the top of Mont Cramont. There the outside temperature was 43° F (6° C) but the temperature inside the box rose to 190° F (88° C). Then he repeated the experiment 4,852 feet (1,480 m) lower down, on the Plains of Cournier. The air temperature there was 77° F (25° C), but the temperature inside the box was almost the same. De Saussure concluded that the Sun shines just as warmly in the mountains as it does on the plains, but that the more transparent mountain air is unable to trap and hold so much warmth.

In 1783, de Saussure invented the HAIR HYGROMETER, which was based on his observation that human hairs increase in length as the HUMIDITY rises and grow shorter as the air becomes drier. Hair hygrometers are still the most widely used instruments for measuring RELATIVE HUMIDITY.

De Saussure married Albertine Boissier in May 1765. In February 1768, in the company of his wife and sister-in-law, he visited Paris, the Netherlands, and England; he returned to Geneva in January 1769. In 1771 he visited Italy, and in the autumn of 1772, with his wife and six-year-old daughter, he toured Italy, visiting Sicily, climbing Mount Etna, and calling at Rome, Rimini, and Venice before returning to Switzerland over the Brenner Pass. He also had an audience with Pope Clement XIV.

These were turbulent times in Geneva and de Saussure became involved in politics. He drew up plans in 1776 for the reform of city institutions and during the troubles of 1782 he was arrested and spent two days in prison. During revolutionary riots in July of the same year he was besieged in his home for several days, suspected of harboring armed men and concealing weapons. The Terror that began in France after the Revolution of 1789 spread to Geneva, and in 1792 Geneva had its own revolution aimed at introducing a measure of democracy. The following year de Saussure was a member of the commission appointed to draft a constitution for the city. This led to an invitation, which he refused, to stand as a candidate for the governing council. The new constitution failed and in 1794 the Terror returned.

In 1787 de Saussure had resigned from his position at the Geneva Academy and spent some time in the south of France, where he could live at sea level and collect measurements of atmospheric pressure that he could compare with those he had taken in the Alps.

His health had begun to deteriorate in 1772, and by 1794 he was a sick man. He was also experiencing financial difficulties and was compelled to return with his family to the country house at Conches where he had been born. News of his poverty spread and he received offers of help from abroad. Thomas Jefferson himself considered offering de Saussure a position at the newly founded University of Charlottesville. It was not to be. De Saussure remained at Conches and died there on the morning of January 22, 1799.

savanna climate The climate that sustains tropical grasslands, where the vegetation resembles open parkland, with tall grasses and scattered trees and shrubs. These are small, umbrella-shaped, and drought-resistant. The number of trees and shrubs is greatest in places where the rainfall is higher than the average for the grassland as a whole. They often extend along river valleys into the drier regions, forming galerias (*galeria* is the Italian for "tunnel"). In the driest areas the tall grasses give way to short grasses. There is a dry season, usually in winter and lasting up to four months. Average temperatures range from about 65° F (18° C) to more than 80° F (27° C).

Savanna (sometimes spelled *savannah*) grassland is found in all the tropical continents. It is bordered on one side by the humid Tropics, at latitude 5°–10° N and S and on the other by the subtropical deserts, at 15°–20° N and S. The name *savanna* is derived from the Spanish *zavana*, which is believed to have come from a word in the Carib language that was once spoken in parts of North America and the southern Caribbean. *Savanna* was originally the name for the tropical grasslands of Central and South America, but its use spread to cover all such grasslands. *See* TROPICAL WET–DRY CLIMATE.

Sc *See* STRATOCUMULUS.

scalar quantity A physical quantity that does not act in a particular direction, as in the case of temperature, or for which the direction of action is unimportant or not specified, as in the case of speed. This is contrasted with a VECTOR QUANTITY, in which the direction of action must be specified.

scale height The thickness the atmosphere would have if its density were constant throughout at its sea-level value of 1.23 kg m^{-3}. The scale height is 8.4 km (5.2 miles).

scanning multichannel microwave radiometer (SMMR) An instrument that was carried on the SEASAT and *Nimbus*-7 satellites (*see* NIMBUS SATELLITES). It first went into service in 1978 and continues to transmit valuable data from *Nimbus*. It carries six RADIOMETERS with 10 channels delivering measurements at five microwave WAVELENGTHS (0.81, 1.36, 1.66, 2.8, and 4.54 cm). The SMMR measures sea-surface temperature, wind speed, WATER VAPOR, clouds and cloud content, snow cover, type of snow, rainfall rates, and different types of ice. It also measures the concentration and extent of SEA ICE. It has provided detailed information about EL NIÑO events since the 1982–83 El Niño and it is used to monitor changes in sea ice.

(You can learn more about the SMMR at podaac. jpl.nasa.gov:2031/SENSOR_DOCS/smmr.html.)

scarlet pimpernel (poor man's weatherglass, shepherd's weatherglass) A small herbaceous plant (*Anagallis arvensis*) that grows on sand dunes and open grassland throughout most of Europe. It is quite common. Its small, scarlet flowers, which appear from June through August, are reputed in weather lore to predict rain reliably. When the weather is to be sunny the flowers open and when rain threatens they close.

scattered cloud Cloud that covers up to half of the sky (4 OKTAS or five-tenths).

scattering The result of the reaction that occurs when VISIBLE LIGHT passes through the atmosphere and collides with air molecules and AEROSOL particles. Molecules and very small aerosol particles, with sizes smaller than the wavelength of the light, absorb the radiation. They are excited by it and reradiate it in all directions.

The size of molecules and particles in relation to the wavelength of light is known as the *size parameter* (*X*) and is given by:

$$X = \pi d/\lambda$$

where *d* is the diameter of the molecule or particle and λ is the wavelength of the light. Air molecules and the smallest aerosol particles are much smaller than the wavelength of light (*d* is smaller than λ) and so *X* is less than 1. The smaller they are, the less efficiently bodies scatter radiation and their efficiency decreases as the difference between their size and the wavelength increases. The efficiency of scattering is inversely proportional to the fourth power of the wavelength (λ^{-4}). This means that the shorter wavelengths are scattered most efficiently.

As light passes through the upper atmosphere, the shortest visible wavelengths of 0.4–0.44 μm, which correspond to violet light, are scattered first. Each time the violet radiation strikes a molecule or particle it rebounds in a random direction. This happens repeatedly, and the amount of scattering increases with the distance the radiation travels through the air. Violet radiation is scattered so thoroughly that it contributes very little to the sky color. The sky is not violet.

Blue light is scattered next. Because the efficiency of scattering is proportional to λ^{-4}, blue light (0.44–0.49 μm) is approximately nine times more likely to be scattered than red light (0.64–0.7 μm). By the time sunlight reaches an observer at the surface, the blue light has been scattered in all directions so that the clear sky appears blue in all directions.

At sunrise and shortly before sunset, the Sun is low in the sky and its light travels through a much thicker layer of atmosphere before reaching the surface. If the sky is clear, the distance is sufficient for all the blue and green light to be scattered, so it disappears, but the longer distance also means there is a greater chance for light at longer wavelengths to be scattered, because the light encounters more air molecules. After the blue and green light have been removed, the scattered light that penetrates to the surface is orange and red. This accounts for the colors of the sky at dawn and sunset.

Bigger particles, for which *X* is greater than 1 (*d* is greater than λ), scatter light much more efficiently. When the RELATIVE HUMIDITY is high, small aerosol particles absorb water vapor and expand. As they grow larger the amount of light they scatter increases. They scatter all wavelengths equally, and so the scattered light is not separated into its constituent colors. The scattered light is white and as scattering increases the sky becomes whiter and hazier. HAZE reduces visibility and it can turn into FOG if the relative humidity reaches 100 percent and water starts to condense onto the particles.

Scattering by molecules and very small aerosol particles was described by LORD RAYLEIGH and is known as *RAYLEIGH SCATTERING*. Scattering by larger particles was described by Gustav Mie and is known as *MIE SCATTERING*.

scatterometer An instrument carried by a satellite that measures the scattering of RADAR waves by the small CAPILLARY WAVES on the ocean surface. The speed and direction of the surface wind can be calculated from these measurements.

scavenging The removal of PARTICULATE MATTER from the air by the action of PRECIPITATION. The natural processes involved are RAINOUT, SNOWOUT, and WASHOUT. These remove most particles within hours or at most days from the time they enter the air (*see* RESIDENCE TIME). The term is also applied to the removal of gaseous pollutants as a result of chemical reactions. These most commonly involve free radicals, such as hydroxyl (OH).

Schaefer, Vincent Joseph (1906–1993) American *Physicist* Vincent Schaefer was born at Schenectady, New York, on July 4, 1906. After leaving school he worked for a time in the machine shop at the General Electric Corporation (GEC) in Schenectady. Then, thinking outdoor work would suit him better, he attended classes at Union College, New York, and enrolled at the Davey Institute of Tree Surgery, from which he graduated in 1928 and became a tree surgeon. He was a keen skier and loved snow but had to abandon tree surgery.

Unable to earn an adequate salary at this profession, in 1933 Schaefer returned to GEC. There he came to the notice of Irving Langmuir (1881–1957). In 1932, Langmuir had become the first American industrial scientist to win the Nobel Prize in chemistry. He was at the peak of his fame, and he recruited Schaefer as a research assistant. Schaefer became a research associate in 1938 and he remained at GEC until 1954.

During World War II, Langmuir and Schaefer studied the problem of icing on the wings and other external surfaces of aircraft. This was extremely dangerous and caused many aircraft to crash, but before remedies could be found the scientists had to discover what was causing ice to form.

Working with his colleague BERNARD VONNEGUT, Schaefer studied the formation of ice and snow using a refrigerated box with an inside temperature that remained at a constant −9° F (−23° C). He hoped to be able to induce WATER VAPOR to be deposited as ice around dust particles. This work continued for some years until July 1946, when there was a spell of unusually hot weather. It became difficult to maintain the temperature inside the box, and so on July 13 Schaefer dropped some dry ice (solid carbon dioxide) into it to chill the air. The result was dramatic. The moment the dry ice entered the air in the box, water vapor turned into ice crystals and there was a miniature snowstorm.

This suggested a way to make PRECIPITATION fall from a cloud that otherwise would not release it. By November 13 Schaefer was ready for a full-scale trial. An airplane flew him above a cloud at Pittsfield, Massachusetts, about 50 miles (30 km) southeast of Schenectady. Schaefer dropped about 6 pounds (2.7 kg) of dry ice into the cloud and started the first artificially induced snowstorm in history. This discovery led to the development of other techniques for CLOUD SEEDING.

Schaefer received the degree of doctor of science (Sc.D.) in 1948 from the University of Notre Dame, and in 1959 he joined the faculty of the State University of New York at Albany, where he founded the Atmospheric Sciences Research Center. He was appointed professor of atmospheric science at the State University of New York in 1964 and held the position until 1976. He was appointed a fellow of the American Academy of Arts and Sciences in 1957, and in 1976 he received a special citation from the American Meteorological Society.

Schaefer died at Schenectady on July 25, 1993.

scharnitzer A cold, persistent northerly wind that blows through the Tyrol, Austria.

sclerophyllous plant A plant that is adapted to prolonged periods of hot, dry weather. Its leaves are evergreen, so they are not all shed at the same time and the plant retains leaves through the year, but they are also small, thick, hard, and leathery. Holly (*Ilex* species), holm (or holly) oak (*Quercus ilex*), and California lilac (*Ceanothus* species) are sclerophyllous plants with broad leaves, as are the gum trees (*Eucalyptus* species) native to Australia. Many pine trees (*Pinus* species) are also sclerophyllous.

(You can read more about sclerophyllous vegetation in Michael Allaby, *Ecosystem: Temperate Forests* [New York: Facts On File, 1999].)

Scotch mist STRATUS cloud that forms suddenly on high ground and is very common in Scotland and on hills and mountains in other high-latitude regions. It is caused by ADIABATIC cooling and CONDENSATION in very moist air that is forced to rise up the hillside, so it occurs only when there is a wind to carry air upward. It resembles a form of liquid precipitation in which MIST and DRIZZLE are mixed. All of the droplets are smaller than 0.02 inch (0.5 mm) in diameter, but they are more closely spaced than in drizzle.

screened pan An EVAPORATION PAN the surface of which is covered by a screen made from wire mesh, usually $1/4$ inch (6 mm) gauge. The mesh reduces INSOLATION and EVAPORATION, producing a PAN COEFFI-

CIENT that is closer to unity than that from an unscreened pan.

scrubber   A device that removes solid and liquid particles and some gases from a stream of gas. It consists of a space into which water is sprayed or that contains wet packing material. The particles and gas molecules adhere to water molecules or dissolve in the liquid water. Scrubbers are used to take samples of gas streams and to remove potential pollutants from waste gases.

scud   Fragments of tattered cloud, most commonly of NIMBOSTRATUS, that lie below the general cloud base.

sea-breeze front (lake-breeze front)   The boundary that is formed when cool MARITIME AIR advances beneath warmer air over land, producing a sea breeze (see LAND AND SEA BREEZES). Fronts of this type occur in summer along seacoasts and the shores of large lakes.

seafloor spreading   *See* PLATE TECTONICS.

sea fog   A type of ADVECTION FOG that forms at sea when warm, moist air moves over an area of much colder water. As the air is chilled to below the DEW POINT TEMPERATURE by contact with the cold water, some of the water vapor it carries condenses. Fog of this type is common in certain areas, especially off the eastern coast of Newfoundland, where air moving northward crosses the warm water of the GULF STREAM and then encounters the cold water of the LABRADOR CURRENT.

sea ice   Ice that forms by the freezing of seawater. When the temperature at the sea surface is below 32° F (0° C) and the sea is calm, snow falling on the sea may settle and accumulate. It is able to do so because snow consists of freshwater, which has a higher freezing temperature than salt water. Ice may also be carried to the sea by rivers or reach it by breaking away from glaciers. Although they may float on the sea, accumulated snow and ice that originated on land are not counted as sea ice. The term is reserved for ice that results from the freezing of the sea itself.

Water freezes at 32° F (0° C) at average sea-level atmospheric pressure, but only if it is pure H₂O. If other substances are dissolved in it, its freezing temperature is lower by an amount proportional to the strength of the solution. The average salinity of seawater, measured as all the dissolved salts but consisting mainly of sodium chloride (NaCl), is 35 grams per kilogram. This is the same as 35 parts per thousand (because 1 g = 1/1,000 kg) and is written as 35‰ (pronounced "per mil"). At 35‰ salinity, the freezing point of water is 28.56° F (–1.91° C). If the sea-surface temperature is between 32° F and 29° F the seawater does not freeze, but snow falling onto it does not melt.

When the temperature falls below freezing, ice crystals start to form. These consist of pure water. The dissolved salt is excluded from the crystal, increasing the salinity of the water adjacent to each crystal, lowering its freezing point still more, but also increasing the density of the water by an amount equal to that of the salt molecules that are added to it. The denser water sinks and less dense water rises from below to replace it, so freezing at the surface increases the rate at which water mixes in the uppermost layer of the sea.

Provided the sea temperature remains below freezing, ice crystals continue to form until they cover large areas of the surface, and as they start to join together the sea is covered with slush. This is known as *FRAZIL ICE*. As the process continues the frazil ice thickens and

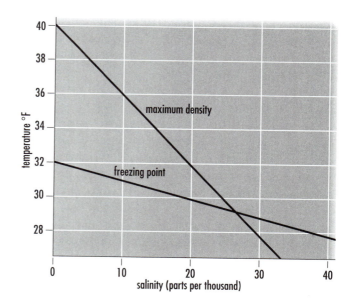

Sea ice. **The freezing point of seawater varies according to the salinity of the water. As seawater freezes the salinity of the surrounding water increases, raising its density.**

then breaks into pieces as a result of the motion of the sea. The pieces jostle against each other and change to a rounded shape. They are then known as *PANCAKE ICE*. As pancake ice forms, salt water becomes trapped between ice crystals, so that although the ice itself comprises only pure water, the pancake ice has salt within it. The amount of salt that becomes trapped in this way depends on the speed with which the ice forms and, therefore, on the air temperature. When the air temperature, while ice is forming, remains at about 3° F (–16° C), the salinity of pack ice is approximately one-fifth that of the seawater, and at –40° F (–40° C) it is roughly one-third.

When the blocks of pack ice unite to form a complete ice cover, they are known first as *YOUNG ICE,* then as *WINTER ICE,* and eventually as *POLAR ICE.*

sea-level pressure   The atmospheric pressure at sea level. This varies, but the average value that is used in the definition of the STANDARD ATMOSPHERE is 101.325 kPa (= 1013.25 mb, 760 mm mercury, 29.9 in mercury, 14.7 lb per square in). Sea-level pressure can be measured directly, but it is usually calculated from station pressures measured at known elevations above sea level. Unless it is stated otherwise, all reported atmospheric pressures are reduced to sea-level values (*see* REDUCED PRESSURE).

Seasat   The first satellite that used imaging RADAR to study the Earth. It was equipped with a SCATTEROME-TER and with a SCANNING MULTICHANNEL MICROWAVE RADIOMETER. *Seasat-A* was launched on an Atlas–Agena rocket from Vandenberg Air Force Base, California, on June 26, 1978, into a slightly elliptical POLAR ORBIT at a height of 482–496 miles (775–798 km). Its orbit carried it over nearly 96 percent of the surface of the Earth every 36 hours. On October 10, when it had transmitted data for 106 days, a short circuit drained all the power from its batteries and the satellite ceased to function.

seasonal drought   One of the types into which DROUGHTS are formally classified. It is less extreme than a PERMANENT DROUGHT but highly predictable, because it occurs in climates where all or most of the precipitation falls during one SEASON. Most plants native to such climates germinate and grow during the rainy season and survive the dry season as seeds or in a dormant state. In Bombay, 94 percent of the annual rainfall falls between June and September.

seasons   In summer, the days are long, the nights short, and the weather is relatively warm. In winter, the opposite conditions prevail: days are short, nights are long, and the weather is relatively cold. These variations in weather conditions define the seasons—the four periods of equal length we know as spring, summer, fall, and winter.

This description is true only in latitudes outside the TROPICS, however, and even there in some places the difference in temperature between one season and another is much less important—or marked—than the difference in rainfall. In these regions, the names *summer* and *winter* are replaced by *rainy season* and *dry season*. In low latitudes, spring and fall are short or barely happen at all. Nor does the change in day length affect all places equally. At the summer SOLSTICE, for example, people at the equator experience 12 hours of daylight, while those at the Pole experience 24 hours. At the winter solstice, people at the equator still experience 12 hours of daylight, but for those at the Pole the Sun does not rise at all and so they have 24 hours of darkness.

There are few places on Earth where no seasonal changes at all occur in day length, mean temperature, or rainfall, but the seasons become more strongly differentiated with increasing distance from the equator. They occur because the amount of solar radiation received at the surface changes through the year as a consequence of the tilt in the rotational axis of the Earth with respect to the PLANE OF THE ECLIPTIC.

Instead of being normal (at right angles) to the plane of the ecliptic, the Earth's axis is at an angle of 66°30' to it, so it is tilted 23°30' from the normal. As the Earth moves along its orbital path, first one hemisphere and then the other is tilted toward the Sun. This produces four clearly defined positions. In two of them, known as the *SOLSTICES,* one hemisphere receives maximum exposure to sunlight and the other receives minimum exposure. In the others, known as the *EQUINOXES,* both hemispheres are equally exposed. Seen from a position on the equator, the Sun at the equinoxes is directly overhead at noon, and at the solstices it is at an elevation of 66°30' above the horizon at noon, or 23°30' from the vertical, dis-

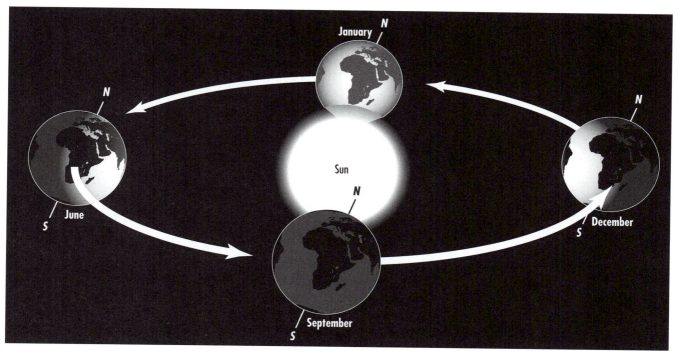

Seasons. Because the rotational axis of the Earth is tilted in respect to the plane of its solar orbit, in June the Northern Hemisphere receives more solar radiation that the Southern Hemisphere and in December the situation is reversed. This produces the seasons.

placed either to the north or to the south. At the solstices, the Sun is directly overhead at one or other of the Tropics. This means that from the other Tropic it is at an angle of 47° to the vertical, displaced in the direction of the equator.

The change in angle alters the ANGLE OF INCIDENCE, which in turn alters the intensity of the radiation that is received at each unit area of the surface. Because the Earth is almost spherical in shape, the angle of incidence increases with latitude and so, therefore, does the intensity of radiation per unit area at the surface. It is this change that causes mean temperatures to be higher in summer than in winter and higher in low latitudes than in high ones.

sea-surface temperature (SST) The temperature of the water at the surface of the sea. This is routinely measured by drifting buoys, ships, and orbiting satellites. Of these, the satellite observations are the most extensive in their coverage and also the most accurate. Buoys can measure only the temperature of the water around them, which may not be representative of the ocean as a whole, and ships measure the temperature

of the water they take on board to cool their engines. Water intakes are located about 16 feet (5 m) below the surface, and so ship measurements must be corrected to give the temperature at the surface.

Sea-surface temperature is climatically important because it affects the temperature of the air immediately above the surface and the EVAPORATION rate of water. This in turn affects air temperature, because WATER VAPOR is the most important GREENHOUSE GAS, as well as affecting cloud formation, ALBEDO, and PRECIPITATION.

Sea-surface temperatures change with the seasons, but they are also subject to other influences. Latitudinally, they change because of the presence of warm and cold ocean currents and they also rise and fall in fairly regular cycles. Some cyclical variations operate with a period of a few years, others of decades or centuries. There is still much to be learned about these cycles.

Eddies in ocean currents can produce local variations in sea-surface temperatures. These are similar to atmospheric CYCLONES and ANTICYCLONES, but they persist for months rather than days.

sea turn A name that is given in the northeastern United States, and especially in New England, to a wind that blows from the sea and often carries MIST.

seawater Water that is found in the seas and oceans and that contains an average of 35 parts per thousand (‰, pronounced "per mil") of dissolved compounds, known collectively as *salts*. The proportion of salts determines the salinity of the water, so if it contains 35‰ of salts, its salinity is 35‰. Salinity varies from 34‰ to 37‰ in coastal areas but may be close to 0‰ where rivers discharge large volumes of FRESHWATER, or as high as 40‰ where a large amount of water is lost by evaporation from a partially enclosed body of water, such as the Persian Gulf and Red Sea. Chlorine (Cl), sodium (Na), sulfur as sulfate (SO_4), magnesium (Mg), calcium (Ca), and potassium (K) account for more than 99 percent of the dissolved matter present in seawater. The minor constituents are bicarbonate (HCO_3), bromide (Br), boric acid (H_3BO_3), and strontium (Sr).

seaweed A plant of any one of several thousand species of multicellular marine algae. Some seaweeds are large. Certain species of *Macrocystis* and *Nereocystis*, found in the Pacific and Southern Oceans, grow to more than 100 feet (30 m) in length. Seaweeds comprise three plant divisions (or phyla), the Rhaeophyta (brown seaweeds), Rhodophyta (red seaweeds), and Chlorophyta (green seaweeds). They grow in coastal waters throughout the world, from the uppermost part of the shore reached by spring TIDES to where the water is about 165 feet (50 m) deep. At greater depths there is insufficient light for PHOTOSYNTHESIS.

Seaweeds have many uses. Some are eaten, others are used to make fertilizer, and some were traditionally used to foretell the weather. These species are adapted to survive the very harsh environment of the upper shore, where they are alternately submerged in seawater and exposed to air and warm sunshine. They survive the dry conditions by shriveling and becoming brittle, but as soon as they detect moisture they start to absorb water and revive. The wracks are the weeds that demonstrate this capacity best. These are brown seaweeds of the genera *Fucus* and *Ascophyllum*, such as bladder wrack (*F. vesiculosus*) and knotted wrack (*A. nodosum*). People used to take them home from the shore and hang them outside the door. If the seaweed was dry and shriveled, the weather would be fine, but if it became flexible and rubbery rain was likely. The method worked, but only up to a point, because by the time the seaweed responded to the rise in HUMIDITY the change in the weather was usually self-evident.

SeaWinds A RADAR instrument that is carried on board the QUICKSCAT satellite. Launched on June 19, 1999, it collects data from a continuous band, 1,800 km (1,118 miles) wide, covering 90 percent of the Earth's surface, and makes approximately 400,000 measurements in a day. SeaWinds has a rotating dish antenna with two spot beams that sweep in a circular pattern radiating microwave pulses at 13.4 gigahertz. It gathers data on low-level wind speed and direction over the oceans and also tracks the movement of Antarctic icebergs. The data are used in scientific studies of global climate change and weather patterns and interactions between the atmosphere and the ocean surface; in tracking of changes in tropical rain forests; and in monitoring of movements at the edge of the Antarctic sea ice and pack ice.

(You can learn more about SeaWinds at winds. jpl.nasa.gov/mission/quikscat/quikindex.html.)

seca The name that is given in northeastern Brazil to a severe drought or very dry wind. Secas are often associated with unusually warm water in the subtropical North Atlantic Ocean.

seclusion An OCCLUSION in which the COLD FRONT first starts to overtake the WARM FRONT some distance from the peak of the WAVE DEPRESSION.

second (s) The SYSTÈME INTERNATIONAL D'UNITÉS (SI) UNIT of time, which is defined as the duration of 9,192,631,770 periods of the radiation that corresponds to the transition of an atom of cesium-133 between two hyperfine levels. The second is also a unit of angle (sometimes called an *arcsecond*), equal to 1/3,600 of a degree, and symbolized by ".

secondary air mass An AIR MASS that has been modified as it passes over a surface that is different from the surface over which it developed its characteristics. For example, in winter, continental polar (cP) air (*see* CONTINENTAL AIR, POLAR AIR) moves outward from Canada and over the North Atlantic. When the cP air passes over the warm water of the NORTH ATLANTIC DRIFT its lower layers become warmer and unstable (*see* STABILITY OF AIR) and their moisture content increases sharply as a result of EVAPORATION. The resulting CONVECTIVE INSTABILITY makes the air turbulent. By the time it reaches the eastern Atlantic the cP air has changed into cool maritime polar (mP) air (*see* MARITIME AIR).

secondary circulation   That part of the GENERAL CIRCULATION of the atmosphere that consists of relatively small-scale, short-lived features that are superimposed on the more permanent features of the PRIMARY CIRCULATION. FRONTAL SYSTEMS, CYCLONES, and ANTICYCLONES dominate the secondary circulation in middle latitudes.

secondary cold front   A front that develops in the cold air behind a FRONTAL CYCLONE when the horizontal temperature gradient is so strong that the cold air starts to separate into two distinct masses, one of which is colder than the other.

secondary cyclogenesis   The development of a second EXTRATROPICAL CYCLONE as the first cyclone becomes occluded (*see* OCCLUSION) and starts FILLING. The new cyclone develops where air is flowing outward from the upper TROPOSPHERE.

secondary cyclone   *See* PRIMARY CYCLONE.

secondary front   *See* PRIMARY FRONT.

secondary pollutant   A polluting substance that is produced in the atmosphere by chemical reactions among PRIMARY POLLUTANTS. A mixture of FOG and SMOKE constitutes SMOG. Smoke is a primary pollutant and the resulting smog is the secondary pollutant. Unburned hydrocarbons, which are primary pollutants released mainly in vehicle exhausts, can be oxidized in a series of steps to form PEROXYACETYL NITRATES. These secondary pollutants may then contribute to the formation of PHOTOCHEMICAL SMOG, a mixture that contains OZONE and NITROGEN OXIDES, both of which are pollutants.

sector   A horizontal plane that is bounded on two sides by the radii of a circle and on the third side by an

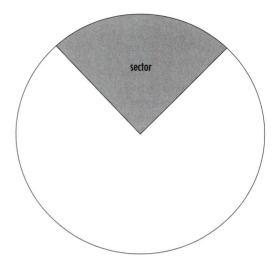

Sector. **A plane surface that is bounded by two radii and an arc of a circle.**

arc that forms part of the circumference of the same circle. A WARM SECTOR is an area that is approximately of this shape and is bounded on two sides by a WARM FRONT and a COLD FRONT, although there is no arc bounding it on the third side.

sector wind   The average direction and speed of the wind that is blowing at the altitude an aircraft plans to fly over one SECTOR of the route it intends to follow. The section of the route is called a *sector* because traditionally it is drawn on a map with two lines drawn at an angle (usually 5°) to each side. This makes it simpler to plot the location of the aircraft and interpret this as the number of degrees it is off its course and, from that, the correction that should be made to the aircraft heading. The sector wind may be either observed or calculated from the broader SYNOPTIC situation.

Seebeck effect   *See* THERMOMETER.

seif dune   A sand dune that is built by winds that blow in two principal directions and that forms in hot deserts. The dune is the extended, tapering, slightly curved arm of a BARCHAN DUNE. Its shape is somewhat reminiscent of a curved sword. Its name is derived from the Arabic word *sayf*, which means "sword."

seistan   A wind that blows almost incessantly from the north or northwest between June and September

Seif dune. **Seif dunes are long, tapering, and slightly curved.**

in eastern Iran and western Afghanistan, but mainly in the Iranian border region of Seistan. This is a depression covering an area of about 7,000 square miles (18,130 km²) centered on latitude 30.5° N, longitude 61° E. The seistan wind is caused by the MONSOON and is also known as the *wind of 120 days*. It can attain speeds of 80 mph (129 km h⁻¹). It is hot and extremely dry and carries large amounts of dust and salt.

selatan   A strong, dry wind that blows over parts of Indonesia from the south during the southeast MONSOON.

selectively permeable membrane   *See* OSMOSIS.

SELS   *See* SEVERE LOCAL STORMS UNIT.

semiarid climate   In the THORNTHWAITE CLIMATE CLASSIFICATION, a climate in which the MOISTURE INDEX is between −67 and −33 and the POTENTIAL EVAPOTRANSPIRATION is less than 5.6 inches (less than 14.2 cm). The climate is designated *D*. In terms of THERMAL EFFICIENCY, this is a frost climate (*E'*).

semidesert climate   In the CLIMATE CLASSIFICATION devised by MIKHAIL I. BUDYKO, a climate in which the RADIATIONAL INDEX OF DRYNESS has a value of 2.0–3.0.

semidiurnal   Occurring twice in every 24 hours.

sensible heat   *See* ENTHALPY.

sensible temperature   The TEMPERATURE that the body feels. This is not always the same as the AIR TEMPERATURE measured by a THERMOMETER, because the sensation of heat or cold is affected by several factors in addition to the actual temperature of the air. These include the wind (*see* WIND CHILL) and the rate at which the body loses heat through CONDUCTION, CONVECTION, and radiation from exposed skin surfaces; from the EVAPORATION of sweat; and from the respiratory tract (which is exposed to inhaled air that is below body temperature). Several of these, and especially the rate of evaporation from the skin, are related to the RELATIVE HUMIDITY (RH) of the air. When the RH is high, the evaporation rate is low, so the body cools itself less efficiently by sweating and the air feels warmer than it is. When the temperature is low the body does not sweat, but heat is lost by conduction to the air in contact with the skin, making the temperature feel lower than it really is. Sensitivity to these effects varies from one person to another, and people acclimatize to the weather conditions to which they are most often exposed but that may feel uncomfortable to a person who is newly arrived.

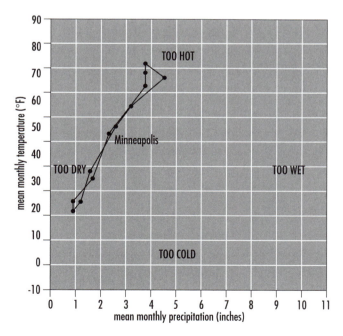

Sensible temperature. **The "comfort chart" plots sensible temperature and humidity and overlays a hythergraph for Minneapolis.**

Sensible temperatures are used in conjunction with RH to define COMFORT ZONES. These can be shown graphically on a chart that plots temperature against PRECIPITATION. When the resulting comfort chart is overlaid with a HYTHERGRAPH it shows at a glance whether or not the climate of a particular place is likely to prove comfortable.

serein Rain that falls from a clear sky. There are several possible explanations for this very rare phenomenon. CLOUD DROPLETS may evaporate after very small RAINDROPS have started to fall. Because of their size the drops take several seconds to reach the ground and by the time they do so the cloud has dissipated. Alternatively, the cloud may move away while the raindrops are falling, so that by the time they reach the ground the cloud is no longer overhead. It may also happen that the wind blows fine rain so that it reaches the ground at a point that is not beneath the cloud. In this case, however, the rain arrives at an angle and its source is fairly obvious.

settled An adjective that describes fine weather conditions that remain unchanged for a minimum of several days and more commonly for a week or more. Settled weather is usually associated with an ANTICYCLONE and is often caused by BLOCKING.

Severe Local Storms unit (SELS) A meteorological center, located in Kansas City, Missouri, where severe weather is monitored and forecasts of storms are issued up to six hours ahead.

severe storm Any storm that damages property or endangers life. The U.S. NATIONAL WEATHER SERVICE defines a severe THUNDERSTORM as one that produces HAIL with HAILSTONES 0.75 inch (19 mm) or more across, or wind GUSTS of 58 mph (93 km h⁻¹) or more, or a TORNADO, or more than one of these.

severe-storm observation A report of a SEVERE STORM that has been positively identified. The report states the time of the observation and the location and direction of movement of the storm.

severe thunderstorm warning A local warning, issued by the NATIONAL WEATHER SERVICE, that violent storms have entered the area. People within the area should take immediate precautions because of an imminent risk of intense hail, possibly with large hailstones, lightning, winds that may gust to 140 mph (225 km h⁻¹), and torrential rain that may cause flooding.

severe thunderstorm watch A preliminary warning issued by the NATIONAL WEATHER SERVICE to areas that may experience violent storms within the next few hours, although such storms have not yet entered the area.

Severe Weather Threat Index *See* STABILITY INDICES.

sferics A word that is derived from atm*ospherics*. Sferics are the electromagnetic disturbances disturbances caused by natural electrical phenomena. They interfere with radio transmissions and can sometimes be heard on a radio as crackling or whistling noises. LIGHTNING causes sferics, which are used to locate the source of the THUNDERSTORM. The device uses two square radio receiver aerials mounted at right angles to each other so that one is aligned north–south and the other east–west. The strength of the sferics signal varies according to the angle at which it approaches the aerials. The signal strength is converted to a direction that is displayed on a screen. A sferics receiver can detect a

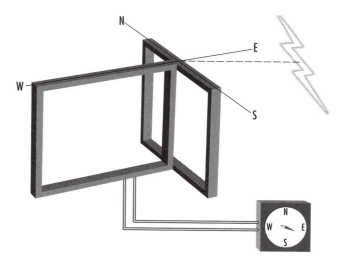

Sferics. A sferics receiver has two square radio aerials at right angles to each other. This arrangement allows the receiver to detect the direction from which a disturbance caused by lightning is coming. The location of the storm can be identified by using two or more sferics receivers.

thunderstorm up to a distance of about 1,000 miles (1,600 km).

When two or more widely separated sferics receivers detect the same thunderstorm they can reveal its location, which is at the intersection of lines drawn on a map from the position of each receiver in the direction it has measured. The position of a storm that is identified in this way is known as a *sferics fix* and a report of a storm that is based on measurements by a sferics receiver is called a *sferics observation*.

sferics fix   *See* SFERICS.

sferics observation   *See* SFERICS.

shade temperature   The AIR TEMPERATURE that is measured inside a STEVENSON SCREEN or other shelter, or anywhere out of direct sunlight. Air is heated almost entirely by contact with the ground and not by direct absorption of solar radiation. A thermometer, on the other hand, absorbs solar radiation directly. This raises its temperature to a level higher than that of the surrounding air and consequently it gives a false reading. For this reason, reported temperatures are always shade temperatures unless stated otherwise.

shallow fog   FOG that does not reduce horizontal VISIBILITY above a height of 6 feet (1.8 m) above the surface. Sometimes shallow fog completely hides the ground but rises no higher than a person's knees and forms when the sky above is blue and cloudless. Fog of this kind is almost always RADIATION FOG.

shamal   A hot, dry wind that blows almost continuously during June and July over Iraq, Iran, and the Arabian Peninsula. It is associated with the flow of air around an area of low pressure centered over Pakistan. The wind seldom exceeds 30 mph (50 km h⁻¹), but it produces large DUST STORMS.

sharki   *See* KAUS.

sharp-edged gust   A GUST of wind that produces an almost instantaneous change in the speed or direction of the wind.

Shaw, William Napier (1854–1945) British *Meteorologist*   An English meteorologist who contributed greatly to the establishment of meteorology as a scientific discipline, he was born in Birmingham and educated there and at Cambridge University, where he also taught. In 1898, he was appointed assistant director of the Cavendish Laboratory at Cambridge University; he resigned in 1900 to take up an appointment as secretary of the Meteorological Council. He was made director of the Meteorological Office in 1905 and remained there until his retirement in 1920.

Shaw introduced the MILLIBAR, in 1909, as a convenient unit of atmospheric pressure. It was adopted internationally in 1929. Some time about 1915, Shaw devised the TEPHIGRAM. He pioneered the use of instruments carried beneath kites and balloons to study the upper atmosphere and wrote several books on weather forecasting, as well as one, *The Smoke Problem of Great Cities* (1925), on air pollution. He received many honors, including the 1910 Symons Gold Medal of the Royal Meteorological Society, and in 1915, he received a knighthood.

shear   A force that acts parallel to a plane, rather than at right angles to it. If two plane surfaces experience a shearing force, one surface is being pushed one way and the other in another (not necessarily opposite) direction. Fluids as well as solids can experience shear. If two bodies of fluid are in motion, the shearing force acting on them may cause a change in their relative speeds rather than direction. When crossing from one body to the other, the shear is evident as an abrupt change in either the direction or the speed of movement, or both.

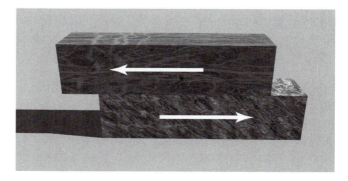

Shear. A shearing force acts parallel to a plane rather than at right angles to it. The arrows represent the direction of shear.

shear instability (Kelvin–Helmholtz instability) The condition in which two layers of a fluid are moving in the same direction but at different speeds. The phenomenon was first recognized and explained in 1868 by the German physicist and physiologist Hermann Ludwig Ferdinand von Helmholtz (1821–94), and in 1871 the Scottish physicist and mathematician William Thomson (later Lord Kelvin, 1824–1907) used Helmholtz's theory to explain the formation of BILLOW CLOUDS. Because of this association, shear instability is often called *Kelvin–Helmholtz instability*.

SHEAR causes the air or water at the boundary between the two layers to roll. This produces a series of small waves at right angles to the direction of the shear. The waves usually extend for a distance equal to several WAVELENGTHS and the entire pattern of waves is carried forward by the wind. The pattern grows and then breaks up; it usually lasts only a few minutes. Shear instability can cause serious CLEAR-AIR TURBULENCE, especially when the waves are starting to break, and it can occur at any height in the TROPOSPHERE.

RADAR can detect shear instability, but it is otherwise invisible unless the RELATIVE HUMIDITY in the waves is very close to SATURATION. In that case clouds may form by the CONDENSATION of WATER VAPOR near the peak of each wave, producing rows of clouds.

shear line A line or narrow belt that marks an abrupt change, or SHEAR, in the direction or speed of the wind.

shear velocity *See* FRICTION VELOCITY.

shear wave A wave that forms where there is strong horizontal WIND SHEAR in stable air (*see* STABILITY OF AIR). The difference in wind speed across a boundary between two layers of air causes TURBULENT FLOW. Air from the lower layer rises, but its stability causes it to sink again, establishing a wave pattern. In a vertical rather than horizontal plane, this is the mechanism that causes a flag to flap in the wind. Air is moving at different speeds on each side of the flag. The wind shear generates a wave pattern that is prevented from breaking down by the cloth of the flag, which acts in the same way as the inherent stability of the air in a shear wave. Where the RICHARDSON NUMBER falls below the critical value of 0.25, the stability of the air is insuffi-

cient to sustain the wave pattern, which breaks down into general turbulence. The breakdown of shear waves plays an important part in transferring energy and transporting materials such as water vapor to and from the ground surface.

sheet lightning LIGHTNING that is seen as a general flash, rather than being precisely located as is FORKED LIGHTNING. Sheet lightning may be forked lightning between two clouds that is seen through an intervening cloud, or it may be a lightning flash between the separated charges inside a cloud. These discharges are thought to be less luminous and to last longer than discharges between clouds or between clouds and the ground.

shelf cloud A layer of cloud that projects like a shelf beneath the anvil (*see* INCUS) of a big CUMULONIMBUS storm cloud. As the storm approaches, it is the anvil that arrives first, as a thin layer of high-level cloud. This quickly thickens and becomes lower. Then a second layer of cloud appears below the anvil. This is the shelf cloud. It marks the region of the storm where warm air is being drawn into the cloud and its WATER VAPOR is condensing as it rises and grows cooler. The GUST FRONT, caused by the inrush of air, is situated beneath the shelf cloud. Heavy PRECIPITATION usually commences near the location where the shelf cloud is attached to the main cloud.

(You can read more about the structure of such storms in Michael Allaby, *Dangerous Weather. Tornadoes* [New York: Facts On File, 1997].)

shelter belt A line of trees that is grown at right angles to the PREVAILING WIND in order to reduce the wind speed on the LEE side as a means of protecting ground or crops that might be damaged by strong winds. Air approaching the trees is forced to rise. This squeezes the STREAMLINES together, accelerating the air as it passes over the tops of the trees, but it decelerates as soon as it has crossed the barrier. The air then separates from the surface of the trees and forms large EDDIES that gradually become smaller in the WAKE of the shelter belt. Finally, the air resumes its former movement and speed. A shelter belt (or wall, fence, embankment, or other obstacle used for the same purpose) affects the flow of air above and in front of the barrier for a distance equal to three times the height of the barrier.

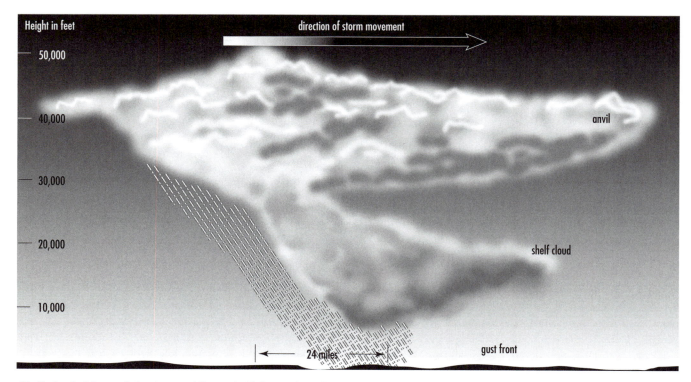

Shelf cloud. A layer of cloud, resembling a shelf, that projects from the main part of a large storm cloud.

The effect on the downwind side of the barrier depends on the density of the barrier. If the barrier is very dense, the wind speed is greatly reduced in the large eddy that forms immediately downwind of it, because air cannot penetrate the barrier. The wind speed recovers quickly to about 90 percent of its former value, however, so the effect is limited to a distance about 10–15 times the height of the barrier. If the barrier is less dense, so it allows some of the air to pass through it, no eddy forms. The reduction in wind speed is smaller immediately behind the barrier, but it extends downwind for a distance that is equal to 15–20 times the height of the barrier. A medium-density barrier performs even better. The wind recovers to 90 percent of its original speed about 20–25 times the barrier height downwind, and some shelter belts provide this amount of protection for a distance equal to 40 times their height.

shelter temperature *See* STEVENSON SCREEN.

shepherd's weatherglass *See* SCARLET PIMPERNEL.

shielding layer A name that is sometimes given to the PLANETARY BOUNDARY LAYER. It refers to the fact that the boundary layer shields the surface from events in the FREE ATMOSPHERE.

shimmer *See* MIRAGE.

SHIP code A version of the international SYNOPTIC CODE that is approved by the WORLD METEOROLOGICAL ORGANIZATION (WMO) for use by VOLUNTARY OBSERVING SHIPS (VOS). Data contained in VOS weather reports are compressed into SHIP code before being transmitted by radio to a shore receiving station, from which they are passed to the WMO Global Telecommunications System.

ship report A weather report that is compiled on board a VOLUNTARY OBSERVING SHIP (VOS) at sea and transmitted to a shore receiving station. At regular intervals, usually at 0000, 0600, 1200, and 1800 GREENWICH MEAN TIME (GMT), officers on the VOS take observations of AIR TEMPERATURE, SEA-SURFACE

Shelter belt. The trees form a barrier that slows the wind. If the barrier is dense the wind is slowed greatly, but soon recovers. A low-density barrier slows the wind less, but the effect extends farther.

TEMPERATURE, WIND VELOCITY, AIR PRESSURE, PAST WEATHER, and PRESENT WEATHER. They may also note the RELATIVE HUMIDITY, cloud cover, and state of the sea. The recorded data are compressed using SHIP CODE, stored in the on-board meteorological logbook, and transmitted to a receiving station on shore. The receiving station then passes the report on to the Global Telecommunications System of the WORLD METEOROLOGICAL ORGANIZATION. When the ship reaches port, its meteorological logbook is handed over to the PORT METEOROLOGICAL OFFICER. Its contents are used to augment the transmitted observations and help to build a long-term picture of the climate over the oceans.

shock wave   A traveling wave that moves through a fluid as a narrow band in which the pressure and/or temperature increases abruptly. Any object that moves through air or water generates a disturbance that propagates as a series of shock waves. Any sudden expansion or movement, such as an explosion or earthquake, generates shock waves. Sound waves are also shock waves.

short-day plant   See PHOTOPERIOD.

short-range forecast   A weather forecast that covers a period of up to two days. The forecast is based partly on SYNOPTIC methods but nowadays more often on NUMERICAL FORECASTING. The synoptic method aims to predict future patterns of high-level pressure distribution and the THICKNESS of the layer between 1000 mb and 500 mb, then to judge the surface conditions that are likely to result. Short-range forecasts are generally fairly accurate, but they are limit-ed. Weather systems may change the speed at which they move, and when the air is moist and unstable it is impossible to predict where and when showers will occur. Consequently, these are forecast somewhat vaguely, as "scattered showers" or "showers with bright periods."

Showalter Stability Index   See STABILITY INDICES.

shower   A short period of PRECIPITATION that falls from a convective cloud such as CUMULUS or, more commonly, CUMULONIMBUS, that is produced when moist, cool, unstable air (*see* STABILITY OF AIR) crosses a warmer surface. Precipitation starts and ends abruptly, varies in intensity but can be very heavy, and is often followed by sunshine and a blue sky.

shrieking sixties   *See* POLAR WET CLIMATE.

Shūrin season   A period of high rainfall that occurs in Japan during September and early October. It is the second wet season of the year; the first is the BAI-U season during June and July. The rains of the Shūrin season are associated with an eastward contraction of the subtropical part of the PACIFIC HIGH that allows low-pressure systems and TYPHOONS to move farther north. Much of the resulting rain is caused by typhoons, but some is produced by the southern sides of DEPRESSIONS that move along the POLAR FRONT, to the north of Japan.

Siberian high   A region of high surface pressure that forms over Siberia in winter. Centered to the south of Lake Baikal, it influences all of Asia north of the Himalayas and extends westward, centered on latitude 50 ° N, across southern Russia and central Europe as far as the Atlantic. It produces a *wind divide*. The midlatitude westerlies prevail to its north and northeasterlies prevail to its south, across the steppes from Ukraine eastward. The northeasterlies carry very dry, CONTINENTAL POLAR AIR. The Siberian high produces the highest pressures known anywhere on Earth. On December 31, 1968, a pressure of 1,084 millibars (32.01 inches [81.31 cm]) was recorded at the town of Agata, Russia (66.83° N 98.71° E). This is believed to be the highest pressure ever recorded.

sidereal day   The time the Earth takes to complete one revolution on its axis so that a point on the surface returns to the same position in relation to a fixed star. This is 4.09 minutes shorter than a mean solar day.

sidereal month   The average time taken by the Moon to complete one orbit of the Earth and return to the same position with reference to the fixed stars. It is 27 days, 7 hours, 43 minutes.

sidereal year   *See* ORBIT PERIOD.

siffanto   A southwesterly wind that blows over the Adriatic Sea and surrounding lands. It is often violent.

sigua   A gale that blows in the Philippines. It is associated with the MONSOON.

silver iodide   A yellow, solid compound of silver (AgI) that melts at 1,033° F (556° C) and boils at 2,743° F (1,506° C). It is used in CLOUD SEEDING because it can be made to form particles the size of FREEZING NUCLEI. A mixture of silver iodide and another compound, commonly sodium iodide (NaI), is dissolved in acetone. The solution is then burned in a propane flame or a flare mounted on an airplane,

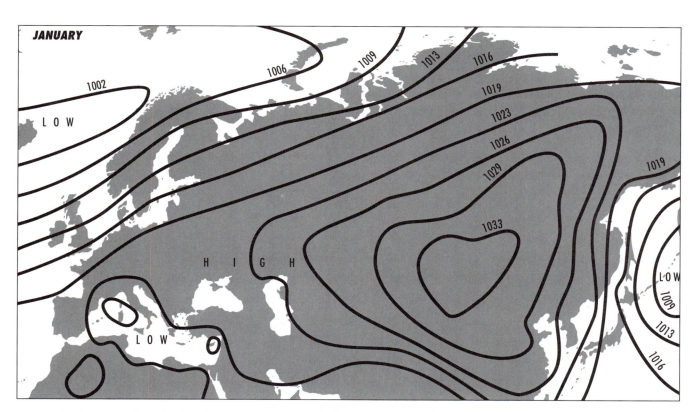

Siberian high. The usual distribution of pressure over Eurasia in January shows a large area of intensely high pressure centered on southern Siberia and Mongolia. This is the Siberian high.

releasing smoke. As the smoke cools the silver iodide solidifies as small crystals. Dropped into a cloud with a temperature of –4° F (–20° C), one ounce of silver iodide produces about 2.8×10^{17} ice nuclei (10^{15} nuclei per gram of AgI).

silver thaw The name that is given to FREEZING RAIN in the Pacific Northwest. It is called *thaw* because it is often followed by warmer weather. This is because the freezing rain is usually associated with an advancing WARM FRONT. Rain produced by the front freezes as it falls through colder air beneath the FRONTAL SLOPE. Air in the WARM SECTOR behind the front soon reaches places on the surface experiencing the "thaw."

simoom A hot, dry wind of the SIROCCO type that blows in spring and summer across the southeastern Sahara and the Arabian Peninsula. It often carries large amounts of dust and sand. It occurs when a DEPRESSION moves along the Mediterranean and air from far to the south is drawn into the CYCLONIC flow.

Simpson, George Clark (1878–1965) British *Meteorologist* A British meteorologist who studied atmospheric electricity and the effect of radiation on polar ice. He also standardized the wind speeds in the BEAUFORT WIND SCALE.

Simpson was born and attended school in Derby. He left school in 1894 to work in his father's business, but his reading of popular books about science aroused his interest and he began attending night school. His father persuaded him to continue his education at Owens College, Manchester. He was coached for the entrance examination, entered the college, and graduated in 1900. He continued his studies at Göttingen University, in Germany, then visited Lapland to study atmospheric electricity. On his return to his college, which by then had become the University of Manchester, he was appointed to head a newly formed meteorology department. He was the first lecturer in meteorology at any British university.

Simpson was also appointed assistant director of the Meteorological Office. He spent some time in India inspecting meteorological stations and in 1910 traveled as meteorologist on Robert Scott's last expedition to Antarctica. Between 1916 and 1920, he worked in the Middle East and Egypt. In 1920, he was appointed director of the Meteorological Office, a post he held

until his retirement in 1938. On the outbreak of war, in 1939, he returned to work, studying electrical storms, and he retired for a second time in 1947.

He was awarded the Symons Gold Medal of the Royal Meteorological Society in 1930. In 1935, he was knighted.

sine wave A curve on a graph that corresponds to an equation in which one variable is proportional to the sine of the other. A point moving along the curve oscillates about a central point so that crests and troughs are of equal amplitude and wavelength (*see* WAVE CHARACTERISTICS).

singular corresponding point A center of high or low AIR PRESSURE that is shown by the pattern of ISOBARS on a CONSTANT-HEIGHT CHART and as an elevated or depressed region on a CONSTANT-PRESSURE CHART and that reappears on succeeding charts.

singularities Certain types of weather that occur fairly regularly at a particular time each year. The INDIAN SUMMER and the sudden advent of ANTICYCLONES at the end of June are probably the best known examples, but there are many more. The JANUARY THAW, affecting the northeastern United States around January 20–23, is a singularity. PRECIPITATION decreases sharply in California in March and April, as a result of an extension of the PACIFIC HIGH, at the same time it increases in the Midwest through an increase in CYCLOGENESIS in Colorado and Alberta and an extension of MARITIME AIR from the Gulf of Mexico.

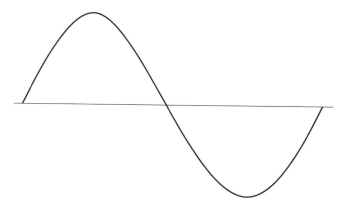

Sine wave. A sine wave oscillates in a very regular manner.

Singularities are major features of WEATHER LORE. Certain months are linked to particular kinds of weather (*see* APRIL SHOWERS, FEBRUARY FILL DYKE). Many of these assumed singularities are imaginary, but others are genuine.

European singularities have been intensely studied and a catalog of them, based on GROSSWETTERLAGE, has been compiled. Spring is often dry over much of northern Europe. This is due to a marked reduction in the frequency of weather systems arriving from the west. Westerly weather increases around the middle of June, producing wet weather. Anticyclones often dominate the weather in the middle of September, broken by stormy weather in late September, and then more fine anticyclonic weather at the end of September and early in October. FOGS and FROST, caused by more anticyclones, are likely in the middle of November. DEPRESSIONS arriving from the west usually make early December a time of mild, wet weather.

Seasonal variations that last rather longer were identified by HUBERT LAMB, who called them *NATURAL SEASONS*.

November 3 is Culture Day in Japan, and according to tradition the weather is usually fine. Naoki Sato and Masaaki Takahashi, who are scientists at the Center for Climate System Research at the University of Tokyo, have studied weather records for Tokyo over 38 years. They have found that the November 3 singularity is real and there are similar singularities in April and to a lesser degree in October.

Singularities also occur during the Asian summer MONSOON. These are associated with CLIMATOLOGICAL INTRA-SEASONAL OSCILLATIONS and produce times when the weather becomes suddenly wetter or drier. For example, the weather is usually dry from August 29 through September 2 in the western North Pacific monsoon region at longitudes 140° E and 15°–20° E.

sink   An area that forms a receptacle for materials that are moving through a system. For example, some atmospheric CARBON DIOXIDE dissolves in seawater, where it is transported into the deep ocean. The ocean therefore acts as a sink for that carbon dioxide. Soil absorbs CARBON MONOXIDE and also solid particles that are removed from the air by RAINOUT, SNOWOUT, and WASHOUT, so the soil is a sink for these substances. A sink represents the endpoint of a transport system that begins with the release of a substance into the environment. The place or process that releases the substance is called its *source*.

sirocco   A hot wind that blows across countries bordering the Mediterranean, most commonly in spring. It occurs ahead of DEPRESSIONS that move from west to east along the Mediterranean, occurring first in the west and then progressively farther to the east; Israel and Lebanon experience it four or five days after Algeria. The CYCLONIC flow of air around the center of low pressure draws air from far to the south, so by the time it approaches the North African coast it has traveled across a long stretch of desert. It reaches the coast as a hot, dry, dusty wind. It may then cross the Mediterranean, cooling and acquiring moisture as it does so, to reach southern Europe as a warm, moist wind that can cause FOG, heavy DEW, and rain—sometimes BLOOD RAIN. As the depression continues to move to the east, the airflow changes and the sirocco disappears, to give way to cold MISTRAL or BORA winds from the north.

SI units   *See* SYSTÈME INTERNATIONAL D'UNITÉS.

skill score   A measure of the accuracy of a weather forecast that is made by comparing the forecast with a reference standard. This standard may be a description of weather conditions that were prevailing when the forecast was compiled, and therefore assuming no change in the weather, or a forecast made by selecting features at random.

Skip   A TYPHOON that struck the Philippines on November 7, 1988. High winds and heavy rain caused mudslides and floods in which at least 129 people died. It was the second typhoon to strike the Philippines in two weeks.

sky color   *See* SCATTERING.

sky cover   The proportion of the sky that is obscured by clouds. The clouds may be partly transparent, so the sky is not completely hidden, and they may extend all the way to the ground. The proportion is expressed in tenths or OKTAS.

sky map   A pattern of light and dark areas that is sometimes seen on the underside of a cloud layer, most commonly in high latitudes when the surface is partly covered by snow or ice. The pattern is made by the different amounts of light reflected from the different types of surface and so it can resemble the surface, like a map.

sky-view factor   The amount of sky that can be seen from a particular point on the surface, expressed as a proportion of the total sky hemisphere.

sleet   Small raindrops that freeze in cold air beneath the base of the cloud in which they formed. They fall as small particles of ice of various shapes. Heavy sleet can reduce visibility. In Britain, sleet is a mixture of rain and snow falling together. All of it falls as snow from the base of the cloud and enters air that is below freezing temperature, but close to the ground there is a layer of warmer air in which some, but not all, of the snowflakes melt.

slice method   A method that is used to study STATIC STABILITY at any level in the atmosphere as an alternative to using the concept of the PARCEL OF AIR. The slice method takes account of air that is moving vertically through the "slice" in both directions.

sling psychrometer   See WHIRLING PSYCHROMETER.

slush   A mixture of melting snow and/or ice and liquid water that is lying on the ground. It has a soft consistency and is very wet. Slush forms when snow and ice start to thaw, when rain mixes with them, or when they are treated, for example, with salt, to lower the MELTING POINT.

small hail   HAIL that consists of HAILSTONES that are less than about 0.2 inch (5 mm) in diameter (smaller than garden peas) but that are sufficiently solid to remain intact until they hit the ground. They are too small to cause any damage or injury. Small hailstones often consist of a nucleus of GRAUPEL surrounded by a layer of clear ice. This gives them a white, frosted appearance. Small hail is often mixed with rain.

small ion   A charged particle of dust or other substance that exists as an AEROSOL. Charged particles can be moved through the air by applying an electric field. Small ions experience less drag than large ions and so they move more rapidly. This difference provides a means for counting the proportion of large and small ions. Usually there are far fewer large ions than small ions, but because of their greater mass the large ions may account for more of the total mass. In 1950, the German atmospheric scientist Christian Junge discovered that the distribution of atmospheric particles is such that for every halving of the diameter of the particles their number increases approximately 10-fold. The smallest particles are AITKEN NUCLEI.

small-ion combination   The removal of SMALL IONS through their reaction with other particles. There are two mechanisms. A small ion may adhere to a neutral AITKEN NUCLEUS. The two then form a LARGE ION. Alternatively, a small ion may combine with a large ion of opposite charge.

SMMR   See SCANNING MULTICHANNEL MICROWAVE RADIOMETER.

smog   A form of air pollution that used to occur frequently in winter in many large industrial cities but was most closely associated with London, where PEA SOUPERS were a familiar feature of winter weather. After the LONDON SMOG INCIDENTS legislation was enacted to prevent the burning of coal in open fires, and with the primary cause of smog removed this type of pollution ceased. Smog of this type is quite different from the PHOTOCHEMICAL SMOG that occurs in cities such as Los Angeles and Athens.

Smog is a contraction of *smoke* and *fog*. The term was first used in 1905, by the atmospheric scientist H. A. Des Voeux, but the phenomenon was far from new. It had been increasing in London since at least 1600, most rapidly in the 17th and 19th centuries.

Like many cities, London is low-lying and beside a large river. Evaporation from the river, as well as from plants in the many parks and open spaces, ensures the air is often moist and dust and other particles from the urban environment provide ample cloud condensation nuclei (CCN). Temperature inversions are also fairly common, and in winter, when the air is cool, the relative humidity beneath an inversion often exceeds 100 percent. Water vapor condenses onto the CCN and the result is a fog.

Coal and wood burn very inefficiently on domestic open fires. When heated, they emit combustible gases, only a proportion of which ignite to produce the fire. The remainder rise up the chimney. As they rise they also cool and condense into solid particles. This is smoke and in 1952 the domestic fires and coal-burning industrial furnaces of london emitted 141,000 tons (143,000 tonnes) of it. Between August 1944 and December 1946, the London suburb of Greenwich had an average of 20 days a month when the visibility was good at 0900 hours. In the city center there were fewer than 15 such days.

When smoke and fog were trapped together beneath an inversion, smoke particles adhered to and mixed with the water droplets. Sulfate particles, produced from sulfur dioxide emitted from the burning of coal with a high sulfur content, dissolved in the water to produce sulfuric acid (H_2SO_4). This made the smog acid, causing damage to buildings by acid deposition. Sulfur dioxide arose mainly from the coal burned in power plants and factories, rather than from domestic fires.

Burning coal also released carbon dioxide and increased the proportion of carbon dioxide to oxygen (the CO_2 partial pressure) beneath the inversion. This made some people breathe faster and more deeply, exacerbating the irritation to the respiratory passages caused by the acid smog itself.

smoke   An AEROSOL that is produced by the incomplete combustion of a carbon-based fuel. It consists of solid or liquid particles, most of which are less than 1 μm (0.00004 inch) in diameter. When coal, wood, or some types of oil are heated certain of their ingredients vaporize. Not all of the vapors ignite. Instead they are carried up the chimney on the rising current of warm air. When they mix with the colder air higher up the chimney or outside, the vapors condense once more. These condensed particles, mixed with fine particles of ash, form SOOT. Soot is readily ignited, producing more unburned vapors that condense once more as they cool in the outside air. Smoke is the mixture of ash and recondensed volatile ingredients from the original fuel.

Smoke particles contain a large proportion of carbon. This gives them a dark color, because of which they absorb radiation and have a warming effect on the air. Smoke can also increase cloud formation and planetary ALBEDO, however, while reducing PRECIPITATION.

Smoke particles are small enough to be active CLOUD CONDENSATION NUCLEI. WATER VAPOR condenses onto them (and becomes acid; see ACID RAIN), but the very small size of the particles produces very small CLOUD DROPLETS. Observations over areas in Central and South America where surface vegetation was being burned during the dry season found that the average size of cloud droplets decreased from 14 μm (0.0006 inch) to 9 μm (0.0004 inch). Small droplets are less likely than large droplets to fall as precipitation. The cloud albedo increased slightly, reducing the amount of sunshine reaching the ground and therefore the surface temperature, but by an insignificant amount.

In colder climates, smoke can mix with FOG to produce SMOG. This has been responsible for many of the most serious incidents of AIR POLLUTION, including the LONDON SMOG INCIDENTS.

smoke horizon   The top of a layer of SMOKE that is trapped beneath an INVERSION and that is seen from above and against the clear sky. The smoke hides the ground and the true horizon, so the boundary between the smoke and the clear air forms a false horizon.

smokeless zone   See CLEAN AIR ACT.

smokes   Clouds of dust and dense, white HAZE and MIST, somewhat similar to the CACIMBO, that form in the morning and evening near the western coast of equatorial Africa during the dry season. They are especially common before the onset of the HARMATTAN.

smudging   Using oil-burning heaters, called *smudge pots,* to protect a delicate farm crop against frost damage. In Florida citrus orchards as many as 70 burners are used to every acre (173 per hectare). They are lit on clear, cold nights when frost is likely and may be used in conjunction with large propellers mounted on tall columns that are used to mix the air and prevent cold air from settling in FROST HOLLOWS. Protection can also be achieved by an alternative version of smudging in which materials are burned in order to produce voluminous amounts of black smoke. The smoke forms a layer above the ground, reducing the amount of heat that is lost by infrared radiation from the surface.

snap   A short period of unusually cold weather that commences suddenly.

sno *See* ELVEGUST.

snow Precipitation that falls as aggregations of ice crystals. If large, these aggregations form SNOWFLAKES. Snow formation begins in clouds with the freezing of water onto FREEZING NUCLEI. The resulting ICE CRYSTALS grow at the expense of supercooled water droplets (*see* SUPERCOOLING) until they are too heavy to remain airborne. Unless the air temperature between the cloud base and the ground is below 39° F (4° C) the snow melts and reaches the surface as rain. Dry snow consists of ice crystals with no liquid water between them. The individual crystals are joined to each other directly or by necks of ice. Wet snow includes 3–6 percent of liquid water held inside the crystal aggregations and in crevices between crystals. Once it has fallen, snow packs together and becomes denser. This happens more quickly with wet snow than with dry snow. Snow melts when sunshine raises its surface temperature above freezing, but if the temperature subsequently falls below freezing again the melted snow refreezes as a sheet of ice across the surface. Dry snow provides good thermal insulation. Small animals survive well beneath it and can move freely through tunnels they excavate.

snow accumulation The depth of snow that is lying on the ground at a particular time. Above the permanent SNOW LINE there is a net accumulation of snow.

snowball Earth Earth when, apart from the highest mountains, its entire surface is covered by ice. The term was coined in 1992 by the American geobiologist Joseph Kirschvink of the California Institute of Technology. This condition is believed to have occurred four times between 750 million and 580 million years ago. Mean temperatures were about −58° F (−50° C) and all of the oceans were frozen to a depth of more than 0.6 mile (1 km). Heat from the crust and core of the Earth prevented the oceans from freezing completely. Dry land then comprised a number of small continents that were formed when a single large landmass broke apart. The small continents were closer to the sea and rainfall over land increased. This washed carbon dioxide from the air. The carbon dioxide reacted with minerals in the rocks and was carried to the sea, so the atmosphere came to contain less carbon dioxide. Temperatures fell and large ice packs formed in high latitudes. These increased the planetary ALBEDO, causing temperatures to fall further (*see* SNOWBLITZ).

Once the oceans were completely sealed, no liquid water was exposed to the air and so precipitation ceased. Volcanoes continued to erupt, however, releasing carbon dioxide into the air. With no precipitation to remove it, the carbon dioxide accumulated. After about 10 million years its concentration was high enough to trigger a huge GREENHOUSE EFFECT. The ice melted in a matter of a few centuries, but this melting did not end the greenhouse warming. Surface air temperatures rose to more than 120° F (50° C). More water evaporated and rainfall became intense. The rain washed carbon dioxide from the air and gradually the climate stabilized.

(You can read more about snowball Earth in Paul F. Hoffman and Daniel P. Schrag, "Snowball Earth" *Scientific American*, January 2000, pp. 50–57.)

snow banner (snow plume, snow smoke) The appearance, when seen from a distance, of snow that is being blown from a mountain top or other exposed high ground.

snowbelt The land lying in a strip up to about 50 miles (80 km) wide parallel to the lee shore of a large lake where winter snowfall is markedly higher than it is on the opposite, upwind side of the lake. Parts of Michigan, New York State, and Ontario, to the lee of the Great Lakes, receive up to three times more snow than regions in the same latitude but a long way from the lakes. This LAKE-EFFECT SNOW is due to modifications in the characteristics of CONTINENTAL AIR as it crosses a large expanse of open, unfrozen water.

snowblink (snow sky) The bright, glaring appearance of the underside of clouds when the ground is covered with SNOW. It is caused by the illumination of the cloud base by light reflected from the snow.

snowblitz A theory, popular in the 1970s, that the Northern Hemisphere could be plunged into a full-scale GLACIATION in a matter of a few decades. This is what is believed to have happened at the onset of the YOUNGER DRYAS and the proposed mechanism is the same.

The first stage requires a large release of freshwater into the northern part of the North Atlantic. Melting of the whole of the GREENLAND ICE SHEET might release sufficient water, but this is extremely unlikely and there is no other large ice sheet in the Northern Hemisphere. The only source for the freshwater would therefore be greatly increased precipitation, perhaps as a consequence of a general rise in temperature.

The second stage involves shutting down the ATLANTIC CONVEYOR. Freshwater is less dense than salt water and would float above it. At first, wave action would mix the two, but as more and more freshwater was added, mixing would become less effective until a layer of freshwater lay permanently above the salt water. The Atlantic conveyor is driven by the sinking of dense surface water, but if the surface water is less dense it cannot sink. This would shut down the Atlantic conveyor.

With the conveyor shut down, conditions would be set for the third stage in the process. Warm water would no longer flow north and the surface of the northern North Atlantic would cool rapidly by releasing its stored heat in the form of infrared radiation. In winter it would freeze, partly because it would be colder than in previous years and partly because freshwater freezes at a higher temperature than salt water. The area of sea ice would expand, increasing the ALBEDO. This would cause further cooling.

The final stage could then follow. Polar continental AIR MASSES that form over North America are very cold in winter. Ordinarily, they warm as they cross the unfrozen Atlantic, but with the Atlantic frozen they would remain cold and would reach Europe essentially unaltered. In summer, when the sea ice melted, the ocean would still be cold, because the NORTH ATLANTIC DRIFT would have disappeared. Consequently air would be cold and moist when it reached Europe, producing reduced precipitation (because of reduced water-holding capacity due to the lower air temperature) but an increase in the proportion of precipitation falling as snow. In winter, increased albedo over both land and sea would produce very low temperatures in Europe. The general movement of air from west to east would spread this cooling to the whole of the Northern Hemisphere.

A year would come when not all the snow that fell on land during the winter melted in summer. The high winter albedo would continue through summer and so the temperature would remain low. More snow would accumulate the following winter and the snow-covered area would increase. Temperatures would stabilize at below freezing, and as more snow accumulated its weight would compress the lower layers into glacial ice. Ice sheets would form in Europe and Canada, and each year their edges would advance farther south until a substantial part of both continents was in the grip of an ice age.

According to the snowblitz theory, the entire process, from the rise in temperature and increased rainfall to the formation and rapid advance of the ice sheets, might take no more than a few decades.

snow chill   The effect of being covered in snow that then melts. In a BLIZZARD or SNOWSTORM, snow itself affords protection from the wind, and it is possible to survive by digging an ice cave and sheltering inside it. If the snow in contact with your clothing begins to melt, however, the LATENT HEAT required to melt the snow is taken from your body, reducing its temperature. At the same time melted snow soaks your clothing, filling all the air spaces between the fibers. This water is close to freezing temperature, and water conducts heat much more efficiently than does air. The combined effects of using body warmth to melt snow and then of wearing clothes soaked in very cold water rapidly reduce the core body temperature.

snow cloud   A cloud from which SNOW falls or appears likely to fall. It differs from a RAIN CLOUD only in being colder. NIMBOSTRATUS may give continuous snow and CUMULONIMBUS may produce snow showers, which may be heavy. ICE PRISMS and SNOW grains may fall from STRATUS. Significant amounts of SNOWFLAKES form if the temperature high in the cloud is below −4° F (−20° C) and it is above 14° F (−10° C) in the lower part of the cloud. Snowflakes melt before reaching the ground unless the air temperature between the CLOUD BASE and the surface is below about 39° F (4° C).

snow cover   All the snow that is lying on the ground, or the proportion of an area that lies beneath snow. Ground is usually said to be snow-covered if more than 50 percent of its area is beneath snow.

snow-cover chart   A SYNOPTIC CHART on which the areas covered by snow are marked and there are contour lines showing the depth of snow.

snowdrift An accumulation of snow that is much deeper than the snow covering adjacent areas. SNOWFLAKES are carried by moving air, and the ability of air to transport them is proportional to its KINETIC ENERGY. If the air loses energy it also loses its ability to keep snowflakes aloft and so they fall to the ground. Drifts occur where moving air has lost a significant amount of its energy.

Air loses energy by friction with the surface. When it encounters woodland or a belt of trees it slows and some of its snow falls, but it falls fairly evenly, so although the depth of snow is likely to be greater inside the wooded area than on open ground to WINDWARD, the difference is not large. The overall effect is to collect snow from the passing air and reduce the amount falling on the LEE side. Isolated plants tend to accumulate small drifts on their lee sides.

Solid barriers, such as walls or high banks on each side of a road, have a much bigger effect. Air approaching below the level of the top of a wall strikes it and is deflected in an EDDY. The resulting loss of

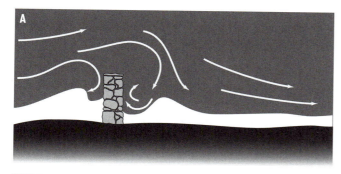

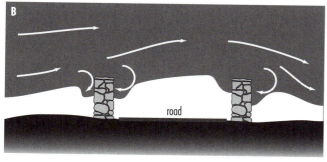

Snowdrift. **A wall (A) produces two snowdrifts, one on each side, with a space between the wall and the drift. High banks lining a road (B) also produce two drifts. In this case the downwind drift behind one bank overlaps the upwind drift in front of the other bank, so the road quickly fills with snow.**

energy causes the air to lose snow, but the eddy also scours snow away from the side of the wall. There is a space between the thin layer of snow lying against the base of the windward side of the wall and the deep drift. On the lee side, air crossing the top of the wall eddies downward. This has a similar scouring effect and a space separates the drift from the wall.

A road that is bordered by high banks fills with snow if the air crosses the road approximately at right angles. The banks have the same effect as a wall, but in this case there are two barriers and the downwind drift caused by one overlaps the upwind drift caused by the other.

A snow fence produces a low drift on its upwind side and a much bigger drift on its downwind side. If the fence is erected to prevent snow from forming drifts on a road the distance between the fence and the road should be equal to 10 times the height of the fence.

snow dune A SNOWDRIFT that is shaped like a sand dune, having formed in the same way.

snow dust Very fine, powdery snow that is driven by the wind.

snow eater A warm, dry wind that removes snow by SUBLIMATION. A CHINOOK wind often removes 6 inches (15 cm) of snow in a day and it can clear 20 inches (50 cm) in a day.

snowfall Precipitation that reaches the surface as snow. The thickness of fallen snow does not provide a direct measure of the amount of precipitation, because SNOWFLAKES vary greatly in size, depending on the temperature of the air through which they fall, and their size determines the amount of air they trap between them as they accumulate. The colder the air, the smaller the flakes and the more air they trap. A report of the thickness of a snowfall provides useful information for road users and anyone who needs to be outdoors, but it conveys very little meteorological information. Consequently, snowfall amounts are always converted into their rainfall equivalents for meteorological purposes. The table gives the snowfall equivalent of a standard amount of water at different surface air temperatures.

Suppose that a fall of snow covers the ground to a depth of 2 inches (5 cm) and that the air temperature is

Snow-to-water ratios		
Temperature		Ratio
°F	°C	
35	1.7	7:1
29–34	−1.7–1.1	10:1
20–28	−6.7 to −2.2	15:1
10–19	−12.2 to −7.2	20:1
0–9	−17.8 to −12.8	30:1
less than 0	less than −17.8	40:1

15° F (−9.4° C). At this temperature the ratio of snow to water is 20:1: that means that 20 in (or 20 cm) of snow is the equivalent of 1 in (or 1 cm) of water. Expressed mathematically in inches, this is

$$20/1 = 2/x$$

where x is the amount of water equivalent to 2 in of snow. Divide the equation by 20 and multiply by x and

$$x = 2/20 = 0.1$$

At this temperature, therefore, 2 in (5 cm) of snow is equivalent to 0.1 in (0.25 cm) of water.

snowfield An extensive, approximately level area that is covered uniformly with fairly smooth snow or ice. Snowfields are found in mountainous areas and high latitudes. A small GLACIER or an accumulation of snow and ice that is too small to be called a glacier is also known as a *snowfield*.

snowflake An aggregation of ICE CRYSTALS that are grouped into a regular six-sided or six-pointed shape between about 0.04–0.8 in (1–20 mm) across. Although all snowflakes have this basic hexagonal form, an extremely large number of variations is possible within it, and consequently every snowflake is unique.

Ice crystals form in clouds where the temperature is below freezing, and once they begin to form they grow at the expense of SUPERCOOLED water droplets. They also fragment as protrusions extending from them are carried away by air currents, so the cloud soon contains splinters of ice that act as FREEZING NUCLEI for the formation of new crystals.

The individual crystals are composed of hexagonal units that are arranged in a number of possible shapes. The shapes depend on the temperature at which the crystals form. As they continue to grow beyond their unit shapes, they often do so by accumulating projecting arms or spikes. When crystals collide, their irregular shapes make it likely that they will lock together. The haphazard way this happens explains why no two snowflakes are identical. Snowflakes grow most readily when the temperature is about 32°–23° F (0° to −5° C). This is because at just below freezing a thin film of liquid water forms on the surface of each crystal and then freezes solid where two crystals make contact.

Tens of individual crystals form a single snowflake, and the flake continues to grow until its fall speed exceeds the speed of the air currents that carry it upward inside the cloud. Then it falls from the cloud. If it falls through air that is above freezing temperature it starts to melt, and if it remains in air at this temperature long enough it reaches the surface as rain rather than snow. It is unlikely to reach the surface as snow if the freezing level is higher than about 1,000 feet (330 m) and the surface air temperature is above 39° F (4° C). Rain is often melted snow and, as Wilson Bentley discovered, the size of raindrops indicates the size of the snowflakes that melted to form them.

Bentley was the most famous student of snowflakes, but he was not the first. In a text called *Moral Discourses Illustrating the Han Text of the "Book of Songs,"* written between 140 and 131 B.C.E., Han Ying described the hexagonal shape of ice crystals and snowflakes. This is the earliest written reference to snowflakes, and their intricate patterns have continued to intrigue scientists ever since. In the 16th century OLAUS MAGNUS referred to them, and in 1611 Johannes Kepler (1571–1630), the German physicist and astronomer, also described them, in the book *A New Year's Gift, or On the Six-Cornered Snowflake*. ROBERT HOOKE (1635–1703), an English physicist and one of the first microscopists, studied snowflakes under the microscope and drew what he saw. He published his drawings and descriptions in 1665, in *Micrographia*. In 1936, Professor Ukichiro Nakaya, a Japanese physicist, was the first person to grow ice crystals in a laboratory, and in 1941 this work led to the establishment of the Institute of Low Temperature Science at the University of Hokkaido. Professor Nakaya became a leading authority on snow. His book on the subject, *Snow Crystals: Natural and Artificial*, was published in 1954 by Harvard University Press.

snow flurry   A light snow shower of brief duration.

snow gauge   An instrument that is used to measure the amount of snow that has fallen over a stated period. It is a modified RAIN GAUGE. In some designs the top edge of the collecting funnel is heated, so that snow falling on it melts and is collected as liquid water. This automatically converts the SNOWFALL amount into its rainfall equivalent. An alternative design has a removable funnel and receiver, so that snow is collected and measured directly. Snowfall can also be measured by taking a core, using an open-ended cylinder pressed vertically all the way through the snow. The snow is then transferred to a measuring beaker. The beaker is calibrated to correct for any difference between its diameter and that of the coring cylinder. Transferring the snow to the beaker inevitably packs the grains together, so it must be melted and the depth read as the water equivalent. The simplest way to measure a snowfall is to use a measuring stick, such as an ordinary ruler, and a thermometer. The ruler measures the thickness of snow, and the air temperature at the time it fell will indicates the water equivalent. Regardless of the method used, snowfall should always be measured in the open, well clear of trees and buildings that might deflect wind and alter the amount of snow reaching the ground.

snow geyser   Fine, powdery snow that is suddenly thrown upward as a result of the disturbance caused when a thick layer of underlying snow settles abruptly.

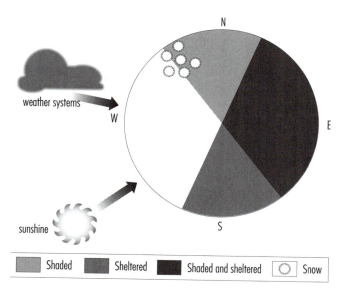

| | Shaded | | Sheltered | | Shaded and sheltered | | Snow |

Snow line. The direction of the noonday Sun and the direction from which most weather systems arrive make it possible to divide an idealized mountain into four sectors. The sectors that are sheltered and both shaded and sheltered receive little precipitation, because of the rain shadow effect. Snow is most likely to fall in the sector that is shaded from the Sun but exposed to the weather.

Such settling is sometimes called a *snow tremor* or *snowquake*.

snow grains (granular snow, graupel)   Very small particles of white, opaque ice. They are flat and less than 0.04 in (1 mm) in diameter.

snow line   The boundary between the area covered by snow, for example, on a mountainside, and the area

Mean snow line					
Latitude	Northern Hemisphere			Southern Hemisphere		
	(feet)	(meters)		(feet)	(meters)
0–10	15,500	4,727		17,400	5,310
10–20	15,500	4,727		18,400	5,610
20–30	17,400	5,310		16,800	5,125
30–40	14,100	4,300		9,900	3,020
40–50	9,900	3,020		4,900	1,495
50–60	6,600	2,010		2,600	793
60–70	3,300	1,007		0	0
70–80	1,650	503		0	0

that is free of snow. The edge of the snow that remains on a mountainside through the summer is also known as the *snow line*. The location of the snow line depends on the temperature, which in turn varies with latitude and with elevation. The location of the snow line on a particular mountain can be determined only by measurement, however. It is extremely difficult to calculate, because every mountain has crags shading the area behind them and deep gulleys. These are places where snow lingers. The side of the mountain that faces the noonday Sun is warmer than the side shaded from it, and precipitation is heavier on the side facing into the direction from which most weather systems arrive. These factors allow an imaginary mountain, which is perfectly conical and smooth, to be divided into four unequal sectors according to whether they are exposed to sunshine, shaded, sheltered from the weather, or both shaded and sheltered. It is possible from this to estimate where snow is most likely to fall. The average snow line elevation at various latitudes is given in the table.

snowout The removal from the air of solid particles that act as FREEZING NUCLEI or CLOUD CONDENSATION NUCLEI and that are carried to the ground in SNOW. The process is identical to RAINOUT, but the PRECIPITATION falls as snow, not rain.

snow pellets (graupel, soft hail) Opaque, white grains of ice that are spherical or less commonly conical in shape and 0.1–0.2 in (2–5 mm) in diameter. They fall in cold weather from CUMULONIMBUS cloud and are brittle. When they strike a hard surface they bounce and shatter.

snow plume *See* SNOW BANNER.

snowquake *See* SNOW GEYSER.

snow ripple *See* WIND RIPPLE.

snow sky *See* SNOWBLINK.

snow smoke *See* SNOW BANNER.

snow stage Condensation that occurs in rising air when the CONDENSATION LEVEL is higher than the FREEZING LEVEL, so that water vapor changes directly into ice by DEPOSITION.

snowstorm A heavy fall of snow is called a *snowstorm,* although there is no precise definition of the term. If more than about 4 in (10 cm) falls, snowplows are likely to be called out to clear roads, so this provides one possible way to define a fall of snow that is severe enough to be called a storm.

The weather conditions required to produce snow are no different from those that produce rain and much of the precipitation that falls as rain begins as snow in the cloud. The heaviest snowfalls occur when the low-level air temperature is between 25° F and 39° F (–4° C and 4° C). This is warm enough for the air to hold a considerable amount of moisture and cold enough to ensure that precipitation falls as snow rather than rain. This is why the heaviest snowstorms usually occur near the beginning and end of winter.

It is possible that snow covering the ground chills the air in contact with it and that this causes air to subside and sustains low pressure high above the surface. These conditions can produce more snowstorms, so one snowstorm may cause another.

Four mechanisms account for most snowstorms. LAKE-EFFECT SNOW falls when cold air crosses relatively warm water and then encounters cold ground. Upslope snowfalls occur when air is cooled by OROGRAPHIC LIFTING. Snow is also produced when warm air is forced to rise over cold air, and by DEPRESSIONS. A particular snowstorm may result from one of these causes or a combination of two or more of them, but depressions cause more storms than any other cause. In North America, lake-effect snow affects cities to the east of the Great Lakes and also Salt Lake City, to the east of the Great Salt Lake. Salt Lake City is a ski resort that receives abundant snow. One storm delivered 27.2 in (69 cm) on October 18, 1984, and on February 2, 1989, the city received 20.9 in (53 cm). Upslope snow is especially common on the western side of the Rocky Mountains. In the month of January 1911, 390 in (991 cm) of snow fell on Tamarack, California, and in the winter of 1998–99 Mount Baker, Washington, received 1,140 in (28.96 m).

snow tremor *See* SNOW GEYSER.

socked in The expression used by aircrews to describe the condition of an airfield that is closed for flying because of bad weather. In the absence of facilities for guiding aircraft to the ground by radar, FOG that severely reduces horizontal visibility and a CLOUD BASE lower than about 200 ft (60 m) close an airfield. They do not prevent aircraft from taking off, provided the pilots are qualified to fly on instruments and can see sufficiently well to taxi onto the runway, but they make it impossible for aircraft to land safely.

soft hail *See* GRAUPEL and SNOW PELLETS.

soft UV *See* ULTRAVIOLET RADIATION.

SOI *See* SOUTHERN OSCILLATION INDEX.

soil air Air that is held in the spaces between soil particles. In the upper layer of a cultivated soil air accounts for about 25 percent of the volume. Soil air is similar in composition to atmospheric air (*see* ATMOSPHERIC COMPOSITION), and most of the constituents are present in similar proportions. The exception is carbon dioxide. Atmospheric air contains about 0.04 percent CO_2, whereas soil air contains about 0.65 percent. The CO_2 in the soil is from RESPIRATION by plant roots and soil organisms, including those engaged in decomposing organic matter. In some soils the proportion of oxygen decreases with increasing depth and that of CO_2 increases by a similar amount. Some of the soil CO_2 dissolves in water and is removed as carbonic acid ($H_2O + CO_2 \rightarrow H_2CO_3$).

soil classification A method for labeling different types of soil that uses principles similar to those used to classify living organisms and clouds (*see* CLOUD CLASSIFICATION). The first attempts to classify soils were made in classical times, but the first steps toward the modern system were made in the latter part of the 19th century by Russian scientists led by Vasily Vasilievich Dokuchaev (1840–1903). Dokuchaev based his classification on the climates in which soils form. It was widely adopted and many of the original Russian names for soil types are still used, although now they are used only informally, because the old system has been replaced. Soil names such as *podzol, chernozem,* and *rendzina* are taken from the early Russian work. Podzols are gray, ashlike, acid soils. Chernozems are black

grassland soils sometimes called *prairie soils,* and rendzinas are brown soils found in humid or semiarid grasslands.

The modern classification was developed by the staff at the Soil Survey of the United States Department of Agriculture. It was published in 1975 and is known as the *Soil Taxonomy.* It divides soils into 12 orders. These are divided further into suborders, great groups, families, and soil series. Some of the names seem strange on first acquaintance and they become stranger at levels below that of the order. The suborders include Psamments, Boralfs, and Usterts; the great groups include Haplargids, Haplorthods, and Pellusterts; and the subgroups include Aquic Paleudults, Typic Medisaprists, and Typic Torrox.

The following are the 12 orders, with a brief description of each.

Gelisols Soils where there is PERMAFROST within 6.5 ft (2 m) of the surface.
Histosols Soils rich in organic matter that are found in bogs and marshes, where the climate is cool and wet.
Spodosols Sandy, strongly acid soils that are found in forests, especially coniferous forests.
Andisols Soils that form from volcanic ash; they are deep and light-textured.
Oxisols Deeply weathered (*see* WEATHERING), acid soils from which most of the plant nutrients have been washed away; they are found in the humid Tropics and subtropics.
Vertisols Clay soils that swell when they are wet and develop deep cracks when they are dry; they are found in climates with marked wet and dry seasons.
Aridisols Desert soils that contain little organic matter; they often have salt layers.
Ultisols Strongly acid, deeply weathered tropical soils.
Mollisols Very dark, grassland soils that are rich in organic matter and highly fertile.
Alfisols Soils that are found mainly in forests where the annual rainfall is 20–50 in (510–1,270 mm); there is a layer of clay beneath the topsoil.
Inceptisol Soils that are found in cold, wet climates; they are at an early stage in their development.
Entisols Soils with little vertical development into layers; they are found on recent floodplains, beneath recently fallen volcanic ash, and as wind-blown sand.

soil heat flux   *See* GROUND HEAT FLUX.

soil moisture   Water that is present in the soil. Since all living organisms require water, the amount of soil moisture affects the rate of biological activity in the soil, including the rate at which organic matter decomposes. Many organisms also require oxygen for RESPIRATION, and they obtain it from air that is trapped in spaces between soil particles. If the soil is completely saturated, these spaces are filled with water and the air is expelled, so too much water is as bad as too little. For most soil organisms, the optimal amount of soil moisture is 50–70 percent of FIELD CAPACITY.

When water arrives at the soil surface, by falling as PRECIPITATION or flowing from adjacent land, it drains downward under the force of gravity. The speed with which water moves depends on the PERMEABILITY of the soil, but even when most of the water has drained away to join the GROUNDWATER, a film of water up to 15–20 molecules thick is left covering all the soil particles.

The amount of water held in a soil is measured by weighing a sample of soil before and after drying it in an oven at 221° F (105° C) or by using an instrument such as a neutron probe to measure the electrical conductivity of the soil between two electrodes. The probe is lowered into a pipe, about 2 in (5 cm) in diameter and of an appropriate length for the conditions being studied. Electrodes are suspended in the pipe at the levels being monitored. The probe contains a radioactive source that emits fast (high-energy) neutrons. When the neutrons collide with hydrogen atoms they change direction and slow down and some of the slow neutrons are deflected back into the probe, where they are counted. The greater the number of slow neutrons the probe detects, the greater is the water content of the soil, because water molecules are the major source of hydrogen atoms.

As water moves through the soil, the hydrogen atoms in its molecules, which carry positive charge, are attracted to oxygen atoms, bearing negative charge, in the molecules of mineral particles and at the surface of the particles. The water molecules attach themselves quite firmly to the mineral particles. This force is called *adhesion*.

Water molecules are also attached to each other by HYDROGEN BONDS. This force is called *cohesion* and because of it other water molecules attach themselves to the molecules coating exposed mineral particles. The resulting film of water is held together by both adhesion and cohesion.

Adhesive and cohesive forces act on surfaces—of mineral particles and water, respectively. Consequently, the amount of water a soil can retain in this way depends on the amount of surface area its particles present. The bigger the soil particles, the smaller their total surface area is.

Soil particles are not spherical, of course, but the ratio of their radius to surface area obeys the same geometrical law. The surface area of a sphere is equal to $4\pi r^2$, where r is the radius. If $r = 2$, then the surface area is 50. If $r = 4$, then the surface area is 201. The volume of each of these particles ($4/3\ \pi r^3$) is 33.5 and 268, respectively, and so any volume of soil contains many more of the smaller particles than of the bigger ones. For example, a soil volume of 10,000 contains 298 of the small particles, but only 37 of the big ones, and the total particle surface area is 14,900 in the soil with small particles and about 7,440 in the other soil. This is why sand, which has big particles, holds much less water than silt or clay, which have very small particles.

Considerable force is needed to remove the film of water from soil particles. Adhesion water moves very little; cohesion water is more mobile. Adhesion water is not available to plants, but molecules of cohesion water are constantly joining and leaving the water flowing through the soil. Water flowing downward through the soil is called *gravitational water*. It is held in the soil with a force that is less than about 30 kPa (300 mb, 4.5 lb in⁻²). Water that can be absorbed by plant roots is called *available water* and is held with a force of about 30–1,500 kPa (0.3–15 bar, 4.5–218 lb in⁻²). Adhesion water is held with a force equal to 1.5–100 MPa (15–1,000 bar, 218–14,500 lb in⁻²).

Water that has drained to below the water table may then move upward again by CAPILLARITY.

solaire   The name that is given to an easterly wind in central and southern France, referring to the fact that easterly winds blow from the direction of the rising Sun (in French, *soleil*).

solano   *See* LEVANTER.

solar air mass   The total OPTICAL MASS of the atmosphere that is penetrated by light from the Sun with the Sun at any given position in the sky.

solar cell   *See* SOLAR ENERGY.

solar collector   *See* SOLAR ENERGY.

solar constant   The amount of energy that the Earth receives from the Sun per unit area (usually per square meter) calculated at a point perpendicular to the Sun's rays and located at the outermost edge of the Earth's atmosphere. The value of the solar constant is not known precisely; the best estimate is 1,367 watts per square meter (1.98 LANGLEYS).

The Sun generates energy by means of thermonuclear reactions in its core. The Sun is made of gas, about 75 percent hydrogen and 25 percent helium, and its mass is 743 times that of the combined masses of all the planets in the solar system. It is 330,000 times more massive than Earth. This mass exerts a gravitational force, pressing material inward. This pressure is so great at the core that the electrons are stripped away from atomic nuclei. These are predominantly hydrogen nuclei, each of which consists of a single proton.

Protons carry positive charge and repel one another, but the pressure at the solar core is sufficient to overcome this repulsion and to force protons together. This causes a reaction known as the *proton–proton cycle* in which four hydrogen nuclei combine to make one helium nucleus.

Later stages in the series of reactions that the proton–proton cycle comprises can follow several different paths, but they all begin with the combination of two hydrogen nuclei to make one deuterium (heavy hydrogen) nucleus with the release of a positron (positive electron, e^+) and a neutrino (ν).

$$^1\mathrm{H} + {}^1\mathrm{H} \rightarrow {}^2\mathrm{H} + e^+ + \nu$$

The deuterium nucleus then captures a third proton. This produces a nucleus of helium-3 ($^3\mathrm{He}$) with the release of a gamma ray (γ).

$$^2\mathrm{H} + {}^1\mathrm{H} \rightarrow {}^3\mathrm{He} + \gamma$$

The final result of the cycle is the production of helium-4 ($^4\mathrm{He}$) with the loss of 0.7 percent of the original mass of the protons. The lost mass is converted into energy according to the Einstein equation $E = mc^2$, where E is energy, m is mass, and c is the speed of light. A little of the energy is carried away by neutrinos, but most is converted into heat.

The temperature inside the solar core is about 15 million K (= $15 \times 10^{6\circ}$ C [=27×10^6 F]). The heat makes the material in the core expand, so it exerts an outward pressure. This outward pressure of expansion precisely balances the inward, gravitational pressure, so the Sun neither expands nor collapses.

The core heats the outer layers of the Sun. The outermost visible layer is called the *photosphere*. Its temperature is 5,800 K (= 5,800° C [= 10,470° F]). Although the temperature is higher in the chromosphere and solar corona, neither of which is visible except during a solar ECLIPSE, these consist of matter that is so tenuous it has little effect on the emission of solar energy. The effective temperature of the Sun is 5,800 K. The chromosphere and corona lie beyond the photosphere. The corona is the source of the SOLAR WIND.

The total amount of energy radiated by the Sun is given by the STEFAN–BOLTZMANN LAW. Earth receives only a small proportion of this energy, about 4.5×10^{-10} of the total. The radius of the Sun (R) is 109 times that of the Earth, but its average distance from the Earth, of 93 million miles (149.5 million km), is equal to $215R$ and it subtends an angle of only 0.5° in the sky. The amount of solar energy intercepted by Earth, the solar constant (S), is equal to $S_O/4\pi R_E^2$, where S_O is the solar output and R_E is the radius of the Earth. It is this calculation that yields the value of 1,367 Wm^{-2}.

Although called a constant, the solar constant varies with changes in the solar output.

A proportion of the solar radiation reaching the top of the atmosphere is reflected by clouds and the surface. This proportion represents the planetary ALBEDO (a) and it must be deducted from the solar constant ($S - a$) to give a value for the amount of solar energy reaching the surface.

solar energy   Energy that the Earth receives from the Sun and that can be used to perform useful tasks, such as heating spaces and water and generating electrical power. Although the concept of solar energy is sometimes extended to include WIND POWER, because wind is produced by weather systems driven by solar energy, the term strictly covers only the direct use of solar radiation as heat or light.

Sometimes the generation of energy by burning biomass is also included. This is the growing of crops, such as fast-growing trees, that can be harvested for fuel or that produce large amounts of sugar, such as corn (maize) and sugar beet, that can be fermented to produce alcohol (ethanol) for use as a liquid fuel. These technologies are included because the crops grow by means of PHOTOSYNTHESIS, which is powered by sunlight.

In principle, a huge amount of energy is available. The surface of the Earth receives about 4×10^{18} J (9.5×10^{17} cal) of energy a year, but the total annual energy consumption of the entire human population of the Earth amounts to only about 3×10^{14} J (7.2×10^{13} cal).

The simplest direct use of solar heat is also the most traditional: south-facing (in the Northern Hemisphere) windows. Add double-glazing to reduce heat loss, and the window allows solar energy to enter the building and warm its interior and contents but prevents warm air and long-wave BLACKBODY radiation from leaving (see GREENHOUSE EFFECT).

The flat-plate solar collector heats water. It consists of a large, shallow box covered with glass and with a base that is painted matte black to absorb radiation. The box contains piping that winds back and forth across the base. Water either is pumped or flows by gravity through the piping. Solar radiation passes through the glass plate and is absorbed by the black base, and heat passes by conduction from the black base to the pipes and the water they contain. The piping carries the heated water inside the building to a heat exchanger in a water tank. Usually a number of solar collectors are mounted as an array on the roof and angled to face the noonday Sun.

Solar collectors are effective provided they are sited in a place where they receive an adequate amount of solar radiation. They are especially useful in latitudes between about 36° N and 36° S. Their disadvantage is that they require warm sunshine. This means they work best during daytime in summer and are much less useful at night, in winter, and when the sky is overcast, the times when heat is most needed.

Mirrors can be used to focus solar radiation onto a central point. Several devices have been built to exploit this property and produce temperatures high enough to heat water to drive generating turbines. Several of these "solar furnaces" are located in the United States, the largest at Albuquerque, New Mexico.

Solar cells, which are also known as *photovoltaic cells*, produce an electric current directly by the action of sunlight. When light strikes certain semiconductor materials some of the light energy frees electrons in the material. An electric field in the material forces all the free electrons to move in the same direction. This constitutes an electric current. Solar cells were developed primarily for use in space, and they supply much of the energy for spacecraft and orbiting satellites. They are also starting to be installed for energy production on the surface. In sunny locations that are too remote to be reached by conventional power lines solar cells may be less costly than alternative forms of generation.

(You can learn more about the uses of solar energy in desert climates in Michael Allaby, *Ecosystem: Deserts* [New York: Facts On File, 2001]. You can find out about solar furnaces at www.wsmr.army.mil/paopage/Pages/solar.htm and www.sandia.gov/Renewable_Energy/solarthermal.furnaces.html, and about solar cells at www.howstuffworks.com/solar-cell.htm.)

solar flare   A sudden increase in the strength of the SOLAR WIND that is associated with SUNSPOT activity. A major flare can double the amount of energy emitted by the Sun over a period of about 0.25 second. The radiation consists mainly of ULTRAVIOLET RADIATION, but substantial amounts of X RAYS and particles are also released. Most flares last for several hours, and the particles, which travel much more slowly than the electromagnetic radiation, take one or two days to arrive at Earth. Radiation and particles from solar flares disrupt radio transmissions and cause AURORAS.

solarimeter   *See* PYRANOMETER.

solar irradiance   The total amount of energy that the Sun emits. The STEFAN–BOULTZMANN LAW relates this to the temperature at the surface of the PHOTOSPHERE, which averages about 10,800° F (6,000 K). At this temperature the total amount of energy radiated by the Sun (its EXITANCE) is about 70 MW m^{-2} and over the entire surface of the photosphere it is about 4.2×10^{20} MW. Emitting energy at this rate, the Sun has so far converted about 0.1 percent of its mass into energy.

The temperature of the photosphere varies slightly, however, and therefore so does the amount of energy it radiates. In 1977, for example, the Kitt Peak National Observatory near Tucson, Arizona, measured a drop of 11 K (19.8° F) in the temperature of the photosphere in a single year. This is very small (a decrease of about 0.18 percent), but if it were sustained for a decade or more it would produce a slight but noticeable cooling of the climate.

During the 1990s the solar output increased. For reasons that scientists do not understand, the amount of cloud over the surface of the Earth is directly proportional to the intensity of COSMIC RADIATION reaching the atmosphere. Cosmic radiation consists of charged particles and these are deflected by the solar wind. Consequently, at times of increased solar output, and a stronger SOLAR WIND, the intensity of cosmic radiation decreases. This reduces cloud formation and with fewer clouds the planetary ALBEDO decreases. More solar energy is absorbed by the surface and the global climate grows very slightly warmer. Some scientists believe that the increased solar output during the 1990s is the principal—or the only—cause of the slight increase in global temperatures recorded and that the effect of changing solar output on the intensity of cosmic radiation and cloud formation is the mechanism by which this occurred.

(You can read the full explanation of this idea in Nigel Calder, "The Carbon Dioxide Thermometer and the Cause of Global Warming" *Energy and Environment*, 10, 1, pp. 1–18, 1999.)

solar radiation cycle The regular increase and decrease in the amount of radiation that is emitted by the Sun. This reaches a maximum every 11 years and is associated with the number of SUNSPOTS visible on the surface of the PHOTOSPHERE.

solar spectrum The full range of electromagnetic radiation that emanates from the Sun. Solar radiation is emitted from the PHOTOSPHERE. The energy the radiation possesses is related to the temperature of the photosphere by WIEN'S LAW. The temperature is about 10,800° F (6,000° C) and at this temperature the Sun radiates at every WAVELENGTH, from GAMMA RAYS (10^{-5}–10^{-8} μm) to radio waves (1–10^9 m). It does not radiate with equal intensity at all wavelengths, however. Solar radiation reaches its maximum intensity at about 0.5 μm, which is the wavelength of green visible light.

When visible white light passes through a prism it divides into its constituent wavelengths. Light of different wavelengths then appears as a band of colors on a screen placed in a suitable position. Violet light, which has the shortest wavelength, appears on the left, and red, with the longest wavelength, on the right. The wavelengths of visible light are

violet 0.4 μm
indigo 0.43 μm
blue 0.46 μm
green 0.5 μm
yellow 0.57 μm
orange 0.6 μm
red 0.7 μm

solar–topographical theory A theory that explains past changes in climate in terms of variations in solar output and the formation and erosion of mountains. Mountains form when part of the Earth's crust is raised by volcanic activity (*see* VOLCANOES) or as the result of a collision between two crustal plates (*see* PLATE TECTONICS). Their height then gradually decreases as material is lost by erosion. The cycle of mountain formation and erosion alters the elevation of the surface. It also affects the flow of air, tending to deflect air to the north or south of a mountain range, and forcing it to rise, and therefore cool and lose moisture, as it crosses high ground. Mountains have a clear climatic influence. So do variations in solar output, because they affect the amount of solar energy reaching the Earth. The theory considers variations of 10–20 percent to each side of the mean to be climatically significant

solar wind A stream of protons, electrons, and some nuclei of elements heavier than hydrogen that flows outward from the Sun. It is generated in the outermost region of the Sun, called the *corona*, where the temperature is so high that particles acquire sufficient energy to escape from the Sun's gravitational field. At the mean distance between the Earth and Sun (known as 1 astronomical unit [AU] and equal to about 93.2 million miles, or 150 million km), the solar wind carries 16–165 protons per cubic inch (1–10 per cubic centimeter) traveling at 220–440 miles per second (350–700 km s⁻¹). The solar wind deflects the tails of comets, so these always point away from the Sun, and compresses the Earth's magnetic field. Interaction between the solar wind and the upper atmosphere produces AURORAS. The intensity of the solar wind varies with the amount of SUNSPOT activity.

solstice   One of the two dates in each year when the difference in length between the hours of daylight and darkness is most extreme. The ARCTIC and ANTARCTIC CIRCLES, at latitudes 66°30' N and S, are defined as the latitudes at which the Sun does not rise above the horizon at the winter solstice and does not sink below it at the summer solstice. In latitudes higher than 66°30', the periods of continuous darkness and daylight are longer than the one day at each solstice. The solstices are also called *midsummer day* and *midwinter day*. At noon on midsummer day, the Sun is directly overhead at the TROPIC in the summer hemisphere, and at noon on midwinter day it is directly overhead at the other tropic. The solstices fall on June 21–22 and December 22–23. The varying length of the day is due to the fact that the rotational axis of the Earth is tilted with respect to the PLANE OF THE ECLIPTIC, and it is this tilt that produces the SEASONS.

solute effect   *See* CLOUD CONDENSATION NUCLEI.

sonde   A package of instruments carried by a free-flying balloon that measure temperature, pressure, and humidity in the air through which they move. When the balloon bursts, the instruments fall to the ground by parachute. Most sondes pass their data to a radio transmitter carried with them. A sonde from which data are transmitted to a ground receiver is known as a *RADIOSONDE*.

sonic anemometer   *See* ANEMOMETER.

sonic boom   The sound, like a clap of thunder, that is heard at the surface when an aircraft or other body flies at a speed greater than the SPEED OF SOUND. At subsonic speeds a moving object disturbs the air, sending out quite gentle waves of pressure that propagate in all directions, traveling at the speed of sound. These disturbances reach the surface continuously, and because they are slight and there is no sudden change in their intensity they make no sound. As the speed of the moving body approaches that of sound, and there-

fore of the pressure disturbances its motion produces, the pressure waves are confined in a decreasing volume. When the body exceeds the speed of sound and also the speed of its own pressure waves, the waves are behind the body and extend horizontally from it in a cone, called a *Mach cone*. The entire pressure field is contained within the Mach cone, where it produces a sudden, sharp increase in pressure, or SHOCK WAVE. An observer on the surface hears this as a loud bang when the edge of the Mach cone passes. Very large aircraft sometimes produce two sonic booms, one from the front of the aircraft and the other from the rear.

soot   Solid particles, composed mainly of carbon, that are emitted when a carbon-based fuel is burned in such a way that not all of the hydrocarbons are fully oxidized. Exhaust emissions from poorly maintained diesel engines are a major source of soot in urban areas. Soot particles are black, but dispersed among water droplets and gases they give smoke its gray color. They vary greatly in size, from less than 1 μm to 3 mm (0.12 in). Most remain airborne for a matter of hours or at most one or two days before they are washed to the ground by rain or snow. If inhaled, however, soot particles smaller than 25 μm in size, and possibly those smaller than 10 μm, are harmful to health.

soroche   A form of mountain sickness that afflicts persons who travel from the coast to the *puna* region of Peru, high in the Andes at an elevation of 10,000–15,000 ft (3,000–4,500 m), where the air pressure is about 675 mb. The soroche causes breathlessness, palpitations, loss of appetite, and sometimes nose-bleeding.

sounder   *See* SOUNDING.

sounding   Any measurement that is made through the column of air between the instrument and the level that is being monitored. An instrument that makes a sounding is called a *sounder*. Satellites usually carry sounders. These measure the amount of radiation being emitted at particular levels in the atmosphere. Weather balloons also take soundings (*see* BALLOON SOUNDING). The word *sounding* was first used at sea to refer to the measurements of the depth of water that were made by using a weighted rope. The word is derived from the French verb *sonder,* which comes from the Latin *sub-*

undare: sub-, meaning "under," and *unda,* meaning "wave."

source   A place at which a particular substance, usually a pollutant, is released into the environment, or the process by which it is released. Road transport is the source of CARBON DIOXIDE, NITROGEN OXIDES, and PARTICULATE MATTER, for example. The opposite of a source is a SINK.

source region   A part of the world where a particular type of AIR MASS originates. A source region must cover an extensive area within which the surface is fair-

ly uniform and where PRESSURE SYSTEMS are stationary most of the time. This allows the air at any height to reach a constant temperature, pressure, and humidity, making the air mass homogeneous. The necessary conditions for the development of an air mass occur where the pressure is high and air flows slowly outward from it. DIVERGENCE prevents outside air from entering. Northern Hemisphere air mass source regions are found over North America, the North Atlantic Ocean, Eurasia, and the North Pacific Ocean. Source regions in the Southern Hemisphere lie over Australia and Antarctica. The air masses that originate in the source regions are classified according to their temperature

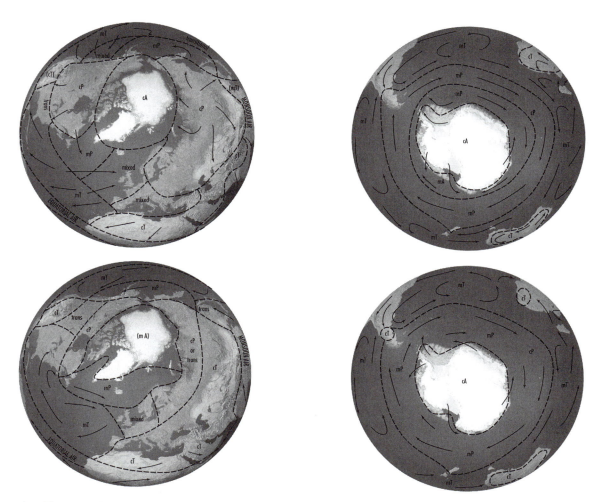

Source region. **The maps show the principal source regions for air masses in the Northern and Southern Hemispheres in winter and in summer. The arrows indicate the directions of the prevailing winds and the letters identify the types of air mass (m = maritime; c = continental; A = arctic; P = polar; T = tropical).**

and the surface over which they develop. Their temperature defines them as ARCTIC AIR, POLAR AIR, or TROPICAL AIR, and the surface over which they form defines them as MARITIME AIR or CONTINENTAL AIR.

South Equatorial Current   Two ocean currents, one in the South Atlantic and the other in the South Pacific, that flow from east to west parallel to and just south of the equator. The current flows within the upper 1,600 ft (500 m) of water at a speed of 0.6–2.5 mph (1–4 km h⁻¹) and is separated from the NORTH EQUATORIAL CURRENT by the EQUATORIAL COUNTERCURRENT.

southerly burster   A strong, cold wind that blows from the south across the eastern part of New South Wales, Australia. It is generated by a TROUGH of low pressure lying between two ANTICYCLONES that are moving in an easterly direction. Sometimes the trough extends almost to the north of Australia. The trough, shaped like an inverted V, is an extension of a DEPRESSION centered over the Southern Ocean. As it passes, the hot air that is drawn down from the north by the anticyclones is replaced by POLAR AIR.

Southern Arch   *See* ARCH CLOUD.

southern circuit   The path that DEPRESSIONS usually follow as they cross the United States from west to east during winter. The path sometimes carries them as far

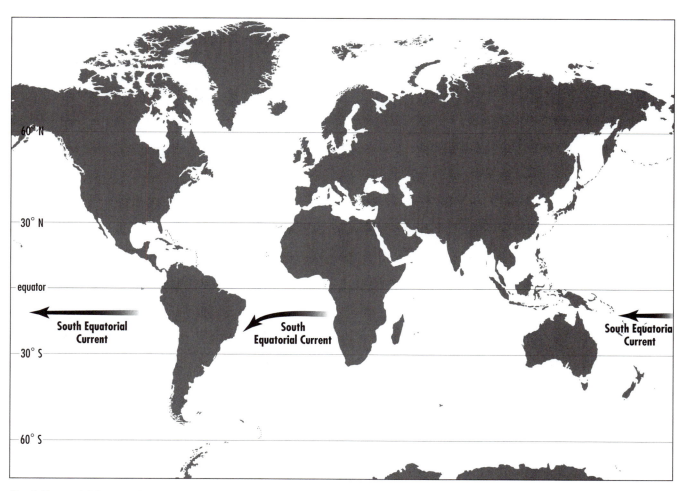

South Equatorial Current. Two ocean currents that flow from east to west in the South Atlantic and South Pacific Oceans, close to the equator and parallel to it. The currents are deflected to the south, where they encounter continents.

south as the Gulf of Mexico. In summer depressions usually follow the NORTHERN CIRCUIT.

Southern Oscillation A change in the distribution of air pressure over the equatorial South Pacific that occurs at intervals of one to five years. Ordinarily, air rises in the region of Indonesia, where the sea-surface temperature is high. It flows at high level from west to east and descends near the South American coast. This produces low pressure over Indonesia and high pressure over the eastern South Pacific. Low-level winds blow from east to west. From time to time this situation is reversed. Pressure is high over Indonesia, low over the eastern South Pacific, and air flows from east to west at high level and from west to east at low level. The change in direction of the surface wind constitutes an EL NIÑO, the change between the two patterns of pressure distribution constitutes a Southern Oscillation, and the two together constitute an ENSO event.

Southern Oscillation Index (SOI) A measure of the difference in sea-level atmospheric pressure between two monitoring stations; those most often used are at Darwin, Australia, and Tahiti, in the central South Pacific. A strongly negative SOI indicates a warming and a strongly positive SOI indicates a cooling in the central and eastern equatorial South Pacific.

South Temperate Zone That part of the Earth that lies between the tropic of Capricorn (see TROPICS) and the ANTARCTIC CIRCLE.

spatial dendrite An approximately spherical ICE CRYSTAL that has branches extending in all directions from a central nucleus. It is a type of DENDRITE.

special observation An observation of a particular weather condition affecting the operation of aircraft that is made because a significant change has occurred since the most recent report. The change might be a lowering or rise in the cloud ceiling, improvement or deterioration in VISIBILITY, or a fall in temperature that causes ice to form on the runway.

special sensor microwave imager (SSM/I) An instrument that is carried on the DEFENSE METEOROLOGICAL SATELLITE PROGRAM (DMSP) *F-8, F-10, F-11, F-12,* and *F-13* satellites. The SSM/I is a passive (*see* PASSIVE INSTRUMENT) MICROWAVE RADIOMETER with

seven channels and operating at four frequencies (19.35 GHz, 22.235 GHz, 37.0 GHz, and 85.5 GHz). It collects linearly polarized microwave radiation and measures the surface brightness over land and sea. This provides data on clouds and other meteorological phenomena, primarily in support of U.S. military operations, which are declassified so they are available to meteorologists. The instruments are carried in a nearly circular, SUN-SYNCHRONOUS, nearly POLAR ORBIT with a PERIOD of 102 minutes, at a height of 534 miles (860 km) and an INCLINATION OF 98.8°.

(You can learn more about the SSM/I at podaac-www.jpl.nasa.gov:2031/SENSOR_DOCS/ssmi.html.)

specific gas constant *See* GAS CONSTANT.

specific heat capacity *See* HEAT CAPACITY.

specific humidity The ratio of the mass of water vapor that is present in the air to a unit mass of that air including the water vapor. It is usually measured as grams of water vapor in one kilogram of air that includes the water vapor. Specific humidity is calculated differently from the MIXING RATIO, but for most practical purposes the two are identical. This is because the amount of water vapor present is rarely more than a very small proportion of the total mass of air, so whether or not it is included in the unit mass of air used for the calculation makes little difference to the outcome.

specific volume The volume that is occupied by a unit mass of a substance.

spectrum *See* RAINBOW, SOLAR SPECTRUM.

spectrum of turbulence The range of FREQUENCIES of the oscillations that make up TURBULENT FLOW. In turbulent flow, air is moving at different local VELOCITIES that vary over different lengths and times. Variation in an oscillation over distance (WAVELENGTH) and time is a variation in frequency. The spectrum at a given point and in a given direction is the ROOT-MEAN-SQUARE velocity of the frequencies contributing to the motion for each bandwidth.

speed of sound The speed with which a sound travels through a medium. Sound propagates as a SHOCK WAVE, and its speed depends on the density and elastic modulus of the medium. The speed of sound through a medium is given by

$$c = \sqrt{(E/\rho)}$$

where c is the speed of sound, E is the elastic modulus, and ρ is the density of the medium.

For a gas, $E = \gamma p$, where γ is calculated from the HEAT CAPACITIES of the principal constituents and p is the pressure. This shows that the speed of sound in a gas is related to the temperature of the gas. This means that the speed of sound through a gas can be expressed as

$$c = c_0 \sqrt{(1 + t/273)}$$

where c_0 is the speed of sound in the gas at 0° C, t is the temperature in degrees Celsius, and dividing by 273 relates the CELSIUS TEMPERATURE SCALE to the KELVIN SCALE.

In the case of a liquid, E is the bulk modulus, which is the ratio of the pressure applied to the medium and the extent to which its volume decreases.

For a solid, the elastic modulus is known as *Young's modulus,* after the British physicist Thomas Young (1773–1829), who proposed it, and it is the ratio between the stress applied to the solid and the change in its length.

Sound travels through air at 68° F (20° C) at 769.5 mph (344 m s⁻¹). It travels through water at 68° F (20° C) at 3,268 mph (1,461 m s⁻¹) and through steel at 68° F (20° C) at 11,185 mph (5,000 m s⁻¹).

spell of weather A period, usually of 5–10 days, during which particular conditions persist. The length of the period must be sufficient to make the spell a notable event, so it must take account of the effect of the weather on the lives of the people who experience it. A spell of FOG might last only two or three days; a spell of warm weather would have to last much longer.

spi *See* SPISSATUS.

spillover PRECIPITATION due to OROGRAPHIC LIFTING that is blown over the top of the mountain by the wind, so that it falls inside the area that is ordinarily in the RAIN SHADOW.

spiral band A pattern on a RADAR screen that is made by echoes from the center of a TROPICAL CYCLONE. The radar reflections are from areas of heavy rainfall and the pattern forms a roughly spiral shape that curves inward toward the storm center and merges in the WALL CLOUD.

spissatus (spi) A species of the cloud genus CIRRUS (*see* CLOUD CLASSIFICATION) that is sufficiently dense to appear grayish when viewed looking toward the Sun.

spontaneous nucleation The formation of water droplets or ICE CRYSTALS by the CONDENSATION of WATER VAPOR in the absence of CLOUD CONDENSATION NUCLEI or FREEZING NUCLEI. Water vapor condenses into liquid spontaneously when the RELATIVE HUMIDITY exceeds about 101 percent (*see* SUPERSATURATION), and ice crystals form spontaneously when the temperature falls below –40° F (–40° C).

spore A microscopically small organic structure that is produced by bacteria, fungi, and some plants, such as algae and ferns. It is capable of giving rise to a new individual without first fusing with another cell, but it does not contain an embryo of the new individual and is therefore quite different from a seed. Some bacteria survive in hostile environments for long periods as spores. Bacterial, fungal, and plant spores can be carried into the air to form part of the AERIAL PLANKTON.

Spörer minimum A period from 1400 to 1510 during which very few SUNSPOTS were observed. The period was identified by the German solar astronomer Friedrich Wilhelm Gustav Spörer (1822–95), who described it in an article published in 1889 that attracted the attention of the English astronomer EDWARD WALTER MAUNDER. Like the MAUNDER MINIMUM, the Spörer minimum was known as the *LITTLE ICE AGE.* Temperatures were abnormally low. Norse settlers were forced to leave Greenland because their crops failed and the sea ice did not thaw, so they could not fish. Famine increased in many parts of the world and in the winter of 1422–23 ice covered the entire surface of the Baltic Sea.

(There is more information in Tim Montfort, "Effects of Sunspots on Earth" at www.cs.usm.maine.edu/~montfort/ast100.htm.)

spot wind The wind that is observed or forecast at a specified height over a specified location.

spray region *See* FRINGE REGION.

spray ring *See* WATERSPOUT.

spring tides *See* TIDES.

sprinkling A very light SHOWER of rain or snow.

squall A sudden, brief STORM in which the wind speed increases by up to 50 percent, then dies away more slowly. For a storm to be described as a squall in the United States, the wind must reach 16 knots (18.4 mph [30 km h^{-1}]) or higher for at least two minutes. The wind speed may reach 30–60 mph (50–100 km h^{-1}) in a severe squall, and speeds of 100 mph (160 km h^{-1}) have been recorded. A series of squalls sometimes occurs along a line known as a *SQUALL LINE.*

squall cloud A roll of dark cloud, often with MAMMATUS, that forms along the leading edge of a SQUALL LINE. It is produced by eddies between the up- and downcurrents.

squall line A series of very vigorous CUMULONIMBUS clouds that merge to form a continuous line, which is often up to 600 miles (965 km) long and advances at right angles to the line itself. Cloud formation begins along a COLD FRONT, where moist air in the WARM SECTOR ahead of the front is being undercut and lifted by the advancing cold air. This produces a number of severe local STORMS. The clouds grow very tall in the unstable air, often penetrating through the TROPOPAUSE. A weak INVERSION in the warm sector restricts the development of CONVECTION CELLS except at the front itself.

Wind speed increases with height, and the speed of the wind in the middle TROPOSPHERE determines that of the clouds. This causes the upper part of each cloud to overtake its base, so the cloud overhangs its own base. Warm, moist air ahead of the cloud is consequently swept up into the convection cell inside it. It is lifted to the FREE CONVECTION level and then rises rapidly to the top of the cloud. There the wind reaches its maximum speed and the cloud is drawn into an anvil. Where the base of the anvil meets the main body of the cloud eddies generated by the vertical air currents produce a characteristic roll of cloud called a *squall cloud,* often with MAMMATUS.

CONDENSATION in the updraft of air produces precipitation. This falls from the trailing edge of the cloud. Some of the water droplets evaporate in the drier air below the cloud. This chills the adjacent air and causes the DOWNDRAFT. The downdraft produces a very local region of slightly raised surface atmospheric pressure. Part of the downdraft flows beneath the cloud and below the updraft. The updraft and downdraft meet at the leading edge of the cloud, where they produce strong GUSTS of wind along a line known as the *GUST FRONT.*

The gust front cuts beneath the warm air ahead of the cloud, lifting it into the updraft. Each individual cumulonimbus contains a single convection cell that lasts for only an hour or two before exhausting its supply of moisture and dissipating. Its gust front, however, scoops up moist air in front and to the right, producing a new cloud. As soon as a cloud attains its maximum development it begins to dissipate, at the same time triggering the development of a new cloud on its right front.

The vigor with which the gust front is able to shovel up warm, moist air causes the line of clouds to move faster than the cold front that initiated their formation. The squall line then becomes detached from the cold front, advancing ahead of it into the warm sector and moving at 10°–20° to the right of the wind direction in the middle troposphere. Behind the local high-pressure band associated with the downdraft there is a region of low pressure, called the *WAKE LOW.*

In a cumulonimbus cloud containing a single convection cell the updrafts and downdrafts conflict. This limits the development of the cloud and contributes to its dissolution. Along a squall line, however, the updrafts and downdrafts augment each other. This allows a squall line to last for several days, rather than the hour or two of an individual cloud, and to produce storms of much greater ferocity. TORNADOES often form along squall lines.

Squall lines are especially common in the central and eastern United States.

sr *See* STERADIAN.

SSI *See* STABILITY INDICES.

SSM/I *See* SPECIAL SENSOR MICROWAVE IMAGER.

SST   *See* SEA-SURFACE TEMPERATURE.

St   *See* STRATUS.

stability chart   A SYNOPTIC CHART that shows the distribution of values given by a particular STABILITY INDEX.

stability index   One of a series of values that are used to summarize the STABILITY OF AIR and the severity of the THUNDERSTORMS, up to and including TORNADOES, that varying degrees of INSTABILITY are likely to generate. Stability indices must be used with caution, because they do not apply to every situation. The most commonly used indices are the K Index (K), Lifted Index (LI), Showalter Stability Index (SSI), Total Totals (TT), SWEAT Index (SI), and Deep Convective Index (DCI). Each index produces a single numerical value that indicates either the degree of instability or the probability of severe storms. The relationship between K values and the probability of severe thunderstorms is shown in the table.

K measures the potential for thunderstorms from the LAPSE RATE and atmospheric moisture. LI and SSI both measure CONVECTIVE INSTABILITY but to different heights. TT has two components, Vertical Totals (VT) and Cross Totals (CT), which measure STATIC STABILITY and DEW POINT temperature, respectively, at particular levels. SWEAT (Severe Weather Threat Index) takes a number of factors into account. DCI combines EQUIVALENT POTENTIAL TEMPERATURE and instability.

The introduction of powerful computers has allowed meteorologists to calculate the more complicated convective available potential energy (CAPE). This has reduced their reliance on stability indices.

K Index	Probability
<15	0%
15–20	<20%
21–25	20–40%
26–30	40–60%
31–35	60–80%
36–40	80–90%
>40	nearly 100%

(You can learn more about stability indices at www.crh.noaa.gov/techpapers/service/tsp-10/10sevweastbind.html.)

stability of air   The tendency of a PARCEL OF AIR to possess neutral BUOYANCY, so that it remains at a constant height. Air with neutral buoyancy is said to be stable and air with positive buoyancy is unstable. Whether air is stable or unstable depends on the difference between the ENVIRONMENTAL LAPSE RATE (ELR) and the ADIABATIC lapse rate. This difference is known as the *POTENTIAL TEMPERATURE GRADIENT*, because it refers to a rate of change, or gradient, in the POTENTIAL TEMPERATURE of the air.

The amount by which the air temperature decreases between the surface and the TROPOPAUSE determines the ELR. This is assumed to remain constant throughout the TROPOSPHERE. The temperature at the tropopause is always about −74° F (−59° C). If, then, the surface temperature is 59° F (15° C), and the height of the tropopause is 36,000 ft (11 km), the ELR is about 3.7° F per 1,000 ft (6.7° C per kilometer). This is an average value for the ELR, based on the mean surface temperature over the whole world, which is 59° F (15° C). The actual ELR varies with every change in the surface temperature. For example, if the surface temperature is 32° F (0° C) the ELR is 2.9° F per 1,000 ft (5.4° C per kilometer), but if the surface temperature is 80° F (27° C) the ELR is 4.3° F per 1,000 ft (7.8° C per kilometer).

Suppose air at 59° F (15° C) is forced to rise. It cools at the DRY ADIABATIC LAPSE RATE (DALR) of 5.5° F per 1,000 ft (10° C per kilometer). This value is greater than the ELR. Consequently, the rising air cools faster than the ELR. At any height the rising air is cooler and denser than the surrounding air and so it sinks back to the level at which it is at the same temperature and DENSITY as the air around it. This air is stable. The most stable condition of all occurs in a layer of air that lies beneath an INVERSION.

If the air is moist it may reach a height at which it becomes saturated and WATER VAPOR starts to condense. CONDENSATION releases LATENT HEAT, reducing the rate of cooling to the SATURATED ADIABATIC LAPSE RATE (SALR) of approximately 3.3° F per 1,000 ft (6° C per kilometer). This is a slower rate of cooling than the ELR and so air cooling at the SALR is always warmer and less dense than the air around it. It therefore continues to rise. This air is unstable.

If the SALR is greater than the ELR rising air always is cooler and denser than the surrounding air, even if the air is saturated. This condition is called *ABSOLUTE STABILITY*. Its opposite is ABSOLUTE INSTABILITY. This exists when the ELR is greater than the DALR. When the ELR is greater than the SALR but smaller than the DALR the condition is known as *CONDITIONAL INSTABILITY*. This is the most common type of instability and is the type illustrated in the preceding example.

Because the ELR is greater when the surface air is warm than it is when the surface air is cold, air is more likely to be unstable in warm weather than it is in cold weather. On really cold days in winter the air is usually very stable, because the ELR is low. When the surface temperature is –10° F (–23° C) the ELR is only 1.8° F per 1,000 ft (3.3° C per kilometer), which is much lower than the SALR.

stable-air föhn   *See* HIGH FÖHN.

stack height (chimney height)   The actual height above the ground of the top of a factory smoke stack (chimney). This determines the height at which gaseous and particulate emissions enter the atmosphere. The more useful measurement, however, is that of the EFFECTIVE STACK HEIGHT, because this takes account of the vertical speed of the material leaving the stack.

stade (stadial)   A period of cold that occurs during a GLACIAL period. It may be marked by a change in the vegetation of ice-free areas or by the advance of ICE SHEETS.

stadial   *See* STADE.

stage   A series of sediments, fossils, or plant remains that are found in a particular place and that indicate the climatic conditions that obtained at the time they were deposited. The stage is named after the place where the evidence for it was first discovered. For example, the HOXNIAN INTERGLACIAL was identified by deposits found at the village of Hoxne and known as the *Hoxnian stage*.

stagnation point   *See* CORNER STREAM.

standard artillery atmosphere   A set of hypothetical values that are used in calculations of the trajectory of missiles through the air (ballistic calculations). These values assume there is no wind, the surface temperature is 59° F (15° C), surface pressure is 1,000 mb (14.5 lb in^{-2}, 29.5 in mercury), surface RELATIVE HUMIDITY is 78 percent, and the LAPSE RATE directly relates air DENSITY to altitude.

standard artillery zone   A layer of the STANDARD ARTILLERY ATMOSPHERE that is of a specified thickness and at a specified altitude.

standard atmosphere (standard pressure)   A unit of pressure that represents the average atmospheric pressure at mean sea level (sea level varies slightly from place to place), assuming the atmosphere to consist of a perfect gas (a gas that obeys all the GAS LAWS) at a temperature of 59° F (15° C [188.16 K]) and the acceleration due to gravity to be 9.80655 m s^{-2}. The standard atmosphere is defined as a pressure of 1.013250×10^5 newtons per square meter. This is equal to 760 millimeters of mercury (29.9213 in mercury) with a density of 13,595 kilograms per cubic meter, 14.691 pounds per square inch, or 1,013.25 MILLIBARS.

standard pressure   *See* STANDARD ATMOSPHERE.

standard temperature   *See* FIDUCIAL POINT.

standard temperature and pressure   *See* S.T.P.

standing cloud   A cloud that remains stationary, usually above or close to a mountain peak or other high ground.

standing waves   *See* LEE WAVES.

stand of the tide   The time of slack water, when the TIDE is about to turn. During this short period the height of the tide does not change and tidal currents slow and then cease before starting to flow again in the opposite direction.

star dune   A sand dune that is built by winds that are highly variable in direction. It consists of a series of ridges that radiate from a central point, making a shape resembling a star.

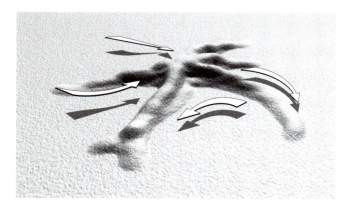

Star dune. Star dunes consist of ridges that radiate from a central point. The arrows indicate the direction of the winds that formed the dunes.

START *See* INTERNATIONAL GEOSPHERE–BIOSPHERE PROGRAM.

state of the sky A full description of the sky that includes the amount, type, and height of all the clouds and the direction in which they are moving.

static electricity An electrical charge that is at rest. One region, such as the upper part of a CUMULONIMBUS cloud, bears a positive charge, and another, such as the lower part of the same cloud, bears a negative charge, but the two areas of charge are separated by an area that is electrically neutral. Consequently, no electrons move from the positive to negative areas. When the charge has accumulated sufficiently to overcome the insulating effect of the neutral region, the static charge is discharged in a spark, in this case a flash of LIGHTNING.

static instability *See* CONVECTIVE INSTABILITY.

static stability (convective stability) The condition in which the atmosphere is stratified so that its DENSITY decreases with height and BUOYANCY increases. The POTENTIAL TEMPERATURE increases with height. If a PARCEL OF AIR rises by CONVECTION it enters a region where it is denser, and therefore heavier, than the surrounding air, so gravity causes it to sink back to its original level. If the parcel should sink, it enters a region where it is less dense, and lighter, than the surrounding air, so it rises to its original level. When the atmosphere is not statically stable it may be neutrally stable (*see* NEUTRAL STABILITY) or convectively unstable (*see* CONVECTIVE INSTABILITY).

stationary front A front at which the air on each side is moving approximately parallel to the front. With no air movement to carry one AIR MASS forward against the other the surface position of the front does not move, or it moves erratically and slowly. It may remain in the same position for several days. A stationary front is shown on weather maps as a line with semicircles on one side and triangles on the other. Stationary fronts can be active (*see* ACTIVE FRONT) or weak (*see* WEAK FRONT). Those that are active produce STRATIFORM cloud with a low base.

station circle *See* STATION MODEL.

station elevation The vertical distance between a weather station and mean sea level. This is measured for a particular point at the station and is used as a reference datum for calculating atmospheric pressure. The station barometers show pressure at the station, but because pressure varies with elevation this is converted to sea-level pressure and it is the sea-level pressure that is reported. Since all stations report sea-level pressure, their reports can be compared directly and ISOBARS can be plotted without a need for further corrections.

station model The formalized diagram that is used to report observations from a weather station. It uses standard symbols to represent cloud, precipitation, and wind direction and speed. Other information is given in numbers. Each item of information occupies a particu-

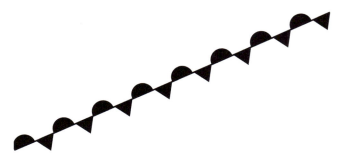

Stationary front. The front is shown on weather maps as a line with red semicircles on one side and blue triangles on the other.

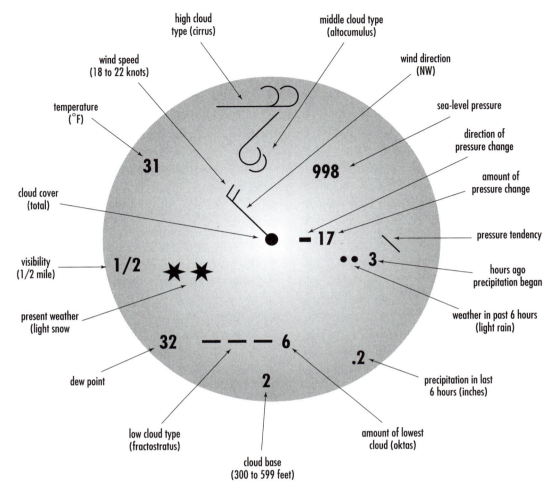

Station model. The formal arrangement of numbers and symbols that is plotted on a weather chart to indicate the conditions reported from a weather station.

lar position around the station, which is represented by a circle at the center.

The central circle, called the *station circle,* is drawn in its geographical position on a weather chart and the other information is positioned around it. Whether the station circle is open or partly or completely black indicates the amount of cloud cover. A line from the station circle indicates the wind direction, and lines or PEN-NANTS at the end of this line indicate the wind speed. CLOUD SYMBOLS indicate cloud types. The station model shows the type of HIGH, MIDDLE, and LOW CLOUD. The CLOUD BASE is reported in a code that translates to a range of heights (in feet) and the amount of the lowest cloud is given in OKTAS. Symbols are also used to indicate the present and past weather.

In addition the model shows the sea-level pressure in millibars. The lowest sea-level pressure ever record-ed (in the eye of TYPHOON TIP) was 870 mb and the highest (recorded on December 31, 1968, at Agata, Siberia, Russia) was 1,083.8 mb. In reporting sea-level pressure, therefore, the thousands and hundreds units (8, 9, or 10) can be assumed. They and also the deci-mal point can then be omitted for the sake of brevity, so that the pressure is always represented by a three-digit number. For example, a pressure of 999.8 mb is reported as 998 (that is, (9)**99**(.)**8**), and a pressure of 1012.4 as 124 (that is, (10)**12**(.)**4**). The model also reports whether the pressure has increased (+) or has decreased (−) over the past six hours, the amount by which it has changed (in tenths of a millibar), and the present BAROMETRIC TENDENCY. Temperature and dew point are reported (in degrees Fahrenheit or Celsius). Horizontal VISIBILITY is given in miles or kilometers. If there is precipitation, the number of hours since it began and the amount of precipitation in the last six hours are reported.

Station models are updated every six hours.

station pressure (surface pressure) The atmospheric pressure that is measured at the STATION ELEVATION.

statistical forecast A weather forecast that is compiled from studies of past weather patterns. These are then compared with present conditions and used to assess the statistical probability of the pattern's repeating. This method can be used fairly reliably to predict a particular climatic feature, such as temperature or precipitation, over a short period. Very general forecasts over longer periods can also be compiled from a detailed knowledge of past patterns, using ANALOG MODELS.

statistically dynamical model A two-dimensional CLIMATE MODEL that deals only with surface processes. It describes the latitudinal (toward and away from the equator) transport of energy and the vertical distribution of energy by radiation and convection.

steam devil A stream of fog that forms an almost vertical column of cloud over the surface of an unfrozen lake in winter. Steam devils develop when cold, dry air crosses the lake and is warmed by contact with the water surface. Water evaporates into the air and the warmed air rises. It cools ADIABATICALLY and its water vapor condenses into streamers of fog. Where warm air is rising especially vigorously, a number of streamers may be drawn into the upcurrents to form a steam devil, which is rather similar to a DUST DEVIL.

steam fog Thin, wispy FOG that forms when cold air crosses a water surface that is warmer. It is often seen over lakes in winter. Because the air is cold, it contains less water vapor than it would if its temperature were the same as that of the water. In the layer of air that is immediately adjacent to the water surface and is warmed by it, the water-vapor content is lower than is needed to produce SATURATION. Evaporation is intense and moist air rises by CONVECTION from the

warm BOUNDARY LAYER. In the colder air above, the relative humidity of the rising air very quickly reaches 100 percent and the water vapor condenses again to form the fine droplets of fog. The fog looks like steam and it forms in the same way as the steam in a bathroom that appears when a hot bath or shower is run when the bathroom is cold and its air temperature is low.

Stefan–Boltzmann constant The amount of radiant energy released by a BLACKBODY. This is represented by σ and is equal to 5.67×10^{-8} W m^{-2} K^{-4} (watts per square meter per kelvin to the fourth power). See STEFAN–BOLTZMANN LAW.

Stefan–Boltzmann law The physical law that relates the amount of radiant energy a body emits to its temperature. The law applies to BLACKBODIES, including the Sun and Earth, and shows the amount of radiation emitted is proportional to the fourth power of the temperature of the body. It is expressed as $E = \sigma T^4$, where E is the amount of radiation emitted, T is the temperature, and σ is the STEFAN–BOLTZMANN CONSTANT. The temperature is in KELVINS and the energy units are watts per square meter (W m^{-2}). The law (and constant) were discovered in 1879 by the Austrian physicist Josef Stefan (1835–93) and at first they were known as *Stefan's law* and *constant*. In 1884, the Austrian physicist Ludwig Eduard Boltzmann (1844–1906), Stefan's former student, showed that the law holds only for blackbodies and his name was added to that of the law and constant. Stefan used the law to make the first fairly accurate estimate of the temperature of the photosphere (visible surface) of the Sun, as 10,800° F (6,000° C).

stem flow Precipitation that falls onto vegetation and reaches the ground by running down the stems of plants. The amount of water reaching the ground as a result of stem flow depends on the total amount of precipitation intercepted by the plants, the amount that evaporates from the leaves, and the amount that evaporates from the stem during stem flow. If the vegetation is dry when precipitation commences, the plants intercept a high proportion of the precipitation, especially if the precipitation is light. Water coating the leaves immediately begins to evaporate, and the rate of EVAPORATION depends on the temperature. In Brazilian forests, for example, about 20 percent of the rain falling onto the leaves evaporates, and almost all the rain falling on them evaporates from the leaves of some forests growing in Mediterranean climates, so there is no stem flow. The amount of stem flow and water dripping from leaves must be measured very precisely in studies of the amount of precipitation that reaches the ground in forests.

steppe climate In the CLIMATE CLASSIFICATION devised by MIKHAIL I. BUDYKO, a climate in which the RADIATIONAL INDEX OF DRYNESS has a value of 1.0–2.0. *See also* MIDDLE-LATITUDE DESERT AND STEPPE CLIMATES.

stepped leader The first stage in a LIGHTNING flash, which carries negative charge away from the cloud as a stream of electrons (negatively charged particles). It occurs when the separation of charge within a CUMULONIMBUS cloud is sufficiently large to break down the insulation of the air. The separated charges produce an electrical field around the cloud and the stepped leader travels where the field is weakest, producing a jagged path that is then established as the LIGHTNING CHANNEL. The stepped leader is almost invisible, but as it approaches close to the region of positive charge it triggers and is met by the much more luminous RETURN STROKE.

steradian (sr) The supplementary SYSTÈME INTERNATIONAL D'UNITÉS (SI) UNIT of solid angle, which is defined as the angle, measured from the center of a sphere, that cuts off an area on the surface of the sphere equal that of a square with sides equal to the radius of the sphere.

Steve A tropical cyclone that struck Cairns, Queensland, Australia, on February 27, 2000. It had winds of 105 mph (169 km h⁻¹). There were no reported injuries or deaths.

Stevenson screen The container that houses the THERMOMETERS and HYGROMETERS or HYGROGRAPHS used at a weather station. It is named after Thomas Stevenson (1818–87), who invented it. Stevenson was an amateur meteorologist, but nevertheless he had a professional interest in the weather. He was a civil engineer, and the Stevenson family firm specialized in build-

ing lighthouses throughout the world. Stevenson was born in Edinburgh, a city that he helped to make into a world center for the science of lighthouse lenses and lights. That is also where his son, the author Robert Louis Stevenson (1850–94), was born.

His screen is a box with louvered walls on all its four sides. The walls are of double thickness; the louver strips form inverted V shapes in cross section. The screen is painted white and stands on legs that raise it so that the bulbs of its thermometers are about 4 ft (1.25 m) above the ground. At this height the thermometers register what is called *air temperature, shelter temperature,* or *surface temperature.* This height is the lowest at which a thermometer gives a reliable reading for the temperature that is experienced by people and that is accurate for a large surrounding area. The temperature at this height is influenced, but not overwhelmed, by variations in temperature closer to

Stevenson screen. **A Stevenson screen is a container of standard construction that is used to house the thermometers used at a weather station. The screen is painted white, has double-louvered sides, and is sited in the open well clear of the ground.**

the ground. Ground-level temperatures often change markedly over short distances, so simply moving the instrument to a different location nearby could alter thermometer readings significantly.

The purpose of the white paint and louvered construction is to shield the instruments from exposure to direct radiation without isolating them from the air they are to monitor. It is because of the shielding it provides that the container is called a *screen*.

Thermometers are sensitive to sunlight, which warms them directly, and they are also affected by radiation rising from the ground. The white color reflects most of the radiation that falls on it and the thick, wooden walls and floor of the screen insulate the air inside. For added protection the screen is positioned so that access to the instruments is from a door on the side facing away from the equator (the northern side in the Northern Hemisphere). This prevents the Sun from shining into the screen while readings are being taken. The louvered construction of the sides allows air to circulate, so the temperature being measured is that of the air outside the screen.

Standardizing both the construction and the siting of Stevenson screens means that all weather stations are obtaining readings under similar conditions. This makes the readings directly comparable.

The disadvantage of the Stevenson screen is that although the double-louvered walls permit air to circulate, the ventilation inside the screen is poor. In hot weather the temperature inside the box can be a degree or more higher than the air temperature outside. Some screens are fitted with fans to improve ventilation.

A Stevenson screen usually contains four thermometers. These are the WET-BULB THERMOMETER and DRY-BULB THERMOMETER used to calculate RELATIVE HUMIDITY and DEW POINT TEMPERATURE and the dry-bulb MAXIMUM and MINIMUM THERMOMETERS. The wet-bulb and dry-bulb thermometers are mounted vertically and the maximum and minimum thermometers horizontally.

(Robert Louis Stevenson wrote an obituary for his father, Thomas; you can read it at www.unibg.it/rls/essays/memport/mp-9.htm.)

Stevin, Simon (1548–1620) Flemish *Mathematician* A Flemish mathematician who published in 1586 the book *Statics and Hydrostatics*. In this, he reported his discovery that the pressure a liquid exerts on the sur-face beneath it depends only on the area of the surface on which it presses and the height to which the liquid extends above the surface. Contrary to what seemed obvious to most people at the time, it has nothing whatever to do with the shape of the vessel holding the liquid. This finding paved the way for EVANGELISTA TORRICELLI, who found a way to "weigh" air and invented the BAROMETER.

Stevin, often known by the Latin version of his name as *Stevinus,* was born in Bruges. He worked as a clerk in Antwerp and then entered the service of the Dutch government; he became the director of the department of roads and waterways and later quarter-master-general to the Dutch army. He devised for military purposes a scheme for opening sluices in the dikes protecting the polders—cultivated land reclaimed from the sea. This would flood the land in the path of any invading army. He also invented a carriage propelled by sails that ran along the beach and could carry 26 passengers, and he is said to have dropped two objects of different weights from a tall tower and found they reached the ground simultaneously—an experiment usually attributed to GALILEO GALILEI. It was Stevin who established the use of decimal notation in mathematics—representing 1/2 as 0.5, for example, and 1/4 as 0.25. He maintained that decimal weights, measures, and coinage would eventually be introduced.

Stevin wrote in Flemish and advised all scientists to describe their work in their own native language. This was unusual, and his works were all translated into Latin later. He died either at the Hague or Leiden, two cities in the Netherlands that are 10 miles (16 km) from each other.

sticky *See* CLOSE.

still well A device that is used to provide an undisturbed water surface for measurements from an EVAPORATION PAN. The standard still well is a brass cylinder 8 in (30.32 cm) tall and 3.5 in (8.89 cm) in diameter, open at both ends, that is mounted over a hole in a galvanized iron base. The base is fitted with leveling screws.

stochastic Obedient to statistical laws. The future behavior of a stochastic system cannot be predicted precisely, but its probable future behavior can be calculated on the basis of its known past behavior. Modern weather forecasters adopt an approach that is partly stochastic and partly deterministic. They recognize that although natural laws certainly determine the behavior of the atmosphere (*see* DETERMINISM), it is impossible to know the condition of the atmosphere in the detail required for an entirely deterministic calculation. Consequently they make the best calculation they can but qualify it by taking account of the way similar conditions developed in the past. This allows them to estimate the probability of a particular weather pattern. A probabilistic forecast makes allowance for the chaotic behavior of weather systems, but because of CHAOS its reliability is inversely proportional to the forecast period.

stochastic resonance The observable effect that results when a STOCHASTIC process acts in the same sense as a natural cycle that is too weak to produce any effect by itself. Stochastic processes ordinarily appear as noise, and scientists try to remove them in order to detect signals. If the signals are extremely weak, however, a stochastic process may reinforce them. The two are then said to *resonate* and because it is the stochastic process that produces the resonance, it is called *stochastic resonance.*

OXYGEN ISOTOPE records from the GREENLAND ICECORE PROJECT indicate that stochastic resonance underlies a 1,500-year climate cycle, the most recent manifestation of which was the LITTLE ICE AGE. Stochastic resonance may also be responsible for the onset of GLACIATIONS and INTERGLACIALS at intervals of approximately 100,000 years. These are driven by the MILANKOVICH CYCLES, but changes in the orbit and rotation of the Earth are believed to be too small to account for ice ages. If stochastic warming and cooling of the climate coincide with the peaks and troughs of the Milankovich cycles, their additional effect would be enough to trigger the observed effects.

Stochastic processes are unreliable and cannot be expected to coincide with every phase in a regular natural cycle. Sometimes resonance should occur and sometimes not. The result of this would be that the observed effect occurs most often at the peak or trough of one cycle but sometimes skips one cycle and more rarely skips two. Very rarely, an event occurs at less than a complete number of cycles since the preceding one. This pattern is what emerges from the ICE CORE record.

In the case of the 1,500-year cycle, this means that most cool episodes resembling the Little Ice Age happen at intervals of 1,500 years, but some occur every 3,000 years and a few every 4,500 years. Similarly, ice ages and interglacials often begin every 100,000 years, but not always. Sometimes 200,000 years can pass without either.

Stokes's law A law describing the factors that determine the magnitude of the friction experienced by a spherical body (such as a raindrop or hailstone) falling by gravity through a viscous medium (such as air). Friction (F) is given by

$$F = 6\pi r \eta v$$

where r is the radius of the body, v is its velocity, and η is the VISCOSITY of the medium.

From Stokes's law it is possible to calculate the TERMINAL VELOCITY (V) of the body as

$$V = 2 g(\rho_p - \rho_a)r^2 \div 9 \eta$$

where g is the acceleration due to gravity (32.18 ft s^{-2} = 9.807 m s^{-2}), ρ_p is the density of the body, and ρ_a is the density of the medium through which it is falling.

A very small water droplet, with a diameter of about 10 μm (0.0004 in), has a terminal velocity falling through air of about 10^{-3} ms^{-1} (0.03 ft s^{-1}). A drop 100 times larger, with a diameter of about 1 mm (0.04 in), falls at about 10 ms^{-1} (33 ft s^{-1}).

The relationship was discovered between 1845 and 1850 by the Irish physicist Sir George Gabriel Stokes (1819–1903).

stoma *See* STOMATA.

stomata (sing. *stoma*) The pores in the surface of plant leaves, especially on the undersides, through which gases are exchanged. In most plants, stomata are open during the day and closed at night. Each stoma can be opened and closed by the expansion and contraction of two guard cells at its mouth. When it is open, air can enter the cell beneath to provide carbon dioxide for PHOTOSYNTHESIS, and oxygen, a by-product of photosynthesis, can leave. Water vapor also escapes while the stomata are open. The loss of moisture in this way is called *TRANSPIRATION*. Plants adapted to dry climates usually minimize the time during which their stomata are open. Some desert plants (called *CAM plants*) open their stomata at night and store the carbon dioxide they absorb until daybreak, after which photosynthesis proceeds with the stomata remaining closed.

storm   In the BEAUFORT WIND SCALE, force 11, which is a wind that blows at 64–75 mph (103–121 km h⁻¹). In the original scale, devised for use at sea, a force 11 wind was defined as "or that which would reduce her to storm staysails." On land, a storm uproots trees and blows them some distance and overturns cars.

storm beach   A linear pile of coarse material, such as pebbles, gravel, and seashells, that has been built on a beach by the action of sea STORMS. During the storm, waves throw the material into a heap, and in the course of many storms they form a distinct ridge or bank. The presence of such a pile of material indicates a beach that is exposed to storms.

storm bed   A layer of sediment that is deposited over a surface by the action of a sea STORM. Shallow waves carry fine-grained material up the shore, where it is precipitated. A storm bed is sometimes called an *event deposit* because it is the result of a single physical event.

storm center   The area of lowest surface atmospheric pressure in a CYCLONE. The term applies to any low-pressure system and not only to a TROPICAL CYCLONE, in which cloud usually clears at the center to form an "eye."

storm detection   The identification of conditions that are leading to the development of a STORM, followed by its observation and tracking. As techniques have improved, the importance of storm detection has increased. Detection involves recognizing particular characteristics, especially wind strength and precipitation, that indicate the type of storm and measuring or calculating the area the storm covers. Balloon SONDES, flights by aircraft equipped with meteorological instruments, RADAR including DOPPLER RADAR, and satellite images are used to detect and then study storms.

storm glass   An instrument that indicates a change in the weather. It is no longer used but was popular in the 18th and 19th centuries. It consisted of a heavy glass tube, tightly sealed to prevent air entering from outside,

that contained a supersaturated mixture of chemical compounds. The precise recipe varied from one instrument to another, but most were based on camphor dissolved in alcohol, with other chemicals. Crystals would form and dissolve inside the glass; the changes apparently were linked to meteorological changes other than simple changes in temperature or air pressure. If the liquid was clear it meant the weather would be fine. Cloudy liquid meant it would rain. If crystals formed at the bottom of the glass in winter there would be frost. Admiral ROBERT FITZROY became very interested in storm glasses. He developed one based on his own chemical mixture that was attached to some versions of the FitzRoy barometer and explained how to interpret the patterns in it in *The Weather Book,* published in 1863.

storm model   A three-dimensional simulation of the way air and water vapor move into, out of, and vertically within a STORM.

storm surge   A rise in sea level, accompanied by huge waves, that is produced by large storms at sea and especially by TROPICAL CYCLONES. Water sweeping inland, often for a considerable distance, causes severe flooding, and in areas struck by a tropical cyclone the storm surge may cause more loss of life, injuries, and damage to property than the wind.

In the SAFFIR/SIMPSON SCALE, a category 1 storm produces a storm surge of 4–5 ft (1.2–1.5 m); category 2 produces 6–8 ft (1.8–2.4 m); category 3 produces 9–12 ft (2.7–3.7 m); category 4 produces 13—18 ft (4.0–5.5 m); category 5 produces a surge of more than 18 ft (5.5 m). Storm surges can be much larger than 18 ft, however. Hurricane GILBERT struck the Mexican coast in 1988 with a storm surge of 20 ft (6 m) that threw onto the shore a Cuban ship that had been several miles out at sea. In 1992 TROPICAL STORM POLLY produced a 20-ft (6-m) surge at the port of Tianjin, China, and in December 1999 CYCLONE JOHN produced one of similar size in Western Australia.

Three factors contribute to produce a storm surge. The first is the drop in surface atmospheric pressure at the center of the storm. A fall in pressure of 1 MILLIBAR (mb) below the mean sea-level pressure of 1,013 mb causes the sea level to rise by about 0.4 in (1 cm). In a tropical cyclone, measuring on the SAFFIR/SIMPSON SCALE, the sea-level rise due to the low pressure in the eye of a category 1 storm is about 14 in (36 cm). In categories 2–4 it is 14.6–20 in (37–51 cm), 20.5–28 in (52–71 cm), and 28.4–37.8 in (72–96 cm), respectively. In category 5, the sea level is raised by more than 37.8 in (96 cm).

The second factor arises from waves driven by the winds. In a severe storm, spray whipped up from the sea surface turns the sea completely white and greatly reduces horizontal visibility. It appears that the sea is boiling, and beneath the white water the waves are huge. Their size, speed, and WAVELENGTH increase in proportion to the wind speed, the distance over which the wind blows (the FETCH), and the length of time the wind has been blowing. Waves reach their maximum size when the wind has been blowing for about 40 hours. A wind of 110 mph (177 km h^{-1}) can raise waves with an AMPLITUDE of 30 ft (9 m) high and a tropical cyclone can raise waves with amplitudes up to 70 ft (21 m). This is close to the maximum size a sea wave can attain, because larger waves fall forward under the weight of water. While sailing through a Pacific typhoon on February 6–7, 1933, the U.S.S. *Ramapo* measured a wave that was 112 ft (34.14 m) high. This is believed to be the biggest wave ever recorded.

At sea, not all waves are the same size. This is because waves interfere with one another in complex ways, which sometimes augment and sometimes diminish them. Of 100 waves passing a fixed point, on average there is one wave 6.5 times bigger than the others. If 1,000 waves pass there is likely to be one that is 8 times bigger.

Where winds drive waves into a confined area, such as a partially landlocked sea, or into shallower water, the water level rises dramatically. Wave amplitude increases as wavelength decreases (*see* WAVE CHARACTERISTICS), so waves grow higher as they approach a shelving coast. Driving water into a confined area causes water throughout the entire basin to slop back and forth as water does in a bathtub when you move your hand back and forth through it. This motion can become very large if the slopping resonates with the natural PERIOD of the sea basin, just as the waves your hand makes in a bathtub grow bigger if they resonate with the size of the tub.

The final component of a storm surge is the TIDE. If the raised sea level and storm-driven waves coincide with a high tide when they reach the coast, the effects of all three factors are added together. It is under these circumstances that a storm surge penetrates farthest inland and causes its greatest damage.

storm track Storms are carried by the prevailing wind systems in the regions where they occur. Generally, therefore, storms in middle latitudes, where the PREVAILING WINDS are from the west, tend to move from west to east. In the TROPICS, the prevailing TRADE WINDS blow from the east, and so storms move from east to west. Storms rarely move in straight lines, however. In North America, a major storm that develops in Alberta is likely to curve southward into Montana and North Dakota before passing to the north of Lake Superior and into Labrador. A storm originating in Colorado may travel northeast and cross the Great Lakes.

TROPICAL CYCLONES start by moving toward the west but soon turn to the northwest in the Northern Hemisphere and southwest in the Southern Hemisphere. This track takes them toward the SUBTROPICAL HIGHS and they curve around the western boundaries of these ANTICYCLONES. They then enter the region of the midlatitude westerlies and their track turns increasingly toward the northeast in the Northern Hemisphere and southeast in the Southern Hemisphere. As they move across cooler water, they weaken and finally dissipate. Although these are the average tracks that tropical cyclones follow, individual cyclones are subject to local influences that cause them to deviate, even to the extent of traveling for a time in the opposite direction. This characteristic makes the task of forecasting their movements very difficult. The speed at which a tropical cyclone moves is determined mainly by the rate at which the warm air above the eye moves. Most travel at 10–18 mph (16–29 km h^{-1}), but some move faster.

storm warning Advance information of the likelihood of severe weather that is issued by the national meteorological service. In the United States, storm warnings are issued by the NATIONAL WEATHER SERVICE using information supplied by the NATIONAL SEVERE STORMS FORECAST CENTER. The warning is of sustained winds, lasting for at least one minute, of 48 knots (55 mph [89 km h^{-1}]) or stronger. A warning is the highest of the three categories of advance notice (the others are advisories and watches) and indicates that the storm is expected within the coming 24 hours.

(You can learn more about the weather warning system at www.spc.noaa.gov/products/wwa.)

stowed wind A wind that is partly blocked by a physical barrier, such as a range of mountains or hills, so it is forced through gaps between them. This increases the speed of the wind.

s.t.p. (standard temperature and pressure) The conditions that are applied when measuring quantities that vary with temperature and pressure and especially when comparing the properties of gases. The conditions are a temperature of 273.15 K (32° F [0° C]) and pressure of 101,326 Pa (1,013.26 mb, 30 in mercury, 760 mm mercury).

str *See* STRATIFORMIS.

Strahler climate classification A CLIMATE CLASSIFICATION that was proposed by Arthur Newell Strahler (born 1918), in his book *Physical Geography* (New York: Wiley, 1969). A. N. Strahler is Emeritus Professor of Geomorphology at Columbia University. His classification is closely related to the KÖPPEN CLASSIFICATION and can be used in conjunction with it.

The Strahler classification is of the genetic type: that is, it is based on the GENERAL CIRCULATION of the atmosphere and relates regional climates to the AIR MASSES and PREVAILING WINDS that produce them. Strahler divided the climates of the world into three main types, or groups, according to the air masses that control them. These are subdivided further into 14 climatic regions, to which he later added highland climates.

His group 1 comprises climates that are controlled by equatorial and tropical air masses and occur in low latitudes: (1) WET EQUATORIAL CLIMATE, (2) TRADE WIND LITTORAL CLIMATE, (3) TROPICAL DESERT AND STEPPE CLIMATES, (4) WEST COAST DESERT CLIMATE, and (5) TROPICAL WET–DRY CLIMATE.

Group 2 comprises middle latitude climates controlled by both tropical and polar air masses. These are (6) HUMID SUBTROPICAL CLIMATE, (7) MARINE WEST COAST CLIMATE, (8) MEDITERRANEAN CLIMATE, (9) MIDDLE LATITUDE DESERT AND STEPPE CLIMATES, and (10) HUMID CONTINENTAL CLIMATE.

Group 3 comprises high latitude climates controlled by polar and arctic air masses. These are (11) CONTINENTAL SUBARCTIC CLIMATE, (12) MARINE SUBARCTIC CLIMATE, (13) TUNDRA CLIMATE, and (14) ICECAP CLIMATE. Group 3 also includes HIGHLAND CLIMATES.

strand line A layer of material that was deposited at the edge of a former lake or sea and that marks the location of a shoreline that has since disappeared. Strand lines indicate higher sea levels or heavier precipitation at some time in the past, and they can therefore be used in the reconstruction of past climates.

strat- A prefix that is derived from the Latin *stratum*, which is the neuter past participle of the verb *sternere*, which means "to strew." It is attached to cloud genera, species, and varieties that form extensive horizontal sheets or layers, as though they were spread (or strewn) across the sky.

stratiform An adjective that is applied to clouds that form extensive horizontal layers.

stratiformis (str) A species of cumuliform clouds (*see* CLOUD CLASSIFICATION) that spread across the sky to form an extensive sheet. The species occurs with the cloud genera CIRROCUMULUS, ALTOCUMULUS, and STRATOCUMULUS.

stratocumulus (Sc) A genus of low clouds (*see* CLOUD CLASSIFICATION) that are composed of water droplets. It is seen as patches, sheets, or a layer of gray, white, or both gray and white cloud. There are always dark areas, shaped as rolls or rounded masses. These sometimes merge into larger masses. They are not fibrous. The smallest elements have an apparent width of about 5°. The cloud is similar to ALTOCUMULUS but heavier and occurs at a lower level.

It often has gaps large enough to allow sunlight to penetrate intermittently. When sunlight shines through gaps in the cloud at around dawn or sunset its converging beams often illuminate dust and other solid particles in the air below the cloud, forming CREPUSCULAR RAYS. These are often described as the SUN DRAWING WATER and regarded as a warning of approaching rain, but stratocumulus does not indicate bad weather.

Stratocumulus develops where rising air encounters a ceiling of warmer air and is flattened against it. In winter it can produce an overcast sky lasting for several days. In summer there is usually sufficient warmth for much of the cloud to evaporate, changing it into fair weather CUMULUS.

Stratocumulus occurs in the species CASTELLANUS, LENTICULARIS, and STRATIFORMIS and in the varieties DUPLICATUS, LACUNOSUS, OPACUS, PERLUCIDUS, RADIATUS, TRANSLUCIDUS, and UNDULATUS.

stratopause   The boundary that separates the STRATOSPHERE from the MESOSPHERE. In summer, it is at a height of about 34 miles (55 km) over the equator and Poles and about 31 miles (50 km) in middle latitudes. In winter, it is at about 30 miles (48 km) over the equator and 37 miles (60 km) over the Poles. The atmospheric pressure at the stratopause is about 1 mb—one-thousandth of the sea-level pressure. Like the TROPOPAUSE, the stratopause is a region where temperature ceases to change with height.

stratosphere   The layer of the atmosphere that lies above the TROPOPAUSE and extends to an altitude of about 31 miles (50 km). In the lower stratosphere, to a height of about 12 miles (20 km), the temperature remains constant with height. Above that, the temperature increases by up to about 2.2° F per 1,000 ft (4° C per kilometer) and sometimes exceeds 32° F (0° C) at the upper boundary, the STRATOPAUSE. The warming is due to the absorption of solar ULTRAVIOLET RADIATION by oxygen and OZONE. The OZONE LAYER is located in the stratosphere. This temperature profile is generally true only in summer, however, and is not the same at all latitudes. In winter, the tropopause is at its maximum height above the equator and the average temperature is –112° F (–80° C). This is also the temperature in the middle stratosphere at high latitudes, but there is a warm region between 50° and 60° N and S, where the temperature averages about –58° F (–50° C). During the winter, the temperature in the stratosphere sometimes rises by as much as 70° F (39° C) over a few days, then falls again. There are strong winds just above the tropopause, but the air is calm at higher levels.

stratospheric aerosol and gas experiment (SAGE) A set of instruments that are carried on the EARTH RADIATION BUDGET SATELLITE and that are used to measure the material injected into the stratosphere by volcanoes. The second set of instruments (SAGE-II) was launched in October 1984. SAGE-III was launched in 1999 on the EARTH OBSERVING SYSTEM satellite. As well as aerosols and volcanic gases such as sulfur dioxide, these versions also measure OZONE, nitrogen dioxide, and water vapor.

stratospheric coupling   Any interaction between disturbances on either side of the tropopause.

stratospheric wind   Wind that blows in the STRATOSPHERE. The little that is known about this air movement has been obtained from satellite observations of the way particles are distributed after a major volcanic eruption. There were two significant eruptions in 1991. MOUNT PINATUBO, located in the Philippines at 15.15° N, erupted in June, and Cerro Hudson, in Chile at 45.92° S, erupted in August. Both volcanoes injected ash and sulfur dioxide into the stratosphere. The cloud from Mount Pinatubo traveled westward, spreading to the north and south as it did so. The cloud from Cerro Hudson traveled eastward and did not widen. This shows that low-latitude stratospheric winds in the Northern Hemisphere blow predominantly from west to east and those in the middle latitudes of the Southern Hemisphere blow from east to west. Very little material from Cerro Hudson reached Antarctica, illustrating the extent to which stratospheric air over the southern polar region is isolated during the late winter (August and September). This isolation from air in lower latitudes prevents air containing OZONE from entering the polar stratosphere and replenishing the depleted OZONE LAYER.

stratus (St)   A genus of low clouds (*see* CLOUD CLASSIFICATION) that form a uniform, gray layer. When stratus forms at the surface it is FOG, and as a cloud it resembles fog that is above the ground. It sometimes delivers DRIZZLE, SNOW GRAINS, or ICE PRISMS, but if it is thin enough for the Sun or Moon to be discernible through it their outlines can be seen clearly. Stratus differs from ALTOSTRATUS only in the height of its base.

It often forms in valleys and as HILL FOG, but also on WARM FRONTS. When CIRROSTRATUS and altostratus are followed by the appearance of stratus, precipitation is very likely. In fine weather, however, stratus often forms overnight, especially over water, and "burns off" as the daytime temperature rises and the cloud droplets evaporate.

Stratus can occur as the species FRACTUS and NEBULOSUS and in the varieties OPACUS, TRANSLUCIDUS, and UNDULATUS.

streak cloud An elongated fragment of fibrous cloud, commonly CIRRUS, that indicates the direction and strength of the WIND SHEAR.

streak lightning A LIGHTNING stroke in which the main channel has branches. It may flash between a cloud and the ground or between a cloud and adjacent air.

streamline The track that is followed by moving air. A wind streamline is shown on a map or diagram as a straight or curved line that is parallel to the wind direction at every point along its path. It may be drawn using information obtained from one or more SYNOPTIC CHARTS showing the wind that was observed and measured at particular times. Streamlines can also be used to show the way air (or water) flows over or around an obstruction. *Streamline* is also used as a verb to describe the inclusion of a design feature in which a body, such as a car, airplane, or boat, is so shaped as to offer the least possible resistance to the flow of air or water around it. Such a shape is said to be streamlined because streamlines show a smooth, laminar flow of fluid around the object. The modification of a design to achieve this objective is called *streamlining*.

Strombolian eruption *See* VOLCANO.

Streamline. **The path that is traced by moving air.**

strong breeze In the BEAUFORT WIND SCALE, force 6, which is a wind that blows at 25–31 mph (40–50 km h^{-1}). In the original scale, devised for use at sea, a force 6 wind was defined as "or that to which a well-conditioned man-of-war could just carry single-reefed topsails and top-gallant sails in chase, full and by." On land, it is difficult to use an open umbrella in a strong breeze.

strong gale In the BEAUFORT WIND SCALE, force 9, which is a wind that blows at 47–54 mph (76–87 km h^{-1}). In the original scale, devised for use at sea, a force 9 wind was defined as "or that to which a well-conditioned man-of-war could just carry close-reefed topsails and courses in chase, full and by." On land, in a strong gale slates and tiles are torn from roofs and chimneys blown down.

Strutt, John William *See* RAYLEIGH, LORD.

stuffy *See* CLOSE.

Stüve chart (adiabatic chart, pseudoadiabatic chart) A THERMODYNAMIC DIAGRAM in which temperature is plotted along the horizontal axis and pressure along the vertical axis, with the highest pressure at the bottom. Pressure is calculated as the atmospheric pressure raised to the power of 0.286. The chart was devised by G. Stüve and has been widely used, although meteorologists now find the TEPHIGRAM more useful.

Sub-Arctic Current *See* ALEUTIAN CURRENT.

Sub-Atlantic period The last of the five ages into which the HOLOCENE EPOCH is divided on the basis of POLLEN ANALYSIS. It began approximately 2,800 years ago, at about the time Iron Age cultures replaced those of the Bronze Age, and it is the period in which we are living today. Summers became cooler and climates became generally wetter than those of the preceding SUB-BOREAL PERIOD. Within the Sub-Atlantic period there have been significant climatic fluctuations. A MEDIEVAL WARM PERIOD was followed by the LITTLE ICE AGE, which ended gradually during the 19th century. During the 20th century climates grew markedly warmer until 1940, then cooler again until the 1970s, after which they remained fairly constant until the 1980s; many climatologists believe they have increased since that time although the evidence for a sustained warming is uncertain (*see* GLOBAL WARMING).

Sub-Boreal period   The fourth of the five ages into which the HOLOCENE EPOCH is divided on the basis of POLLEN ANALYSIS. It lasted from about 5,000 years ago until 2,800 years ago and was a time of colder, drier, more CONTINENTAL CLIMATE than the preceding ATLANTIC PERIOD, although summers were warmer than those of today.

subcloud layer   Some or all of the air that lies below the CLOUD BASE. The term is sometimes used to describe all the air between the cloud base and the surface. At other times it describes the shallow layer of stable air that occurs beneath the base of clouds produced by CONVECTION.

subgeostrophic   An adjective that describes a wind that blows with less force than the GEOSTROPHIC WIND. The GRADIENT WIND around a CYCLONE is subgeostrophic.

subgradient wind   A wind that blows with less force than the GRADIENT WIND that would be expected from the PRESSURE GRADIENT and latitude (which determines the magnitude of the CORIOLIS EFFECT).

subhumid climate   *See* GRASSLANDS CLIMATE.

sublimation   The change of ice into water vapor, without passing through a liquid phase. Sometimes the reverse process, in which water vapor changes directly into ice, is also called sublimation, but it is more correctly known as *deposition*. The two processes can be seen happening in winter when patches of snow and ice disappear from the ground or frost appears while the temperature remains well below freezing. They also occur in a freezer, when ice cubes left there for too long dwindle in size and eventually disappear, and when the sides of the freezer become frosted.

sublimation nucleus   A solid particle present in the air onto which water vapor solidifies by deposition (also called *sublimation*) to form an ice crystal. Sublimation nuclei are believed to play an important part in the formation of ice in clouds containing SUPERCOOLED water droplets. Nuclei are necessary for the formation of ice crystals in the air. Sublimation nuclei trigger the formation of ice crystals by direct deposition from air adjacent to supercooled droplets. The ice crystals then act as FREEZING NUCLEI, and the proportion of ice particles in that part of the cloud increases at the expense of supercooled water.

subpolar glacier   *See* COMPOSITE GLACIER.

subpolar low   A belt of low atmospheric pressure that lies between latitudes 60° and 70° in both hemispheres. This is where the polar easterly and midlatitude westerly winds converge. The strong contrast in temperature between the TROPICAL AIR arriving from one side and POLAR AIR from the other gives rise to frequent DEPRESSIONS and storms that are carried in a westerly direction by the prevailing winds on the low-latitude side of the POLAR FRONT. The ALEUTIAN LOW and ICELANDIC LOW are the most prominent parts of the subpolar low.

subpolar region   The part of the world that lies between the low-latitude margin of land occupied by tundra and the high-latitude margin of lands with cool temperate or desert vegetation. Winters are long and cold, summers short, and the climate is fairly dry throughout the year. Coniferous forest is the vegetation most typical of subpolar regions.

subsidence   A general sinking of air over a large surface area. Subsidence produces high surface AIR PRESSURE and DIVERGENCE. As air subsides, more air is drawn into the column at a high level, producing high-level CONVERGENCE. Subsiding air is compressed by the increasing weight of air above it and warms in an ADIABATIC process. Subsidence occurs on a global scale in the HADLEY, FERREL, and POLAR CELLS that are part of the GENERAL CIRCULATION of the atmosphere (*see* THREE-CELL MODEL).

subsidence inversion   *See* ANTICYCLONIC GLOOM.

substratosphere   A name that is sometimes given to the uppermost part of the TROPOSPHERE, lying immediately below the TROPOPAUSE. Air in this layer is cold, dry, and similar to the air above it in the lower STRATOSPHERE.

subsun A bright spot of light that is seen on the top of a layer of cloud by an observer who is flying in an aircraft or standing high on a mountain. The light is caused by the reflection of sunlight from the horizontal upper surfaces of ICE CRYSTALS at the top of the cloud.

subsynoptic An adjective that is used to describe conditions that are seen to cover an area that is large, but not so large as in the picture presented in a SYNOPTIC view. Satellite images of cloud patterns over the ocean are subsynoptic in extent. They are able to show features the size of a TROPICAL CYCLONE in an area that is approximately 1,000 miles (1,600 km) square, but a SYNOPTIC CHART would cover a much bigger area.

Subtropical Convergence Zone *See* SUBTROPICAL FRONT.

subtropical cyclone A CYCLONE that occurs in the subtropics. It develops when the southern tip of a POLAR TROUGH in the upper atmosphere becomes cut off from the main part of the trough. The isolated pocket of low pressure, with cold air at its center, may then extend downward to the surface, where it forms a very symmetrical cyclone. The center of the cyclone is up to 100 miles (160 km) in diameter and the highest rainfall occurs about 300 miles (480 km) from the center. Some subtropical cyclones, such as the KONA CYCLONES of Hawaii, are an important source of rain. Subtropical cyclones are very persistent and are often absorbed into new polar troughs, rather than dissipating. Subtropical cyclones can also develop from a TROPICAL CYCLONE that moves across land. This process prolongs the rainfall from the decaying tropical cyclone. This transformation can also occur in the other direction, as a subtropical cyclone turns into a TROPICAL STORM. This happens if warm, moist air is drawn strongly toward the center of the subtropical cyclone. Rainfall intensifies around the center and the temperature at the center rises until the cyclone has the warm core typical of a tropical cyclone.

subtropical desert One of the deserts that lie approximately between latitudes 20° and 40° in both hemispheres. They occur in all the continents. Examples include the Sonoran Desert of the United States and Mexico, the Atacama Desert of Chile, the Sahara Desert, the Namib Desert of southwestern Africa, and the deserts of the Near East, Middle East, and Australia.

They are caused by the SUBSIDENCE of air on the descending sides of the HADLEY CELLS. Air in the Hadley cells rises by CONVECTION and loses its moisture as it ascends and cools in an ADIABATIC process. It descends again as very dry air that heats adiabatically during its descent. It reaches the surface as hot, dry air and its subsidence produces high surface pressure and an outward flow of air. The outward flow prevents moister air from entering the high-pressure region. Consequently, the climate is extremely arid.

The deserts tend to occupy the western sides of the continents, although the Sahara extends across the whole of Africa. This is because the flow of air is ANTICYCLONIC and the centers of the ANTICYCLONES are over the eastern sides of the oceans. This flow moves cool, dense air toward the equator on the western sides of the continents, intensifying the subsidence and high pressure there. In some places the offshore winds that are part of the general outward flow of air produce UPWELLING. This produces cool surface water and inhibits convection that might otherwise produce clouds and PRECIPITATION, some of which might benefit the coastal belt. Areas that experience a WEST COAST DESERT CLIMATE are very dry, but often foggy.

(You can read more about these and other deserts in Michael Allaby, *Ecosystem: Deserts* [New York: Facts On File, 2001].)

subtropical easterlies index A value that is calculated for the strength of the easterly winds in the subtropics between latitudes 20° N and 35° N. The index is based on the difference in the average sea-level atmospheric pressure between these latitudes. It is expressed in meters per second as the east–west component of the GEOSTROPHIC WIND.

subtropical front The boundary that separates the cold water of the Southern (Antarctic) Ocean around Antarctica from the warmer, subtropical waters farther north. The front extends from the coast of Antarctica to about 40° S. Within the front the temperature of the water rises to approximately 39° F (4° C) and the salinity increases by about 0.5 parts per thousand (‰), although in summer it can be as low as 33 ‰. The boundary was originally known as the *Subtropical*

Convergence Zone and was renamed the *Subtropical Front* in the 1980s.

subtropical high A semipermanent ANTICYCLONE that lies in the subtropics. There are several subtropical highs. They are located over the ocean, are most developed in summer, and strongly influence the climates to the east of them by BLOCKING or diverting DEPRESSIONS traveling from west to east in middle latitudes. The AZORES HIGH is a subtropical high that affects the climates of western Europe. The intensity of the subtropical highs varies from time to time (*see* NORTH ATLANTIC OSCILLATION).

subtropical high-pressure belt One of the two regions of generally high atmospheric pressure that surround the Earth in the subtropics, centered at about latitude 30° N and 30° S. The high pressure is caused by the SUBSIDENCE of air on the descending side of the HADLEY CELLS.

subtropical jet stream The JET STREAM that is closest to the equator in both hemispheres. It is at about 30° N and S latitude throughout the year and is produced by a TEMPERATURE GRADIENT that occurs only in the upper TROPOSPHERE (*see* THERMAL WIND). The subtropical jet stream is more persistent than the POLAR JET STREAM. Consequently, references to the jet stream usually relate to the subtropical jet stream and it is this jet stream that is most often shown on maps. Its strength varies between summer and winter with the changing temperature gradient. In the Northern Hemisphere its speed in winter is greatest over the Pacific Ocean and it is greatest over North America in summer.

subtropics The two belts surrounding the Earth in both hemispheres that lie between the TROPICS and the TEMPERATE BELT. The subtropics are not sharply defined, but they are bounded by the Tropics on the side nearer the equator and by approximately latitude 35°–40° on the side nearer the Pole. This boundary represents an average. The actual boundary is farther from the equator on the western sides of continents and closer to the equator on the eastern sides. This difference is due to the ANTICYCLONIC circulation in the

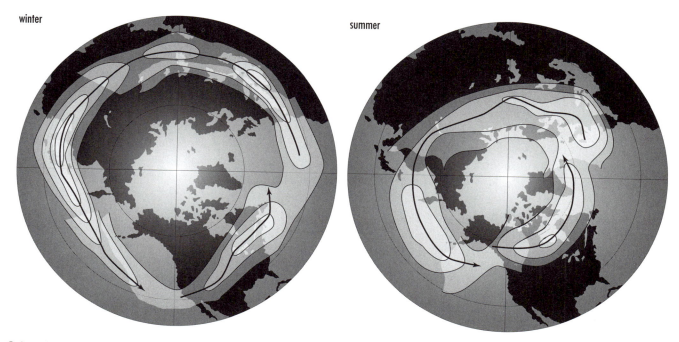

winter   summer

Subtropical jet stream. **The maps show the location of the Northern Hemisphere jet stream in winter and summer. The shading indicates the wind speed, with the darkest shading signifying the greatest speed.**

SUBTROPICAL HIGH-PRESSURE BELT, which carries cooler air toward the equator on the western sides of the continents and warmer air toward the equator on the eastern sides.

suction scar An approximately circular mark that is left on open ground by a SUCTION VORTEX. As it passes, the intense updraft of the suction vortex draws in loose dirt that is scattered from the top of the vortex. The vortex is so short-lived that hardly has it made its "scar" before it dies down. The resulting mark is a shallow depression a few feet across. At one time, some people believed suction scars were the footprints of giants.

suction vortex A small column of spinning air that rotates about its own axis and also moves in a circle around the main vortex of a TORNADO. A major tornado may generate two or more suction vortices. They are often hidden in the dust cloud that surrounds the base of the tornado, but they are responsible for many of the freakish effects a tornado sometimes produces. A suction vortex has been known to destroy half of a house but leave the other half unscathed and vanish before it reached the house next door. Suction vortices may spin in either direction, but where one rotates in the same clockwise direction as the main tornado, its own ANGULAR VELOCITY is added to the angular velocity of the main tornado and its MESOCYCLONE and to the forward speed with which the tornado is moving over the ground. This makes the winds in some suction vortices up to 50 percent faster than the wind in the tornado itself. A tornado with winds of 200 mph (322 km h⁻¹) may be surrounded with suction vortices spinning at 300 mph (483 km h⁻¹). Few suction vortices are more than 100 ft (30 m) in diameter and some are no more than 10 ft (3 m) across. They are very short-lived. Few last longer than about three minutes and few survive long enough to complete even one full orbit of the main tornado.

(You can read more about tornadoes and their suction vortices in Michael Allaby, *Dangerous Weather: Tornadoes* [New York: Facts On File, 1997].)

sudestada A strong southeasterly wind that blows in winter along the Atlantic coasts of Argentina, Uruguay, and southern Brazil. Sudestadas create heavy rain and FOG as well as stormy seas. They are caused by PRIMA-RY CYCLONES, many with long TROUGHS extending from them, that cross South America from west to east below about latitude 40° S, moving at up to 30 mph (50 km h⁻¹) in summer and 40 mph (64 km h⁻¹) in winter. The sudestadas are the Southern Hemisphere counterparts of the Northern Hemisphere NOR'EASTERS.

sugar snow *See* DEPTH HOAR.

suhaili A strong southwesterly wind that blows across the Persian Gulf in the wake of a DEPRESSION. It often follows a KAUS wind. The suhaili generates thick cloud, rain, and sometimes FOG.

sukhovei A hot, dry, easterly wind that blows across southern Russia and the European steppe. It occurs most often in summer and it can damage crops and produce DROUGHT.

sulfate *See* SULFUR DIOXIDE.

sulfur cycle The pathways by which the element sulfur moves among the rocks, air, and living organisms. Fairly small amounts of sulfur enter the air from volcanic eruptions (*see* VOLCANO), but the principal sources are the WEATHERING of rocks and the atmosphere, which contribute approximately equal amounts. Sulfur is an essential ingredient of proteins and protein–carbohydrate complexes and so all living organisms require it. Whether it is released through weathering, dissolves into surface waters from the air, or is incorporated in living tissue, sulfur eventually reaches rivers that carry it to the sea. Some sulfur is deposited as a variety of compounds in the airless muds of estuaries and bogs where BACTERIA obtain energy by reducing it and releasing gases as a by-product of the chemical reactions involved. Hydrogen sulfide (H_2S) is the gas released in the largest amounts by sulfur-reducing bacteria. In the oceans several species of phytoplankton (very small plants) release DIMETHYL SULFIDE, some of which enters the air, where it is oxidized eventually to sulfate (SO_4) AEROSOL. Sulfate aerosol particles act as CLOUD CONDENSATION NUCLEI, returning sulfur to the surface.

The burning of fuels now adds significantly to the natural cycle. Depending on its quality, coal contains an average of 1–5 percent sulfur and oil contains 2–3 percent. When these fuels are burned the sulfur enters the air as SULFUR DIOXIDE unless it is removed from the waste gases. Averaged over the whole world, the amount of sulfur entering the air as a consequence of burning of fuel is approximately equal to the amount entering the air in the course of the natural cycle. The difference is that the burning of fuel is concentrated in the industrialized regions of the world.

sulfur dioxide The product of the oxidation of sulfur, sulfur dioxide (SO_2) is released into the atmosphere by the natural oxidation of reduced compounds such as hydrogen sulfide (H_2S) and DIMETHYL SULFIDE ((CH_3)$_2$S) that are emitted by living organisms. These natural processes maintain a background atmospheric concentration of about 0.0001 parts per million of SO_2. SO_2 is also released through the burning of fuels (see SULFUR CYCLE) and smelting of metal ores that contain sulfur, and it is emitted by pulp and paper mills. When the atmospheric concentration of SO_2 rises significantly above the background level the gas becomes extremely irritating and causes damage to respiratory and other plant and animal tissues.

Once it is airborne, SO_2 continues to oxidize, often to sulfur trioxide (SO_3). In the presence of catalysts this reacts with water droplets to form sulfuric acid ($SO_3 + H_2O \rightarrow H_2SO_4$). This reaction increases the acidity of CLOUD DROPLETS, FOG, and MIST and contributes to the problems associated with ACID RAIN.

SO_2 also reacts with other substances to form sulfate (SO_4) particles. These reflect sunlight, and they can be carried long distances. Sulfate particles are less than 1 μm (0.00004 in) across and act as CLOUD CONDENSATION NUCLEI. They therefore increase cloud formation, but the resulting clouds are composed of CLOUD DROPLETS that are so small they fall very slowly, so PRECIPITATION can be reduced.

Sullivan winter storm scale A 5-point scale for classifying winter storms that was devised in 1998 by Joe Sullivan, a meteorologist employed by the NATIONAL WEATHER SERVICE in Cheyenne, Wyoming. The scheme ranks storms as (1) minor inconvenience, (2) inconvenience, (3) significant inconvenience, (4) potentially life-threatening, and (5) life-threatening.

sultry An adjective that informally describes the uncomfortable conditions that result when a high air temperature coincides with high RELATIVE HUMIDITY and still air.

sumatra A strong SQUALL that crosses the Malacca Strait between Malaysia and Sumatera (Sumatra) during the southwest MONSOON. The squalls are generated along a FRONT that is aligned from northwest to southeast and is sometimes more than 100 miles (160 km) long, and they advance from the southwest. Sumatras originate as sudden surges of wind that are intensified by a katabatic effect (see KATABATIC WIND) as they descend from the mountains. They often last for several hours and produce winds that occasionally exceed 30 mph (48 km h^{-1}) with gusts of up to 50 mph (80 km h^{-1}). Winds of this force are classed as 7 or 8, moderate gale or fresh gale, on the BEAUFORT WIND SCALE. Sumatras sometimes reach the Malaysian coast but penetrate only a short distance inland.

summation principle A method that is used in United States practice to report the SKY COVER at any specified level. The principle is that the sky cover is equal to the total sky cover at each cloud layer from the lowest to the level specified. This means that at no level can the cloud cover be less than the cover at lower levels and the total cloud cover cannot exceed 10/10 (or 8 OKTAS).

sun cross A rare optical phenomenon that consists of a vertical band of white light in the sky with a horizontal band across it. It is seen when a SUN PILLAR and SUN DOG both appear at the same time and part of the circle of the sun dog intersects the sun pillar.

sun dog (mock sun, parhelion) A bright patch that is sometimes seen to one side of the Sun, and usually slightly below it. It is caused by the refraction of light by ice crystals and occurs when ice crystals are falling very slowly. Sun dogs are usually associated with a 22° HALO, in which case they occur as a pair, one to each side of the halo, but they can occur singly. Moon dogs, or mock moons, form in the same way but are less common.

Sun drawing water See CREPUSCULAR RAYS and STRATOCUMULUS.

Sun photometer An instrument that measures the intensity of direct sunlight. It is held in the hand and pointed directly at the Sun. The VOLZ PHOTOMETER is widely used to monitor AIR POLLUTION.

(You can learn more about Sun photometers at www.concord.org/haze/ref.html.)

sun pillar An optical phenomenon in which a bright shaft of light extends upward from the Sun. A pillar may also extend below the Sun, but this condition is much rarer. It is caused by the reflection of light from the undersides of the types of ICE CRYSTALS known as *plates* and *capped columns*, when these are falling very slowly.

sunshine recorder An instrument that measures the intensity and duration of INSOLATION. There are several designs. The CAMPBELL–STOKES SUNSHINE RECORDER is one of the oldest and is still widely used. PYRANOMETERS measure solar radiation at all WAVELENGTHS and over a complete hemisphere. A pair of pyranometers, one facing upward and the other downward, are used to measure surface ALBEDO. PYRHELIOMETERS measure direct solar radiation perpendicular to a surface. DIFFUSOGRAPHS, which are pyranometers modified by being surrounded by a shade ring, measure diffuse sunlight. Possibly the simplest device is the ACTINOMETER, which consists of two thermometers, one with a blackened bulb. This thermometer absorbs more heat than does the thermometer with a clear-glass bulb and the difference between the two temperatures they record can be used to calculate insolation.

sunspot A dark patch on the visible surface (photosphere) of the Sun, up to 31,070 miles (50,000 km) across, where the temperature is about 2,700° F (1,500° C) cooler than the surrounding area. Sunspots are caused by intense magnetic fields and strongly affect the strength of the SOLAR WIND: the greater the number of sunspots, the more intense the solar wind. The number of sunspots increases and decreases over an approximately 11-year cycle that coincides with climatic changes on Earth; the first such correlation to be recognized is now known as the *MAUNDER MINIMUM*. The solar wind affects the intensity of COSMIC RADIATION that reaches the Earth's atmosphere, which in turn influences cloud formation. Sunspot activity is also linked to the intensity of ULTRAVIOLET RADIATION. This, through its absorption by OZONE, affects the temperature in the upper atmosphere, which in turn affects the JET STREAM and the temperature in the lower atmosphere. Some scientists believe GLOBAL WARMING can be wholly attributed to variations in solar output indicated by changes in the number of sunspots.

(*See* Michael Allaby, *Deserts* [New York: Facts On File, 2001]. Also see Fred Pearce. "Sunnyside Up." *New Scientist*, July 11, 1998, pp. 44–48.)

sun-synchronous orbit A satellite orbit in which the satellite remains in the same position relative to the Sun. The satellite orbits at a height of about 560 miles (900 km), which is equal to one-seventh of an Earth radius (of 3,959 miles [6,371 km]). It passes close to the Poles, but at an angle to the meridians (lines of longitude), and takes about 100 minutes to complete one orbit. The satellite crosses the equator about 15 times each day and passes over every part of the surface of the Earth at the same time each day.

superadiabatic *See* ENVIRONMENTAL LAPSE RATE.

supercell The type of CONVECTION CELL that sometimes develops in a very massive CUMULONIMBUS storm cloud. A supercell storm is capable of producing TORNADOES.

Three conditions are needed for the growth of a supercell storm. There must be a warm, moist AIR MASS in which the air is conditionally unstable (*see* CONDITIONAL INSTABILITY). There must also be an advancing COLD FRONT to lift the warm air. Finally, there must be a strong WIND SHEAR at a high level to carry away the rising air.

Under these conditions, warm, moist air is first lifted by the cold air moving beneath it and then drawn upward by CONVECTION. It cools in an ADIABATIC process and its WATER VAPOR condenses, releasing LATENT HEAT that warms the air and increases its INSTABILITY. CONVERGENCE at a low level is accompanied by DIVERGENCE at high level, which is due to the wind shear. This accelerates the rising air. When the storm is fully developed the upcurrents may travel at 100 mph (160 km h^{-1}).

PRECIPITATION falls as HAIL, SNOW, or RAIN from the upper part of the cloud. As it falls, the precipitation chills the air around it. This produces cold downcurrents. In most storm clouds the precipitation and its cold air fall into the upcurrents. They chill the rising air and within an hour or so they suppress the upcurrents altogether. Then the cloud dies.

A supercell is different, because its upcurrents rise at an angle to the vertical. Instead of falling directly into the rising air, the precipitation falls to the side of it, and so the upcurrents are not suppressed. This allows the cell to continue growing. The biggest supercell clouds break through the TROPOPAUSE and reach a height of 60,000 feet (18.3 km). A supercell is also different for another reason. In ordinary storm clouds there are usually several convection cells. They share the energy of the storm, the size of each individual cell is limited, and the cold downcurrents of one cell suppress the warm upcurrents of the adjacent cell. In a supercell storm there is only one cell. It is huge and occupies the whole of the interior of the cloud. The resulting storm releases very heavy rainfall and hail consisting of large HAILSTONES.

Supercell storms last much longer than ordinary storms do, but they seldom survive more than about three hours. This does not mean that the storm dies after a maximum of three hours, however, because the downdraft emerging at the base of the cloud can lift more conditionally unstable air and trigger the formation of a new cloud. This is what happens in a SQUALL LINE, as each storm produces a new storm as it dies.

supercooling The chilling of water to below its freezing temperature without triggering the formation of ice. Ordinarily, pure water freezes when its temperature falls to 32° F (0° C). If the water contains dissolved impurities, such as salt, it freezes at a slightly lower temperature. In the air, however, a cloud droplet that falls slowly through very cold air can be chilled to well below freezing temperature without freezing. It is then said to be *supercooled,* and clouds often contain supercooled water droplets at temperatures as low as –20° F (–29° C). At temperatures of about –5° F to –13° F (–15° C to –25° C) water droplets freeze, but only if FREEZING NUCLEI are present. In the absence of freezing nuclei, water droplets have been cooled under laboratory conditions to temperatures as low as –40° F (–40° C). Below this temperature ice starts to form by HOMOGENEOUS NUCLEATION.

supergeostrophic An adjective that describes a wind that blows with greater force than the GEOSTROPHIC WIND. The GRADIENT WIND around an ANTICYCLONE is supergeostrophic, although the effect is small.

supergradient wind A wind that is stronger than the GRADIENT WIND predicted by the balance of the PRESSURE GRADIENT FORCE and the CORIOLIS EFFECT.

superior image *See* MIRAGE.

supernumerary bow A RAINBOW that appears inside the primary bow with the colors in the same order as those in the primary (the secondary and occasionally tertiary bows appear on the outside of the primary with the colors reversed). It is seen when the water droplets are very small. Supernumerary bows are often superimposed on the primary bow. They may broaden particular bands or cause them to disappear. If the droplets are smaller than about 0.004 in (0.1 mm) in diameter the superposition of colors may produce a rainbow with pale colors at the edges but an almost white central band. These droplets are similar in size to those of fine drizzle and the bow they produce is similar to a FOGBOW. It is sometimes called *Ulloa's ring.*

super outbreak A chain of 148 tornadoes that occurred on April 3 and 4, 1974. The tornadoes were produced by three separate SQUALL LINES that developed simultaneously and moved eastward. Together they extended from the southern shore of Lake Michigan to Alabama and at one point from Canada to the Gulf of Mexico. More than 300 people were killed.

supersaturation The condition of air in which the RELATIVE HUMIDITY (RH) exceeds 100 percent. The RH of air inside clouds is usually between 100 percent and 101 percent, so the supersaturation is said to be between 0 and 1 percent (RH – 100). Supersaturation occurs because water vapor condenses only with difficulty in the absence of CLOUD CONDENSATION NUCLEI. A droplet of clean water with a diameter of 0.2 μm can exist with a supersaturation of 1 percent (RH 101 percent), but one with a diameter one-hundredth of that size, of 0.002 μm, requires a supersaturation of 220 percent (RH 320 percent). This is a result of the THOMSON EFFECT.

supertyphoon A TROPICAL CYCLONE in the Pacific Ocean that covers an area very much larger than the area covered by most HURRICANES or TYPHOONS. The area of a supertyphoon can be 3 million square miles (8 million km²). For comparison, the area of the United States is about 3.7 million square miles (9.5 million km²). Fortunately, supertyphoons are rare.

surazo A cold, dry southerly or southwesterly wind that blows across the mountain ranges and high plateau of Peru, often with great force. It produces clear skies and temperatures below freezing.

surface analysis The study of a SYNOPTIC CHART of surface conditions by which meteorologists abstract the information they need to identify AIR MASSES, FRONTAL SYSTEMS, and other features and to plot their locations.

surface boundary layer *See* FRICTION LAYER, PLANETARY BOUNDARY LAYER.

surface inversion (**ground inversion**) A temperature INVERSION that begins at ground level, so that in the lowest layer of air the temperature increases with height.

surface melting *See* FREEZING.

surface pressure *See* STATION PRESSURE.

surface pressure chart A SYNOPTIC CHART that shows the distribution of STATION PRESSURE. This is the atmospheric pressure that is measured at the surface rather than the REDUCED PRESSURE, and it is indicated by ISOBARS.

surface shearing stress The force exerted on a surface by air that passes across it. Because the force acts on a surface area, surface shearing stress is expressed as a pressure, measured in pounds per square inch or, in SYSTÈME INTERNATIONAL D'UNITÉS (SI) units, in PASCALS. The pressure exerted as surface shearing stress is equal to the opposing force of DRAG that is exerted by the surface on the moving air. This slows the air, but its immediate effect is felt only in the layer of air in immediate contact with the surface.

surface temperature The temperature of the air or sea measured close to the surface of land or water (*see* SEA-SURFACE TEMPERATURE). *See* STEVENSON SCREEN.

surface tension A property of a liquid that makes it seem that a thin, flexible skin covers its surface. The phenomenon is caused by the mutual attraction of molecules in the liquid. Below the surface, each molecule is attracted equally in all directions. The forces acting on it are balanced and it can move in any direction. Molecules at the surface are attracted by molecules to each side and beneath them, but there are no molecules beyond the surface to balance this attraction. Consequently, surface molecules are held firmly by molecular attraction from the sides and below. This attraction draws the surface molecules into a spherical shape, in which the forces are most evenly distributed. It is surface tension that draws small volumes of liquid into drops and the surface of liquids contained in tubes or vessels into a convex shape called a *meniscus*.

A raindrop, falling through the air, is surrounded on all sides by air and the water molecules are subjected only to the attraction of other molecules within the drop. A raindrop is therefore able to assume a spherical shape that is then distorted into a teardrop shape by the force of gravity. Bubbles, in which air is contained by water, are spherical because of surface tension, but ordinarily the surface tension of water is so high that

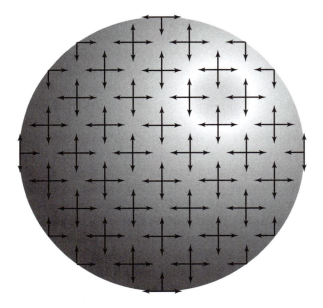

Surface tension. **The arrows indicate the forces acting on the molecules of the liquid. Inside the mass of liquid the forces act equally in every direction. At the surface there are no molecules to exert an attraction from outside, so the molecules are held by forces acting to the sides and toward the center.**

they survive for only a very short time before the water molecules move toward each other to form a drop, excluding the air. Bubbles can be made to last longer by adding detergent to the water. Detergent reduces the surface tension that causes the bubble to collapse.

A drop of liquid that is lying on a solid surface also experiences attraction to molecules in the solid material. This produces *adhesion tension*. If the liquid is held in a container, adhesion tension draws it up the sides, so the surface of the liquid assumes a concave shape. CAPILLARITY is caused by adhesion tension. If the liquid is in contact with another liquid, the surfaces of both liquids generate forces of attraction. This is called *interfacial tension*.

Surface tension (symbolized by γ) is measured in thousandths of a NEWTON per square meter (mN m^{-2}). It decreases with increasing temperature. For water in contact with air, γ = 74.2 mN m^{-2} at 50° F (10° C), 72.0 mN m^{-2} at 77° F (25° C), and 62.6 mN m^{-2} at 176° F (80° C).

Water has a much higher surface tension than many other liquids, but that of mercury is higher (γ = 472 mN m^{-2} at 68° F [20° C] and 456 mN m^{-2} at 212° F [100° C]). The alcohol used in some thermometers (*see* MINIMUM THERMOMETER) has γ = 23.6 mN m^{-2} at 50° F (10° C) and 21.9 mN m^{-2} at 86° F (30° C).

surface visibility The horizontal VISIBILITY measured by an observer on the ground, rather than in an airfield control tower.

surface weather observation An observation or measurement of weather conditions that is made at ground level or on the surface of the sea. Surface observations have been made and recorded for many years and so they provide the longest continuous series of data. Surface data from the past are difficult to interpret, however. At one time the methods of observing, measuring, and recording were not standardized, so it is unwise to draw any conclusions from comparisons of data from different stations.

It is not safe even to compare data from the same station at different times unless the full history of the station is available. Most weather stations have been moved at one time or another, and a change in location of only a few yards may alter the winds, temperatures, pressures, and humidity to which the instruments are exposed. A station may remain in the same place, but

its surroundings may change. Urban developments may envelop it, but the change may be subtler than that. The closure of a factory upwind or the installation of equipment to remove pollutants from its chimney emissions may alter the temperature of the air or intensity of INSOLATION at the weather station. Even the removal of trees can affect wind patterns.

It is impossible to correct for all these influences and dangerous to assume that they tend to cancel one another because for any factor an increase in one place is compensated for by a decrease somewhere else. Surface weather observations made today under standard conditions are essential for the compilation of SYNOPTIC CHARTS and the preparation of weather forecasts. They can also be used, with caution, to detect climatic changes over recent years, but they provide an unreliable base for estimates of climatic change since earlier times.

surface wetness gauge *See* DEW GAUGE.

surface wind The speed and direction of the wind that blows at the surface of land or sea. In order to minimize the deflection of the wind by trees, buildings, and similar obstacles, the instrument used to measure the surface wind is mounted on a pole or tower at least 33 ft (10 m) tall.

surge line A line just ahead of a local group of THUNDERSTORMS where there is a sudden change in the wind speed and direction.

surge of the trades *See* TRADE WINDS.

Surtseyan eruption *See* VOLCANO.

swamp model A MODEL of the GENERAL CIRCULATION in which the oceans are treated as though they were permanently wet land—in fact, a swamp. The wet surface in a swamp model has no HEAT CAPACITY or active mechanisms for transferring heat. Swamp models were constructed because of the difficulties and uncertainties involved in describing the thermal behavior and movement of the real ocean mathematically. Modern models are better able to do this and so swamp models are no longer used.

SWEAT Index *See* STABILITY INDICES.

sweeping A mechanism by which RAINDROPS are believed to grow. It is a variant of the COALESCENCE and COLLISION THEORY that is based on the fact that the TERMINAL VELOCITY of a raindrop is proportional to its size. Large drops therefore fall faster than small drops. They collide with some slower, smaller drops, which merge with them (collision and coalescence). Other small droplets are swept into the wake of the bigger drops and absorbed by them in that way. This is sweeping.

swinging-plate anemometer *See* ANEMOMETER.

Sybil A TROPICAL STORM that struck the Philippines on October 1, 1995. It triggered floods, landslides, and volcanic mudflows, causing damage in 29 provinces and 27 cities. More than 100 people were killed.

synoptic An adjective that indicates that something is based on a general view of conditions over a large area at a particular time. The word is derived from the Greek word *sunoptikos,* from *sun,* meaning "with," and *optikos,* meaning "seen." In synoptic meteorology, data gathered from many places are assembled to provide a picture of atmospheric conditions over a large area.

synoptic chart A map that shows a general picture of the weather conditions over a large area at a particular time (*see* SYNOPTIC). Synoptic charts are produced at regular intervals, usually of six or 12 hours, and are based on reports from WEATHER STATIONS. These are plotted as STATION MODELS, and ISOBARS are drawn to link stations where the AIR PRESSURE is the same. Small differences are smoothed out and the isobars then indicate the distribution of pressure and wind. The isobars show the REDUCED PRESSURE (sea-level pressure) or the contours at several CONSTANT-PRESSURE SURFACES. The chart shows the CLOUD AMOUNT, surface AIR TEMPERATURE, pressure, BAROMETRIC TENDENCY, wind direction, and wind strength. It also shows the surface position of COLD FRONTS, WARM FRONTS, and OCCLUSIONS.

synoptic climatology The branch of CLIMATOLOGY that deals with the influence of atmospheric circulation patterns on regional CLIMATES. It is concerned with regional climates up to the scale of a hemisphere, but not with global climates. Studies of ENSO events are classed as synoptic climatology.

synoptic code One of the recognized codes that are used to transmit SYNOPTIC weather data and observations. It is used widely throughout the world and is the code officially approved by the WORLD METEOROLOGICAL ORGANIZATION. It is one of the codes used by the U.S. NATIONAL WEATHER SERVICE (NWS). The NWS also uses the METAR code, and the SYNOP code is also used extensively. All of these codes translate observations and data into groups of numbers. This procedure compresses the data and improves the reliability of transmission and reception.

(You can learn more about the synoptic and METAR codes at www.met.fsu.edu/Classes/Common/ sfc.html, about the SYNOP code at www.zetnet. co.uk/sigs/weather/Met_Codes/codes.html, and about other weather codes at www.usatoday.com/weather/ wpcodes.htm.)

synoptic meteorology The study of simultaneous weather conditions over a large area (that is, synoptically).

synoptic report Any weather report that is based on SYNOPTIC WEATHER OBSERVATIONS that is encoded by an authorized code and transmitted to a weather center.

synoptic scale (cyclonic scale) The scale of weather phenomena that can be shown on a SYNOPTIC CHART. These events extend horizontally for about 600 miles (1,000 km) and vertically for about 6 miles (10 km), and they last for about one day. These are the approximate dimensions of a CYCLONE. In the classification of meteorological scales that scientists use, events of these dimensions are said to occur on a *meso-α scale,* both spatially and temporally.

synoptic weather observation A set of observations and measurements of surface weather conditions that is made at a WEATHER STATION at one of the times specified by the WORLD METEOROLOGICAL ORGANIZATION using standard instruments, calibrations, and methods. The observations should include REDUCED PRESSURE, TEMPERATURE, DEW POINT TEM-

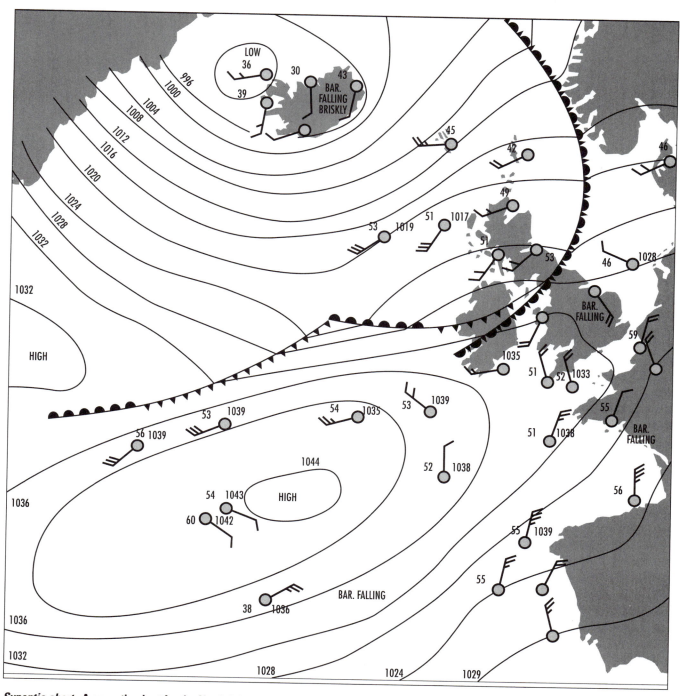

Synoptic chart. **A synoptic chart for the North Atlantic on a day in April.**

PERATURE, PRECIPITATION, SKY COVER, VISIBILITY, WIND SPEED, and WIND DIRECTION (*see* STATION MODEL), as well as any other details that may be relevant.

synthetic aperture radar (SAR) A type of RADAR in which the instrument moves on board an aircraft or satellite. The SAR transmits a continuous-wave signal at a precisely controlled frequency, so the signal is coherent, and it stores the reflected signals in a memory. This allows it to process a large number of echoes at one time. The effect is similar to that of an instrument with a large antenna, although only a small antenna is used. SAR is used to map the surface of Earth or other planets in great detail.

Système International d'Unités (SI units) The units of measurement that are used throughout the

SI units and conversions

Unit	Quantity	Symbol	Conversion
Base units			
meter	length	m	1 m = 3.2808 feet
kilogram	mass	kg	1 kg = 2.205 pounds
second	time	s	
ampere	electric current	A	
kelvin	thermodynamic temperature	K	1 K = 1° C = 1.8° F
candela	luminous intensity	cd	
mole	amount of substance	mol	
Supplementary units			
radian	plane angle	rad	$\pi/2$ rad = 90°
steradian	solid angle	sr	
Derived units			
coulomb	quantity of electricity	C	
cubic meter	volume	m^3	1 m^3 = 1.308 $yards^3$
farad	capacitance	F	
henry	inductance	H	
hertz	frequency		
joule	energy	J	1 J = 0.2389 calorie
kilogram per cubic meter	density	$kg\ m^{-3}$	1 kg m^3 = 0.0624 lb ft^{-3}
lumen	luminous flux	lm	
lux	illuminance	lx	
meter per second	speed	$m\ s^{-1}$	1 m s^{-1} = 3.281 ft s^{-1}
meter per second squared	acceleration	$m\ s^{-2}$	
mole per cubic meter	concentration	$mol\ m^3$	
newton	force	N	1 N = 7.218 lb force
ohm	electric resistance	Ω	
pascal	pressure	Pa	1 Pa = 0.145 lb in^{-2}
radian per second	angular velocity	$rad\ s^{-1}$	
radian per second squared	angular acceleration	$rad\ s^{-2}$	
square meter	area	m^2	1 m^2 = 1.196 $yards^2$
tesla	magnetic flux density	T	
volt	electromotive force	V	
watt	power	W	1 W = 3.412 Btu h^{-1}
weber	magnetic flux	Wb	

Prefixes attached to SI units alter their value.

Prefixes used with SI units

Prefix	Symbol	Value	Prefix	Symbol	Value
atto	a	$\times 10^{-18}$	deci	d	$\times 10^{-1}$
femto	f	$\times 10^{-15}$	deca	da	$\times 10$
pico	p	$\times 10^{-12}$	hecto	h	$\times 10^{2}$
nano	n	$\times 10^{-9}$	kilo	k	$\times 10^{3}$
micro	µ	$\times 10^{-6}$	mega	M	$\times 10^{6}$
milli	m	$\times 10^{-3}$	giga	G	$\times 10^{9}$
centi	c	$\times 10^{-2}$	tera	T	$\times 10^{12}$

world for scientific purposes. Their values are defined by the General Conference of Weights and Measures, comprising delegates from more than 40 countries, which meets every few years. Its central office and laboratory, the International Bureau of Weights and Measures, is at Sèvres, near Paris, France, and is managed by the International Committee of Weights and Measures. This has 18 members, each from a different country, and it operates under the auspices of the General Conference. It can make recommendations to the General Conference for changes to the system, but they can be adopted only with the authority of the General Conference.

The system comprises seven base units, two supplementary units, and a large number of derived units.

syzygy The approximate alignment of the Earth, Sun, and Moon, or of the Earth, Sun, and another planet. The Earth–Sun–Moon syzygy causes spring TIDES.

T

T *See* TESLA.

Tad A TYPHOON, rated category 1 on the SAFFIR/SIMPSON SCALE, that struck central and northern Japan on August 23, 1981, killing 40 people and leaving 20,000 homeless. It had winds of up to 80 mph (128 km h⁻¹).

TAFB *See* TROPICAL PREDICTION CENTER.

taiga climate The climate that is typical of northern Russia, northern Scandinavia, northern Canada, and the interior of Alaska. These are the regions that support coniferous forest mixed with some broad-leaved species, especially birch (*Betula* species), alder (*Alnus* species), and willow (*Salix* species). This type of forest is known as *boreal forest* in North America and as *taiga* in Russia. Winters are long and cold and the hours of daylight are short. Summers are short and warm, with long hours of daylight. There is a large seasonal range in temperature. At Fairbanks, where the climate is typical of inland Alaska, the average July daytime temperature is 72° F (22° C) and the average January daytime temperature is –2° F (–19° C), a difference of 74° F (41° C). Verkhoyansk, Russia, has an even wider range, of 120° F (67° C), from a daytime average of 66° F (19° C) in July to –54° F (–48° C) in January. The climate is fairly dry and in winter the low temperatures cause the ground to freeze, so water is not available for plants. PERMAFROST is found over much of the region, although it occurs in patches, discontinuously.

tailwind (following wind) A wind that blows in the same direction as that in which a body is traveling. If the body is traveling along the surface of the land or sea, the assistance provided by a tailwind reduces the amount of fuel that must be consumed to maintain a given speed and accelerates wind-powered transport such as sailing ships and yachts. Aircraft measure their speed as the speed at which they move through the air. This is known as their *airspeed*; their speed in relation to the surface is known as their *groundspeed*. A tailwind does not alter the airspeed, because it is the air surrounding the aircraft that moves, carrying the aircraft with it. The speed at which the aircraft moves through the air is unaltered, but its groundspeed is increased to a value equal to the sum of the airspeed and the speed of the tailwind, or of the tailwind component of a wind blowing at an angle to the direction of flight. The opposite of a tailwind is a headwind.

taino A local name for a TROPICAL CYCLONE that occurs in some parts of the Greater Antilles.

talik *See* PERMAFROST.

TAO *See* TROPICAL ATMOSPHERE OCEAN.

tarantata A strong northwesterly breeze that blows in the Mediterranean region.

Tay Bridge disaster   A devastating catastrophe that occurred in the 19th century on the rail bridge that crosses the Firth of Tay, Scotland, linking Fife to the city of Dundee on the northern side of the river. The present bridge is the second (and a road bridge also spans the Firth). The first Tay Bridge was opened on June 20, 1877. It was rather more than 1 mile (1.6 km) long and the engineers who designed it believed it was strong enough to withstand any weather. On the evening of Sunday, December 28, 1879, a train departed as usual from Edinburgh with six carriages carrying passengers bound for Dundee. There were between 75 and 90 persons on the train as it began to cross the Tay Bridge. The weather was stormy, with gale-force winds blowing at up to 75 mph (120 km h^{-1}). At about 7:l5 P.M., when the train was about halfway across, the bridge collapsed and the train fell into the river below. There were no survivors. The subsequent investigation concluded that the bridge had not been properly built and maintained and that its design failed to allow adequately for wind loading. Some scientists now believe it was destroyed when two TORNADOES struck it simultaneously.

(You can read more about the Tay Bridge disaster at www.tts1.demon.co.uk/tay.html and about tornadoes in Britain in Michael Allaby, *Dangerous Weather: Tornadoes* [New York: Facts On File, 1997].)

Technical Support Branch   *See* TROPICAL PREDICTION CENTER.

teeth of the wind   An old nautical term that means the direction from which the wind is blowing; "in the teeth of" means "in face of opposition." Sailing into the teeth of the wind meant sailing directly to WINDWARD.

tehuantepecer   A cold, northerly wind that blows almost constantly in winter over the isthmus of Tehuantepec, Mexico. It carries heavy rain to the northern coast, but is dry by the time it reaches the Pacific. The tehuantepecer is a southerly extension of the NORTHER and NORTE.

Teisserenc de Bort, Léon Philippe (1855–1913) French *Meteorologist*   The scientist who discovered the STRATOSPHERE was born in Paris on November 5, 1855, the son of an engineer. His career began in 1880 when he went to work in the meteorological department of the Central Bureau of Meteorology, in Paris. He undertook three expeditions to North Africa, in 1883, 1885, and 1887, to study the geological features and geomagnetism. During the same period he was growing increasingly interested in the distribution of atmospheric pressure at an altitude of about 13,000 feet (4,000 m).

In 1892 he was made chief meteorologist at the bureau, but he resigned in 1896 in order to establish a private meteorological observatory. This was located at Trappes, near Versailles, and Teisserenc de Bort used it primarily to study clouds and the upper air. He was one of the pioneers in the use of balloons to take soundings of the upper atmosphere; he described their use and the results he had obtained with them in a paper he published in 1898 in the journal *Comptes Rendus*. As well as working from his own observatory, Teisserenc de Bort conducted investigations in Sweden, over the Zuider Zee and Mediterranean, and over the tropical Atlantic.

His balloon soundings revealed that the air temperature decreased with height as expected, but only to an altitude of about 7 miles (11 km). Above this height the temperature remained constant as far as the greatest altitude his balloons could reach. In 1900 he proposed that the atmosphere comprises two layers. He called the lower layer, in which temperature decreases with height and air is constantly moving, the *troposphere* (the Greek word *tropos* means "turning"). He said the upper boundary of the troposphere is marked by a layer he called the *tropopause*.

Above the tropopause there is an isothermal layer. Teisserenc de Bort believed that because the temperature in this layer remains constant, there is no mechanism such as CONVECTION to make the air move vertically. Consequently, he thought the air might separate into its constituent gases, which would then form layers, with the heaviest gases at the bottom and the lightest at the top. Immediately above the tropopause there would be a layer of oxygen, above that a layer of nitrogen, then helium, and hydrogen at the top. This stratified arrangement led him to name the isothermal layer the *stratosphere*. The stratosphere is not, in fact, layered in the way Teisserenc de Bort supposed, but his name for it has survived.

Teisserenc de Bort died at Cannes on January 2, 1913.

teleconnections Linked atmospheric changes that occur in widely separated parts of the world. The SOUTHERN OSCILLATION INDEX is a typical example. When the sea-level atmospheric pressure rises above normal at Darwin, Australia, it falls by an approximately similar amount at Tahiti, in the central South Pacific, thousands of miles away. EL NIÑO produces dry weather to northeastern Brazil and in the western Mediterranean, the Sahel region of Africa, northeastern China, and Australia, but wet weather in much of the United States, Israel, and northwestern Europe.

telegraphy The transmission of information by means of electric pulses that travel along a wire cable from a sender to a receiver. The word is derived from two Greek words, *tele,* which means "far," and *graphein,* which means "to write."

Until the invention of telegraphy, information could be communicated over a long distance no faster than a horse could gallop. Weather forecasting was impossible, because assembling detailed information about conditions over a sufficiently large area took so long that by the time the data had been analyzed the weather system described had disappeared.

Scientists knew by the middle of the 18th century that an electric current would travel a considerable distance along a metal wire if the wire were connected to the earth to complete the circuit. The first practical idea for a telegraph was suggested in 1753, in an article in the *Scots Magazine* by someone identified only as "C.M." C.M. proposed a separate insulated wire for each letter of the alphabet. At the receiving end, each wire was to be attached to a ball that hung above a piece of paper with a letter written on it. When the current reached the ball, the paper would jump up to it, and so the message could be spelled out letter by letter. Alternatively, each wire could end at a bell that would be struck by a ball when a current traveled along the wire. Various other inventors suggested similar systems and they were tried out.

These early methods were slow and cumbersome. A breakthrough occurred in 1819, when the Danish physicist Hans Christian Oersted (1777–1851) discovered that an electric current produces a magnetic field. Other scientists, including André Marie Ampère (1775–1836) and PIERRE LAPLACE, developed this idea, and in 1825 the English inventor William Sturgeon (1783–1850) enclosed a needle within a coil of wire.

The needle became a magnet when a current passed through the wire and its polarity changed when the current changed direction. Sturgeon called his device an *electromagnet.*

In 1831 JOSEPH HENRY made a signaling apparatus that used an electromagnet. It consisted of a magnetized steel bar that was pivoted in a horizontal position and could be attracted by an electromagnet. When it was, the end of the bar struck a bell. Messages could be conveyed by a sequence of sounds. In 1835 Henry invented the relay, a series of similar circuits in which each circuit activated the next. It overcame the diminution in the signal passing through a length of wire that is caused by resistance in the wire itself. SAMUEL MORSE devised a code suitable for telegraphic use and the MORSE CODE is still used today.

The first telegraph line in the world linked Baltimore and Washington, D.C., and was opened in 1844. The first message, sent in Morse code, was "What hath God wrought?" Telegraph lines were soon being installed in many countries and between them. The first submarine cable was laid in 1850 to link England and France. An attempt to lay a transatlantic cable was made in 1857, but the cable broke and the project had to be abandoned. The first successful transatlantic cable was laid in 1866.

Joseph Henry had been elected secretary of the Smithsonian Institution in 1846, and he used his position to establish a network of weather observers throughout the United States. The network became operational in 1849, and the observers used the telegraph to send data to the Smithsonian. This system formed the basis on which the U.S. WEATHER BUREAU was formed in 1891. The first national network of meteorological stations linked to a central point by telegraph opened in France in 1863.

(You can read more about the history of weather reporting and forecasting in Michael Allaby, *Dangerous Weather: A Chronology of Weather* [New York: Facts On File, 1998].)

Television and Infrared Observation Satellite (TIROS) The first weather satellite, which was launched by the United States on April 1, 1960. By 1965, nine more TIROS satellites had been launched, several of them into POLAR ORBIT rather than equatorial orbits. *TIROS-8* carried the first AUTOMATIC PICTURE TRANSMISSION equipment. The first ADVANCED

VERY HIGH RESOLUTION RADIOMETER was carried on *TIROS-N,* launched in October 1978. TIROS satellites are now known as *NOAA-class satellites.* These satellites travel in polar orbits, scanning the entire surface of the Earth over a 24-hour period. Their instruments are sensitive to visible light and infrared radiation and they scan to the sides of the orbital path, covering an area 1,864 miles (3,000 km) wide and 1.2 miles (2 km) high. At one time, for example, they would be able to observe the entire area from southern Florida to Hudson Bay and from the Atlantic to the Great Lakes. They transmit a constant stream of data by automatic picture transmission or HIGH RESOLUTION PICTURE TRANSMISSION. As well as monitoring the weather, the satellites carry search and rescue transponders that are used to help locate ships and aircraft in distress.

(There is more information about U.S. weather satellites at http://ecco.bsee.swin.edu.au/chronos/metsat /weather.html.)

temperate belt   The region that lies approximately between latitudes 25° and 50° in both hemispheres. The lower latitude is close to the TROPICS and the higher latitude is about at the 50° F (10° C) ISOTHERM of mean SEA-SURFACE TEMPERATURE. The belt lies in the middle latitudes and its climates are often described as *middle latitude* (or *midlatitude*). The climates of the temperate belt correspond to category C in the KÖPPEN CLIMATE CLASSIFICATION, with temperatures in the coldest month between 26.6° F (−3° C) and 64.4° F (18° C) and in the warmest month higher than 50° F (10° C).

temperate glacier (warm glacier)   A GLACIER in which the temperature of the ice is above the PRESSURE MELTING point except during winter. The glacier flows because pressure melting produces a thin layer of liquid water at the base of the ice, allowing the overlying ice to slide.

temperate rain forest   *See* RAIN FOREST.

temperate rainy climate   A middle latitude climate that corresponds to category Cf in the KÖPPEN CLIMATE CLASSIFICATION. Temperatures in the coldest month are between 26.6° F (−3° C) and 64.4° F (18° C), and in the warmest month they are higher than 50° F (10° C). There is at least 2.4 inches (60 mm) of rain in the driest month. *See* TROPICAL WET–DRY CLIMATE, HUMID SUB-

TROPICAL CLIMATE, MARINE WEST COAST CLIMATE, and MEDITERRANEAN CLIMATE.

Temperate Zone   *See* MATHEMATICAL CLIMATE.

temperate zone   The regions of the Earth that lie between the TROPICS (latitude 23.5° N and S) and the ARCTIC CIRCLE and ANTARCTIC CIRCLE (66.5° N and S).

temperature   A measure of the relative warmth of an object or substance that allows it to be compared to another object or substance (one is warmer or cooler) or to a standard (so many degrees). Temperature and heat are not the same: heat is a form of energy, temperature the effect it produces.

All objects and substances, including the air and our own bodies, are made from atoms and molecules. Atoms and molecules move. If they are in the form of a gas they move freely and rapidly. In a liquid they move more slowly and have less freedom. In a solid they are unable to move around, but they vibrate. How fast they move or vibrate depends on the amount of energy they have. Their KINETIC ENERGY (energy of motion) can increase or decrease. One form of energy can be converted into another (*see* THERMODYNAMICS, LAWS OF). When an atom or molecule absorbs heat (one form of energy) the heat is converted to kinetic energy (another form of energy).

The kinetic energy is that of the atoms or molecules, and it is measured as motion in relation to the center of mass of the object or substance. When moving or vibrating atoms or molecules strike another object a proportion of their kinetic energy is transferred. An appropriate sensor can detect this transferred energy. Nerve endings in our skin are sensors that detect the impact of fast-moving or vibrating atoms and molecules. The message that the nerves send to the brain is interpreted as temperature. We feel that something is hot or cold, either in relation to our own skin temperature or in an absolute sense if the skin is exposed to a temperature so high it will burn or so cold it will freeze the tissues.

A THERMOMETER is an instrument that absorbs kinetic energy from impacting atoms and molecules and converts it to a reading against a scale. Three temperature scales are used. Scientists usually prefer the KELVIN SCALE, in which the temperature is written in the unit K (for kelvin), without a degree sign. The CEL-

SIUS TEMPERATURE SCALE is the most widely used everyday scale; it is sometimes still called the *centigrade scale*, because there are 100 of its degrees between the freezing and boiling points of water. Celsius temperatures are written as ° C. Its name was officially changed from *centigrade* in 1948, at the Ninth General Conference on Weights and Measures. The FAHRENHEIT TEMPERATURE SCALE is more often used in the United States and Britain (where it is being replaced by the Celsius scale). Its temperatures are written as ° F.

A fourth scale, the RÉAMUR TEMPERATURE SCALE, is used in very few places today. It was devised in 1730 by the French physicist and naturalist René Antoine Ferchault de Réamur (1683–1757). Réamur measured the expansion of a mixture of water and alcohol as its temperature increased. The liquid was held in a bulb at the base of a tube, as in any thermometer. When it was at the freezing point he marked the point it reached on the tube as zero. He then graduated the remainder of the tube into units, each of which was equal to one-thousandth of the volume of the liquid in the bulb and tube when it was at freezing. When the liquid reached boiling point he found its length had increased to 1,080 units, so it had risen 80 units (or degrees). Consequently his scale ran from 0° R at freezing to 80° R at boiling point.

temperature belt The area that lies between two lines on a graph that show the daily maximum and minimum temperatures for a particular place. The belt indicates the temperature range for that place.

temperature gradient The rate of temperature change over a horizontal distance.

temperature–humidity index (comfort index, discomfort index, heat index, THI) A numerical value that relates the temperature and humidity of the air to the conditions that make a sedentary person wearing ordinary indoor clothes feel comfortable (*see* COMFORT ZONE). It is calculated by

$$THI = 0.4(T_a + T_w) + 15$$

where T_a and T_w are the DRY-BULB TEMPERATURE and WET-BULB TEMPERATURE, respectively, measured in degrees Fahrenheit. If the temperatures are measured in degrees Celsius the equation is

$$THI = 0.4(T_a + T_w) + 4.8$$

temperature range The difference between the highest and lowest mean temperatures that have been recorded for a particular place. If the annual range is required, only daytime temperatures should be used. The DIURNAL RANGE compares the mean daytime and mean nighttime temperatures for a month, season, or year. The mean range uses only mean temperatures, but the absolute range takes account of the highest and lowest temperatures ever recorded by day or night. Chicago, for example, has a mean annual temperature range of 49° F (27° C), but an absolute temperature range (measured over 75 years) of 128° F (71° C).

TEMPO *See* COOPERATIVE HOLOCENE MAPPING PROJECT.

temporale A strong southwesterly or westerly wind that blows onto the Pacific coast of Central America during the summer. It brings hot, humid air and weather similar to that of the Asian MONSOON. The wind is caused by the deflection of the southeasterly TRADE WINDS in the eastern South Pacific Ocean.

tendency The rate of change of a VECTOR QUANTITY at a specified place and time.

tendency interval The period of time that elapses between the measurements that are used to determine the TENDENCY of a meteorological factor. The interval is usually three hours.

tephigram (TΦgram) A THERMODYNAMIC DIAGRAM on which the temperature and humidity of the air are plotted against pressure. This reveals the entire structure of a column of air. The actual temperature is plotted against one axis and the logarithm of the POTENTIAL TEMPERATURE against the other. Vertical lines then represent ISOTHERMS and horizontal lines are isotherms of potential temperature. These are also dry ADIABATS and the distance between them decreases as the potential temperature increases. Saturated adiabats appear as curved lines. In the lower TROPOSPHERE, they cross the dry adiabats at about 45°, but the angle becomes smaller with increasing altitude. ISOBARS are slightly curved lines running diagonally across the diagram from lower left to upper right and almost bisecting the right angles at which the isotherms intersect.

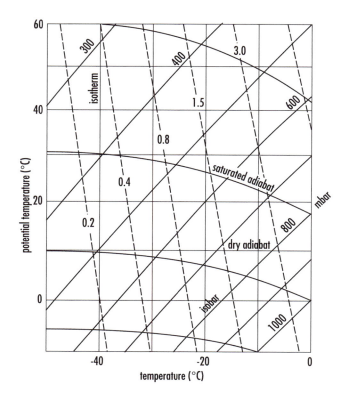

Tephigram. **A graph on which temperature and humidity are plotted against pressure to illustrate the structure of an entire column of air. The actual temperature is plotted against the horizontal axis, the logarithm of the potential temperature against the vertical axis, pressure by the solid diagonals, and saturation specific humidity by the dotted lines.**

ISOTHERMS, showing the saturation SPECIFIC HUMIDITY, are shown as dotted lines that make a small angle with the vertical isotherms. When the tephigram has been constructed, it is rotated clockwise by about 45° so that the isobars lie horizontally. They then correspond to altitude and can be labeled as such, in addition to being labeled in units of pressure (usually millibars). The conventional symbol for potential temperature is the Greek letter phi (Φ), so the tephigram is a *t* (for temperature) *phi* (for potential temperature) *-gram*. The tephigram was devised by SIR WILLIAM NAPIER SHAW.

terdiurnal   Happening every three days.

Teresa   A TYPHOON that struck Luzon, Philippines, on October 23, 1994. It killed 25 people.

terminal velocity (fall speed)   The maximum speed that a falling body can attain. Once the body has accelerated to its terminal velocity it continues its descent at that constant speed. The terminal velocity of a body is therefore proportional to its weight and to the drag exerted on it as a result of the resistance to its passage offered by the medium through which it is falling. That drag is proportional to the surface area of the body, because it is the surface against which resistance acts, and the weight is proportional to its volume.

If the body is very small, the flow of air around it is dominated by the VISCOSITY of the air. For larger bodies, the downward force acting on the body is equal to the weight of the body minus the weight of the air it displaces. In the lower TROPOSPHERE, the terminal velocity (V) of a falling body the size of a small raindrop is given by $V = 8 \times 10^3 r$, where r is the radius of the body. The airflow around large raindrops and HAILSTONES is much more turbulent and their terminal velocity is given by $V = 250r^{1/2}$. *See also* STOKES'S LAW.

terral   A land breeze (*see* LAND AND SEA BREEZES) that blows with great regularity along the western coasts of Chile and Peru. It complements the VIRAZON. Northwesterly land breezes that occur in Brazil and in Spain, sometimes producing squalls, are called *terral levante*.

terral levante   *See* TERRAL.

terrestrial radiation   The BLACKBODY RADIATION that is emitted by the surface of the Earth. This is approximately equal to the radiation emitted by a BLACKBODY at a temperature of 255 K (−0.67° F [−18.15° C]). It consists of INFRARED RADIATION in the WAVE BAND 4–100 µm, with a strong peak at a WAVELENGTH of about 12 µm, which is consistent with WIEN'S LAW.

Tertiary   The subera of the CENOZOIC era of geological time (*see* GEOLOGICAL TIME SCALE) that began about 65 million years ago and ended about 1.64 million years ago. It includes the PALEOGENE and NEOGENE periods.

tertiary circulation   The circulation of air that takes place on a very local scale, such as a THUNDERSTORM, local wind, FÖHN WIND, or TORNADO.

tesla (T) The derived SYSTÈME INTERNATIONAL D'U-NITÉS (SI) UNIT of magnetic flux density, which is the amount of magnetism per unit area of a magnetic field measured at right angles to the magnetic force. The tesla is equal to one WEBER per square METER (1 T = 1 Wb m^{-2}). The unit was adopted in 1954 and was named in honor of the Croatian-born American physicist and electrical engineer Nikola Tesla (1856–1943).

Testing Earth System Models with Paleoclimatic Observations See COOPERATIVE HOLOCENE MAPPING PROJECT.

tetrachloromethane See CARBON TETRACHLORIDE.

thaw A warm SPELL OF WEATHER in winter or early spring during which snow and ice melt.

thawing index The cumulative number of DEGREE-DAYS when the AIR TEMPERATURE is above freezing (32° F [0° C]). Thawing indices are used to estimate the distribution of PERMAFROST and the depth to which the ground has thawed. The indices are available for the entire world, broken into areas measuring 0.5° longitude by 0.5° latitude.

(You can learn more about thawing indices at nsidc. org/NSIDC/CATALOG/ENTRIES/nsi-0063. html.)

thawing season The time that elapses between the lowest point and the succeeding highest point on the time curve of cumulative DEGREE-DAYS when the AIR TEMPERATURE is above and below freezing (32° F [0° C]).

Thelma Two TYPHOONS, the first of which, rated category 4 on the SAFFIR/SIMPSON SCALE, struck Kaohsiung, Taiwan, on July 25, 1977. It generated winds of up to 120 mph (193 km h^{-1}). Nearly 20,000 homes were destroyed and 31 people died. The second Typhoon Thelma struck South Korea on July 15, 1987. It caused floods, landslides, and mudslides in which at least 111 people died.

thematic mapper (TM) A sensing device that is carried on the *LANDSAT 4* and *LANDSAT 5* satellites in SUN-SYNCHRONOUS ORBITS. It detects reflected visible and NEAR-INFRARED RADIATION and obtains information about the surface of the Earth with a resolution of 100 feet (30 m), producing images that are detailed enough to show individual fields. All TM transmissions from *Landsat 4* ended in August 1993 as a result of failure of the equipment and some *Landsat 5* transmissions ended in February 1987, also after failure.

(You can learn more about Landsat and TM at edcwww.cr.usgs.gov/glis/hyper/guide/landsat_tm.)

theodolite A surveying instrument that is used to track PILOT BALLOONS. A theodolite consists of a telescope with crosshairs in the eyepiece that is used for sighting and focusing on the balloon. The telescope is mounted on a tripod fitted with a spirit level to indicate when the instrument is horizontal, and it can be rotated in both the horizontal and vertical planes. The instrument is tilted until the telescope is focused accurately on the balloon. The horizontal and vertical angles between the instrument and the balloon are then read from graduated circles that are seen through a second eyepiece.

Theophrastus See WEATHER LORE.

thermal A CONVECTION current that forms locally, where the ground has been heated strongly by the Sun. As the warm air rises it expands to the sides and drifts downwind. If it reaches its LIFTING CONDENSATION LEVEL, CUMULIFORM cloud develops. The condensation of WATER VAPOR releases LATENT HEAT, which warms the air and generates another thermal that produces CASTELLANUS cloud. Many birds use thermals to gain height effortlessly, and glider pilots seek them for the same purpose. They must seek the thermals, because in addition to swaying this way and that with the wind, they soon dissipate (see EROSION OF THERMALS).

thermal belt A fairly well defined area on many mountainsides in middle latitudes where nighttime temperatures are higher than those at higher and lower elevations. Below the thermal belt cold air subsides in a KATABATIC process at night to produce low temperatures and often a FROST HOLLOW in the valley bottom. Above the thermal belt the ADIABATIC decrease in temperature with height also produces cold air.

thermal capacity See HEAT CAPACITY.

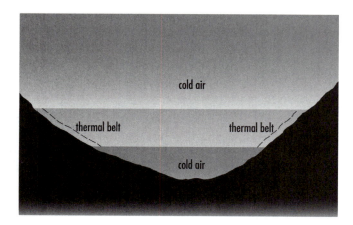

Thermal belt. **The thermal belt is a region on a mountainside where the nighttime temperature is higher than it is at higher or lower elevations.**

thermal climate   A climate that is defined only in terms of temperatures.

thermal conductivity   The rate at which heat passes through a substance. It is measured as the amount of heat that is transmitted in unit time over a unit distance in a direction perpendicular to a surface of unit area under conditions in which the transfer of heat depends only on the temperature gradient. It is measured in joules per second per meter per kelvin ($J\ s^{-1}\ m^{-1}\ K^{-1}$). 1 $J\ s^{-1}\ m^{-1}\ K^{-1}$ = 5.598 Btu $s^{-1}\ ft^{-1}\ °F^{-1}$. The thermal conductivity of dry sand is 0.3 $J\ s^{-1}\ m^{-1}\ K^{-1}$ (1.68 Btu $s^{-1}\ ft^{-1}\ °F^{-1}$), of wet sand 2.2 $J\ s^{-1}\ m^{-1}\ K^{-1}$ (12.32 Btu $s^{-1}\ ft^{-1}\ °F^{-1}$), and of water 0.561 $J\ s^{-1}\ m^{-1}\ K^{-1}$ (3.14 Btu $s^{-1}\ ft^{-1}\ °F^{-1}$).

thermal diffusivity   The rate at which heat penetrates a material. This depends on the THERMAL CONDUCTIVITY of the material, its density, and its specific HEAT CAPACITY. Thermal diffusivity (κ) is given by

	k
quartz	0.0000044
dry sand	0.0000023
wet sand	0.0000074
clay	0.0000015
still water	0.00000014
still air	0.00002

$$\kappa = k/\rho C$$

where k is the thermal conductivity, ρ is the density, and C is the specific heat capacity. Thermal diffusivity is measured in square meters per second.

thermal-efficiency index   A value for the amount of energy, as heat, that is available for plant growth in the course of a year. The concept was devised by CHARLES W. THORNTHWAITE and forms one of the bases of the THORNTHWAITE CLIMATE CLASSIFICATION. It is calculated from measurements of the amount by which the mean temperature in each month is above or below freezing. For each month the thermal efficiency is $(t - 32)/4$ and the thermal-efficiency index is the sum of the thermal efficiencies for each month through the year. A value of 0 indicates what is called a *frost climate* and a value of more than 127 indicates a tropical climate.

thermal equator   The belt around the Earth where the temperature is highest. Its location changes with the SEASONS between 23° N and between 10° S and 15° S. The mean location of the thermal equator is about 5° N.

thermal equilibrium   The condition of two or more bodies that are at the same temperature and therefore possess the same amount of KINETIC ENERGY (*see* THERMODYNAMICS, LAWS OF). Unless some outside process intervenes, energy is not exchanged between bodies in thermal equilibrium.

thermal high   An area of high AIR PRESSURE that is produced by the cooling of air that is in contact with a cold surface. The cold air contracts and becomes denser, drawing in more air higher in the air column and increasing the surface pressure.

thermal inertia   A measure of the rate at which a particular material responds to a change in the amount of energy it absorbs. Substances with a high HEAT CAPACITY, such as water, respond very slowly. Those with a lower heat capacity, such as rocks and dry sand, respond more rapidly. The composition of land surfaces can be inferred from measurements made by REMOTE SENSING of the changes in temperature during the day.

thermal-infrared radiation INFRARED RADIATION that has a WAVELENGTH in the middle of the infrared WAVE BAND, of about 3–15 μm.

thermal low (heat low) An area of low AIR PRESSURE that is produced by the warming of air that is in contact with a warm surface. The warm air expands and becomes less dense, and the expansion extends throughout much of the column of air above the warm surface. Thermal lows occur in the Persian Gulf, Spain, northern India, and southwestern United States, which are regions dominated by SUBTROPICAL HIGHS. The lows are caused by intense surface heating. They draw in air that is sometimes moist and produces rain. A thermal low develops over southwestern Arizona every summer. As a result, Phoenix receives an average of 1 in (25 mm) of rain in both July and August, when the average daytime temperatures are 104° F (40° C) and 101° F (38° C),respectively, compared with 0.1 in (2.5 mm) in both May and June, when the average temperatures are 91° F (33° C) and 101° F (38° C), respectively.

thermal pollution The release into the environment of air or water that is markedly hotter than the air or water into which it is discharged. Raising the temperature of water can harm aquatic organisms, because they depend for RESPIRATION on oxygen that is dissolved in water, and the amount of dissolved oxygen water can hold is inversely proportional to the water temperature. Discharging hot air can produce a HEAT ISLAND effect that radically alters the local climate.

thermal resistivity A measure of how poorly a material conducts heat. This is the opposite of THERMAL CONDUCTIVITY.

thermal soaring Circling in the rising air of a THERMAL in order to gain height with the minimum expenditure of energy. Many birds, including some gulls, eagles, hawks, swallows, and and swifts, practice thermal soaring, and glider pilots imitate them.

thermal steering The movement of an atmospheric disturbance in the direction of the nearest THERMAL WIND. The THICKNESS of the layer used to calculate the thermal wind is usually taken to extend from the sur-

face to the middle TROPOSPHERE, and thermal steering is equivalent to movement along THICKNESS LINES.

thermal tide A change in AIR PRESSURE that is produced by the heating of the surface by sunshine. It follows the progress of the Sun and therefore progresses around the Earth in a similar fashion to that of the gravitational TIDES.

thermal wind A wind that is generated when the air temperature changes by a large amount over a short horizontal distance. The JET STREAM and the EASTERLY JET are the most important examples.

Warm air is less dense than cool air. Where warm and cool air lie adjacent to one another, therefore, atmospheric pressure decreases with height more rapidly in the cool air than in the warm air, because the cool air is more compressed. Consequently, an ISOBARIC SURFACE (a surface across which the atmospheric pressure is the same everywhere) slopes upward from the cool, dense air to the warm, less dense air, and the THICKNESS of each atmospheric layer increases along a gradient from the dense to the less dense air. This gradient increases with height, because the thickness of a layer depends on the degree to which the air is com-

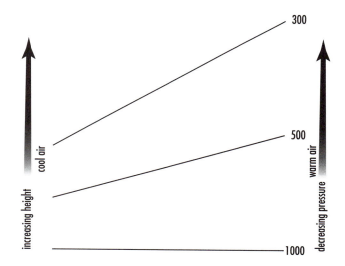

Thermal wind. Pressure decreases with height more rapidly in cool, dense air than it does in warm air, which is less dense, and the thickness of each layer is proportional to the temperature. Where warm and cool air lie adjacent to one another this produces a temperature gradient that increases with height. It is the gradient that generates the thermal wind.

pressed and compression decreases with height more slowly in dense air than in air that is less dense.

The speed of the GEOSTROPHIC WIND is proportional to the PRESSURE GRADIENT or, to put it another way, to the slope of the isobaric surface. This means that if the slope of the isobaric surface changes with height, so must the speed of the geostrophic wind. The thickness of a layer that is bounded by two isobaric surfaces is proportional to the mean temperature in the layer. It therefore follows that the change in the speed of the geostrophic wind across the layer is proportional to the temperature gradient across the layer. This means the layer must be BAROCLINIC. The relationship between the geostrophic WIND SHEAR and baroclinicity is known as the *thermal wind relation.*

Since the temperature gradient increases with height, so does the wind speed. It blows with the cool air to its left in the Northern Hemisphere and to its right in the Southern Hemisphere; that is why the POLAR FRONT and subtropical jet streams blow from west to east in both hemispheres.

thermal wind relation   *See* THERMAL WIND.

thermistor   A THERMOMETER in which the principal component is an electronic device that resists the flow of an electric current by an amount that varies with the temperature. As the temperature rises, the resistance increases and the current flowing through the device decreases. As the temperature falls the resistance decreases and the current increases.

thermocline   Literally, a change of temperature that occurs along a gradient between two places. More specifically, the thermocline is a layer in the ocean where the temperature decreases with depth much more rapidly than it does in the water above or below. The depth of the thermocline varies from place to place and with the seasons, but it may commence as little as 33 feet (10 m) or as much as 660 feet (200 m) below the surface and end at depths between 500 feet (150 m) and 5,000 feet (1,500 m). In Arctic and Antarctic waters, there is usually no thermocline, because the sea surface is covered by ice during the winter and there is only slight warming by solar radiation in summer. The strongest thermocline is in the Tropics.

Water at the ocean surface is warmed by solar radiation. In the Tropics, the sea-surface temperature com-

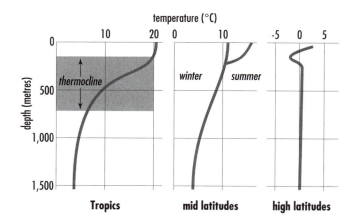

Thermocline. **The ocean layer in which temperature decreases most rapidly with depth. The tropical thermocline is of much greater vertical extent than the high-latitude thermocline. In mid-latitudes the thermocline almost disappears in winter.**

monly exceeds 68° F (20° C) and can reach 80° F (27° C). This is probably a maximum because the rate of evaporation increases with temperature to a point at which the latent heat of vaporization cools the surface layer sufficiently to prevent the temperature from rising any higher.

Radiant heat does not penetrate very deeply, and the ocean loses heat to the atmosphere by long-wave radiation, but winds and currents mix the upper waters and it is this mixing that carries warm water to a greater depth.

At about 13,000 feet (4,000 m), which is the average depth of all the oceans, the water temperature is between 34° F and 36° F (1° C and 2° C). This temperature remains constant throughout the year, regardless of latitude. The deep ocean is the most unchanging environment on Earth.

Mixing produces an upper layer of water in which the temperature decreases only very slightly with depth. Below the mixed layer, water temperature begins to decrease sharply with depth, and by about 3,300 feet (1,000 m) it has fallen to approximately 40° F (4.4° C). From there it decreases much more gradually.

The layer in which temperature decreases rapidly is the thermocline, and it is most strongly marked in the Tropics because there the temperature must fall from about 68° F (20° C) to about 40° F (4.4° C). In midlatitudes, where the water in the mixed layer is cooler, the temperature gradient is shallower, especially in winter, when the surface temperature is about 54° F (12° C). In summer, when the sea is warmer, there is a more sharply defined summer thermocline very close to the surface. In latitudes higher than about 50° N and 50° S there is no thermocline.

Solar radiation is absorbed by the oceans, but it warms only the water above the thermocline. When warmed, the oceans warm the air in contact with them. Cooler water from the thermocline that becomes incorporated into the mixed layer is immediately replaced by cold water from a higher latitude. It is partly through the coupling of oceans and atmosphere that heat is transferred by ADVECTION from low to high latitudes.

A thermocline also develops in many lakes in summer.

thermocouple   *See* THERMOMETER.

thermodynamic diagram   A diagram that summarizes the factors affecting the temperature and pressure

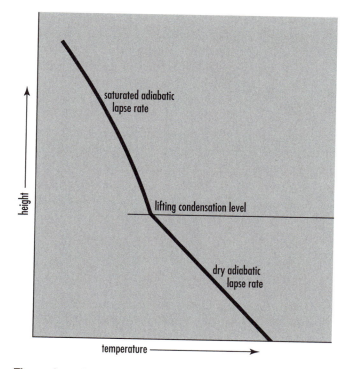

Thermodynamic diagram. **A diagram that illustrates the amount of energy a parcel of air possesses. In this example, temperature is plotted against height to show the lapse rate. The lifting condensation level marks the height at which the dry and saturated adiabatic lapse rates meet.**

of a PARCEL OF AIR. A point on the diagram then refers to the thermodynamic (energy) state of the air in that location. A simple thermodynamic diagram measures altitude along its vertical axis, temperature along its horizontal axis. A line showing the change of air temperature with height corresponds to the LAPSE RATE and indicates the height of the LIFTING CONDENSATION LEVEL, marking the boundary between the dry and saturated adiabatic lapse rates and the height of the cloud base. The most widely used types of thermodynamic diagram are the STÜVE CHART and TEPHIGRAM.

thermodynamics, laws of   Thermodynamics is the scientific study of the ways in which energy changes from one form into another, the way it is transmitted from one place to another, and its availability to do work. It is based on the idea that in any isolated system, anywhere in the universe, there is a measurable amount of internal energy. The internal energy is the sum of the KINETIC ENERGY and POTENTIAL ENERGY of all the atoms and molecules within the system. This total amount of internal energy cannot change without intervention from outside the system, in which case the system ceases to be isolated. This principle can be described by four laws.

The first law of thermodynamics was suggested in the 1840s by the German physicist Julius Robert Mayer (1814–78, *see also* TYNDALL, JOHN) and was verified in 1843 by the English physicist James Prescott Joule (1818–89). Lord Kelvin (1824–1907), the Scottish physicist, and the German physicist Hermann Ludwig Ferdinand von Helmholtz (1821–94) also made important contributions to the development of the law. The law states that energy can be neither created nor destroyed, but it can be changed from one form into another. This is sometimes called the law of *conservation of energy*.

There was a difficulty with this law. Joule had measured the mechanical equivalent of heat, which is the change of energy from one form (heat) to another (kinetic energy). Heat engines, such as steam and internal combustion engines, exploit this transformation, and the French theoretical physicist Nicolas Léonard Sadi Carnot (1796–1832) had shown that the efficiency of such an engine depends only on the difference in temperature between the source of heat and the sink into which the heat is finally discharged. Some heat is lost, however, as energy flows through the engine. This loss was explained in 1850 by the German theoretical physicist Rudolf Clausius (1822–88), who asserted that heat does not pass spontaneously from a colder to a hotter body. For example, if you leave a cup of hot coffee to stand in a cold room, the coffee becomes colder by losing its heat, and not hotter by absorbing heat from its cold surroundings. In 1851 Lord Kelvin arrived at the same conclusion.

This is the second law of thermodynamics. There are several ways it can be expressed. Clausius summarized it in two ways: heat cannot be transferred from one body to a second body at a higher temperature without producing some other effect; the ENTROPY of a closed system increases with time. It was Clausius who coined the word *entropy*. The second law means that most physical processes are irreversible.

Entropy increases and so there must be a point at which it reaches a maximum and can increase no further. This point is described by the third law of thermodynamics, which was discovered in 1905–06 by the German physical chemist Hermann Walther Nernst (1864–1941). The third law states that in a perfectly crystalline solid there is no further increase in entropy when the temperature falls to ABSOLUTE ZERO. This also means that it is impossible to cool any substance to absolute zero (although substances have been cooled to a tiny fraction of a degree above absolute zero).

This should have completed the list of laws of thermodynamics, but there is a further principle that is more fundamental than any of the others. It had been well known for centuries and was taken for granted until the English physicist Sir Ralph Fowler (1889–1944) drew attention to it. It could not be called the fourth law, because it is the principle that underlies the second law (and it can be derived from it). Nor could it be called the first law, because that would mean renumbering the existing laws, and that would cause endless confusion. So Fowler proposed that it be called the *zeroth law* (law number 0). It states that if two isolated systems are each in thermal equilibrium with a third, then they are in thermal equilibrium with each other. This law makes it possible to use a THERMOMETER to measure the temperature of a substance or body. The thermometer is an isolated system that is brought into equilibrium with the system being measured, and the temperature scale marked on the thermometer is derived from a third system that was used to calibrate the instrument.

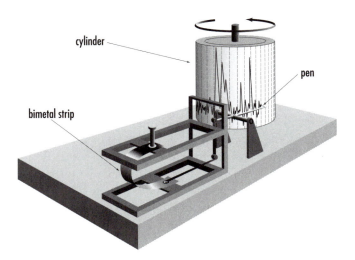

Thermograph. An instrument that provides a continuous record of temperature as a line drawn by a pen on a chart attached to a cylindrical drum.

thermograph An instrument that provides a continuous record of temperature. It consists of a component that changes shape with changes in temperature. This is connected by a system of levers to an arm that terminates in a pen held against a calibrated chart mounted around a cylinder. The cylinder rotates at a constant speed, so the pen traces around the chart a line that rises and falls with temperature changes. The principal component may be a bimetal strip or a Bourdon tube. A bimetal strip consists of two pieces of different metals that expand and contract by widely different amounts when heated and cooled. The two strips are bonded together, so that their differential expansion causes the combined strip to bend. A Bourdon tube is a curved container made from phosphor bronze and filled with alcohol. Like a bimetal strip, it bends in response to changes in temperature.

thermohaline circulation The exchange of surface and deep water that takes place in the oceans, but only in high latitudes, due to differences in temperature *(thermo-)* and salinity *(-haline)*. Over most of the oceans, the surface layer remains warmer than the deep water at all times, because it absorbs solar radiation. This means the surface layer, being warmer and therefore less dense, mixes only slightly with the water below it. The warmer the surface layer, the greater is the difference in density between the surface and deep water. This means that very little mixing can occur between the surface and deep waters in the tropical oceans, but rather more in midlatitude oceans. In Arctic and Antarctic waters, however, the surface layer loses so much heat by radiating it toward the sky that it becomes colder than the water beneath it. As it freezes, salt is expelled from the ice crystals and into the water adjacent to the ice. Water near the ice is therefore colder than the subsurface water and also more saline. Both factors increase its density, so it sinks and deep water rises to take its place, establishing a thermohaline circulation.

thermometer An instrument that is used to measure the TEMPERATURE of a substance. Thermometers used in meteorology measure the temperature of air and water.

There are several ways to measure changes in temperature. The most common method is based on the fact that many substances expand when they are warmed and contract when they are cooled. If this property is to be used to measure small changes in temperature, the substance chosen must expand and contract by the largest amount possible. Air was the first to be tried, in the AIR THERMOSCOPE that was invented by GALILEO GALILEI, It was very inaccurate, however. Liquids were better. Alcohol was the first to be tried and in 1714 DANIEL FAHRENHEIT made a thermometer using mercury. This was the first thermometer to give reliable, accurate readings and Fahrenheit is credited with having invented the thermometer.

Today, both alcohol and mercury are used. Alcohol expands and contracts more than mercury does, but an alcohol thermometer is less accurate than a mercury thermometer. This is because the rate at which substances expand with increasing temperature varies slightly across the temperature range, and the variation, although small, is greater for alcohol than for mercury. Both alcohol and mercury remain liquid at temperatures encountered in the lower atmosphere. A dye is added to alcohol to make it visible. Most alcohol thermometers use a red dye, but some use blue.

Alcohol and mercury thermometers consist of a narrow capillary tube (*see* CAPILLARITY), sealed at one end and blown into a bulb at the other end. The bulb acts as a reservoir for the liquid. As the liquid expands and contracts, the change is exaggerated by the narrowness of the tube in which it moves.

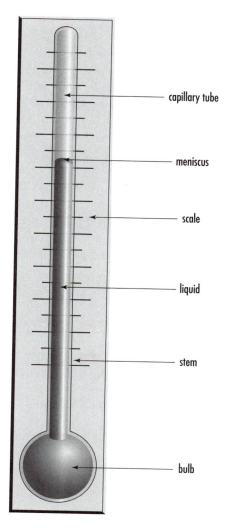

Thermometer. The most widely used thermometer measures the expansion of a liquid held in a reservoir, called the *bulb*, opening into a capillary tube. The expansion and contraction of the liquid are read against a scale.

The thermoscope failed because Galileo did not know that the volume of a body of air changes with the atmospheric pressure, as well as the temperature. Account can now be taken of this factor and modern gas thermometers are the most accurate of all thermometers based on the changing volume of a fluid. The gas used is not air, but either nitrogen, hydrogen, or helium. It is held in a vessel, so its volume remains constant. As the temperature changes, the pressure it exerts also changes according to the GAS LAWS, and this is the change that is registered. Gas thermometers are used in industry.

The amount by which the volume of a substance changes with changing temperature is called the *coefficient of thermal expansion* of that substance. This varies greatly from one substance to another. As the temperature changes, two metals with different coefficients of thermal expansion increase and decrease in length by different amounts. If strips of these metals are securely bound together into a bimetal strip, as the length of one changes more than that of the other, the bimetal strip curls by an amount proportional to the change in temperature. This property provides the basis for a thermometer using a bimetal strip. It is often used to operate thermostats. Its principal meteorological use is in THERMOGRAPHS. A bimetal strip is less accurate than an alcohol or mercury thermometer.

The properties of dissimilar metals can also be used to measure temperature electrically. A thermocouple consists of two wires or rods of different materials, each of which is made into a half loop. The two half loops are welded together at their ends to make a circuit. If one of the joints is at a different temperature from the other, an electric current flows through the circuit. This phenomenon was discovered in 1821 by the Estonian–German physicist Thomas Johann Seebeck (1770–1831) and is known as the *Seebeck effect*. The first thermometer to use it was made in 1887 by the French physical chemist Henri Louis Le Châtelier (1850–1936), using platinum and rhodium as the two metals.

The electrical resistance of a metal increases as the temperature of the metal rises. This property is exploited to make resistance thermometers, which commonly use platinum, nickel, tungsten, copper, or alloy wires. Resistance thermometers are accurate to within 0.2° F (0.1° C).

Ceramic semiconductors have a similar property. They are used to make THERMISTORS.

Once a thermometer has been made, it must be calibrated. This is done by means of two reference points, commonly the freezing and boiling temperatures of pure water under standard sea-level atmospheric pressure. The distance between these two points is then divided into a convenient number of gradations, called *degrees*. Three calibration systems are in common use. The FAHRENHEIT TEMPERATURE SCALE is still popular in the United States and Britain, but it is being replaced by the CELSIUS TEMPERATURE SCALE, which is the one used in the rest of the world. Scientists use the KELVIN SCALE.

thermosphere The layer of the atmosphere that lies above the MESOPAUSE. Its lower boundary is about 56 miles (90 km) above sea level. It is the uppermost layer of the atmosphere and has no precise upper margin. Although the gases composing it are extremely rarefied they do exert drag on spacecraft at heights of more than 155 miles (250 km). The lower thermosphere consists mainly of molecular nitrogen (N_2) and molecular (O_2) and atomic (O) oxygen. Above 125 miles (200 km) there is more atomic oxygen than molecular or atomic (N) nitrogen. Temperature increases with height, mainly as a result of the absorption of ULTRAVIOLET RADIATION by atomic oxygen and probably rises to more than 1,800° F (1,000° C). These temperatures refer only to the estimated speed of particles, however, because the atmosphere is so tenuous that satellites orbiting in it are not warmed by the air, although they are warmed by the solar radiation they absorb. The region of the atmosphere between 50 and 250 miles (80 and 400 km) is sometimes called the *IONOSPHERE*.

thermotidal oscillations *See* ATMOSPHERIC TIDES.

thermotropic model A model of the atmosphere that is used in NUMERICAL FORECASTING. The model aims to forecast the height of one CONSTANT-PRESSURE SURFACE, usually that at 500 mb, and the height of one temperature, usually the mean temperature between 1,000 mb and 500 mb. The THERMAL WIND is assumed to remain constant with height. Given these two parameters, it is possible to construct a forecast surface chart.

THI *See* TEMPERATURE–HUMIDITY INDEX.

thickness The difference in altitude between two ISOBARIC SURFACES. This varies with the temperature of the air, because the warmer air is, the less dense it is, so that a given mass of warm air occupies a greater volume than a similar mass of cold air. Consequently a layer of warm air bounded by two isobaric surfaces is thicker than a layer of cold air bounded by the same surfaces. The resulting gradient is responsible for generating the THERMAL WIND.

thickness chart A SYNOPTIC CHART that shows the changing THICKNESS of a particular atmospheric layer.

thickness line (relative isohypse) A line drawn on a map that joins places where the THICKNESS of a given atmospheric layer is the same.

thickness pattern (relative hypsography) The pattern that is made by the THICKNESS LINES on a THICKNESS CHART.

Thomson effect A water molecule at the surface is more tightly bound in a body of liquid that has a plane (level) surface than it is in a spherical droplet, and the smaller the droplet the easier it is for the molecule to escape. The effect was first described mathematically by the Scottish physicist William Thomson (1824–1907), who later became Lord Kelvin. (It should not be confused with the Thomson effect in thermodynamics, which was also discovered by Lord Kelvin.) The equation describing the Thomson effect is

$$\rho_r/\rho_s = \exp{(A/rT)}$$

where ρ_r is the EQUILIBRIUM VAPOR density, ρ_s is the density at SATURATION of the layer adjacent to it, A is a constant for the liquid, r is the radius of curvature of the droplet, T is the ABSOLUTE TEMPERATURE, and exp indicates that the relationship is EXPONENTIAL.

The forces that bind water molecules are exerted in all directions. Molecules at a plane surface are attracted from the sides and from below, but there are no molecules to attract them from above. A molecule at a curved surface is attracted from below and from the sides, but because of the curvature, molecules to the sides are also a little below it and so the lateral attraction is reduced. This means it is easier for a molecule to escape into the air from a curved surface than from a plane surface. The smaller the radius of curvature, the greater the reduction in the lateral attraction.

The ratio of ρ_r to ρ_s increases as the radius of curvature (r) decreases and the droplet becomes smaller. Consequently, the smaller the droplet, the more water molecules that escape from the equilibrium vapor around it into the air beyond and droplets can survive only if ρ_s increases until it is greater than ρ_r. In other words, the air surrounding the droplet must be supersaturated. Water molecules have a definite size, which limits the minimum size it is possible for a droplet to

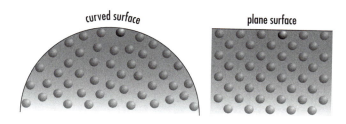

Thomson effect. A molecule at the surface of a spherical droplet is bound to the liquid less strongly than a molecule at a plane surface.

be. When this is taken into account, it is found that HOMOGENEOUS NUCLEATION of liquid droplets can occur only when the RELATIVE HUMIDITY (RH) reaches about 300 percent. Air is never this humid. The fact that water vapor is able to condense so readily demonstrates the importance of CLOUD CONDENSATION NUCLEI to the process of CONDENSATION.

The exponential nature of the relationship between ρ_r and ρ_s is critical. As soon as r begins to increase, by only the smallest amount, condensation accelerates rapidly. Once a droplet has a radius of 0.15 μm (0.000006 in) an RH of 101 percent is sufficient for it to grow.

The Thomson effect can be modified by RAOULT'S LAW, when the droplet is not of pure water, but is a solution. Where a mass (m) of a solute is dissolved in a droplet of water the overall effect is modified by a constant (B) that is determined by the composition of the solute. The equation is then

$$\rho_r/\rho_s = 1 + (A/rT) - (Bm/r^3)$$

Thornthwaite, Charles Warren (1899–1963) American *Climatologist* C. W. Thornthwaite was one of the most eminent climatologists of his generation, with an international reputation. He held many important positions, but his enduring fame rests on the CLIMATE CLASSIFICATION he devised. This remains in widespread use, especially among agricultural scientists, because of its emphasis on THERMAL EFFICIENCY and PRECIPITATION EFFICIENCY, two concepts that Thornthwaite introduced.

Major climate classifications do not appear all at once, in their complete and final form. Their authors revise and amend them over the years, and Thornthwaite was no exception. The first version of his scheme applied only to North America. It appeared in October 1931, in "The Climates of North America According to a New Classification" in *Geographical Review* (21, pp. 633–55). He expanded the classification to cover the world in another article, "The Climates of the Earth," which appeared in *Geographical Review* in July 1933 (23, pp. 433–40). In January 1948 he published a second version of his classification, in which he introduced the concept of POTENTIAL EVAPOTRANSPIRATION. He described the new scheme in a third article, "An Approach toward a Rational Classification of Climate" (*Geographical Review* 38, pp. 55–94).

Charles Warren Thornthwaite was born on March 7, 1899, in Bay County, Michigan. He graduated as a science teacher in 1922 from Central Michigan Normal School and received his doctorate from the University of California, Berkeley, in 1929. Thornthwaite held faculty positions from 1927 to 1934 at the University of Oklahoma, from 1940 to 1946 at the University of Maryland, and from 1946 to 1955 at the Johns Hopkins University. He headed the Division of Climatic and Physiographic Research of the U.S. Soil Conservation Service from 1935 to 1942; from 1946 until his death, he was the director of the Laboratory of Climatology, at Seabrook, New Jersey, and professor of climatology at Drexel Institute of Technology, Philadelphia.

From 1941 to 1944, he was president of the Section of Meteorology of the American Geophysical Union. In 1951, he became president of the Commission on Climatology of the World Meteorological Organization, a post he held until his death. He died of cancer on June 11, 1963.

Thornthwaite climate classification A scheme for classifying climates that was devised by CHARLES WARREN THORNTHWAITE. The first version of the scheme, published in 1931, applied only to North America, but in subsequent years it was expanded to cover the entire world.

Like the KÖPPEN CLIMATE CLASSIFICATION, it is generic, in that it uses quantitative criteria or temperature and precipitation to define the boundaries of climatic types. It differs from the Köppen classification in its use of the concepts of precipitation efficiency and thermal efficiency, but its most important contribution was part of the 1948 revision of the scheme, in which Thornthwaite introduced the concept of POTENTIAL

Moisture provinces

Climate type		Moisture index
A	perhumid	100 or more
B_4	humid	80–100
B_3	humid	60–80
B_2	humid	40–60
B_1	humid	20–40
C_2	moist subhumid	0–20
C_1	dry subhumid	–20–0
D	semiarid	–40– -20
E	arid	–60– -40

Temperature provinces

Climate type		Potential evapotranspiration	
		inches	centimeters
E'	frost	5.61	14.2
D'	tundra	11.22	28.5
C'_1	microthermal	16.83	42.7
C'_2	microthermal	22.44	57.0
B'_1	mesothermal	28.05	71.2
B'_2	mesothermal	33.66	85.5
B'_3	mesothermal	39.27	99.7
B'_4	mesothermal	44.88	114.0
A'	megathermal	44.9	114

EVAPOTRANSPIRATION and a moisture index. Precipitation efficiency is calculated as the sum of the ratios of precipitation (*P*) to evaporation (*E*) for each month through the year. This yields a PRECIPITATION-EFFICIENCY INDEX (P-E index). Thermal efficiency similarly relates temperature (*T*) to evaporation to yield a THERMAL-EFFICIENCY INDEX (T-E index). Because temperature and evaporation are so closely linked, the *T-E* value is shown in the table of climate types as the potential evapotranspiration.

From these calculations, Thornthwaite recognized nine moisture provinces based on the P-E index, and nine temperature provinces based on the T-E index. These are given names and also designated by letters and numerals. These are related to vegetation, and the moisture provinces correspond to rain forest (A), forest (B), grasslands (C), steppe (D), and desert (E).

Moist climates (A, B, C$_2$)	Aridity index
r little or no water deficiency	0–16.7
s moderate water deficiency in summer	16.7–33.3
w moderate water deficiency in winter	16.7–33.3
s$_2$ large water deficiency in summer	more than 33.3
w$_2$ large water deficiency in winter	more than 33.3
Dry climates (C$_1$, D, E)	**Humidity index**
d little or no water surplus	0–10
s moderate water surplus in winter	10–20
w moderate water surplus in summer	10–20
s$_2$ large water surplus in winter	more than 20
w$_2$ large water surplus in summer	more than 20

In addition, the classification adds code letters that qualify these main categories by referring to the amount and distribution of precipitation associated with them. This brings to 32 the total number of climate types recognized in the scheme. These additional qualifications are based on an aridity index for moist climates and a humidity index for dry climates.

three-cell model   A description of the GENERAL CIRCULATION of the atmosphere that is widely used by meteorologists and climatologists. It is known to be an oversimplification, but it nevertheless provides a good approximation of the way the atmosphere behaves.

The general circulation transports warm air from the equator to the Poles and returns cool air to the equator. This is achieved by a system of vertical cells. These are of three types, and hence the name of the model.

The cell responsible for the equatorial circulation is a modified version of the cell described in 1753 by GEORGE HADLEY and it is known as the *HADLEY CELL.* Air is warmed strongly by contact with the surface (mainly ocean surface in equatorial regions). It rises, cools in an ADIABATIC process, loses most of its moisture, and moves away from the equator at the level of the TROPOPAUSE. CONVERGENCE produces a belt of low AIR PRESSURE at the surface.

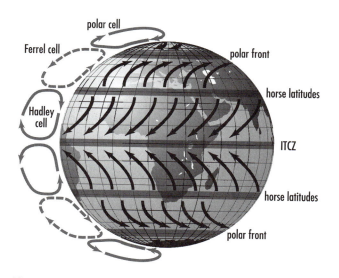

Three-cell model. **The general circulation of the atmosphere can be described by three vertical cells, the Hadley, Ferral, and polar cells. The arrows indicate the direction of surface winds.**

At about latitude 30° in both hemispheres the air descends, warming adiabatically as it does so. By the time it reaches the surface the air is hot and dry. The subsiding air produces a belt of high surface pressure in the subtropics. This reinforces the aridity of the climate by producing an outward flow of air at the surface, which prevents moister air from moving into the region.

Over the Poles, where the land and sea surfaces are extremely cold, the air is chilled at the surface and contracts. Air descends, producing high surface pressure and an outflow of air away from the Poles. This constitutes one side of the POLAR CELLS.

Both the polar and Hadley cells are thermally direct, which is to say they are driven directly by the surface temperature. They are separated by midlatitude cells that were first discovered by WILLIAM FERREL and are known as *FERREL CELLS*. The Ferrel cells are indirect, in that they are driven by the Hadley and polar cells to each side of them.

When subsiding air reaches the surface on the poleward side of the Hadley cells it divides. Some returns to the equator and some moves toward higher latitudes. The subsiding air of the Hadley cells and the poleward movement of air near the surface form one side and the base of the Ferrel cells. Near the POLAR FRONT, the air moving toward the Pole meets air moving away from the Pole in the polar cells. Convergence makes the air rise and produces low pressure at the surface. At the height of the tropopause the air divides. Some flows toward the Poles, to complete the polar cells, and some flows toward the equator, to complete the Ferrel cells.

As air moves toward or away from the equator it is deflected by the CORIOLIS EFFECT. Near the surface this produces only a slight deflection in low latitudes, where the magnitude of the Coriolis effect is small (it is zero at the equator), but winds are affected by VORTICITY. This produces the low-level northeasterly and southeasterly TRADE WINDS. The belt where the trade winds from both hemispheres meet is called the *INTERTROPICAL CONVERGENCE ZONE* (ITCZ).

High-level winds in the Hadley cell are westerly above about 25,000 feet (7,600 m) over the equator and above about 5,000 feet (1,500 m) over the subtropics. Winds in midlatitudes are from the west at all altitudes. These are the PREVAILING WESTERLIES. They include the westerly JET STREAM.

At the surface, there is a belt where the prevailing westerlies flow away from the equator, as southwesterlies in the Northern Hemisphere and northwesterlies in the Southern Hemisphere, and the trade winds flow toward the equator. This produces the HORSE LATITUDES, where winds are light and variable.

Surface winds in the polar cells, flowing toward the equator, are easterlies. At high level, winds flowing away from the equator are westerlies.

Although the three-cell model provides a useful general outline, it should not be taken too literally. There is not one Hadley cell in each hemisphere, for example, but several. The winds predicted by the model should be GEOSTROPHIC. This would produce easterly winds at high level in midlatitudes, but the prevailing midlatitude winds are westerlies at all levels, although they are dominated by ROSSBY WAVES and the variations in them through the INDEX CYCLE. The simple description of the indirect Ferrel cell is also inaccurate. In midlatitudes, heat is transferred mainly by large-scale horizontal waves and smaller disturbances embedded within them.

three-front model   A MODEL of the distribution of AIR MASSES over North America that is used to analyze the BAROCLINIC structure of DEPRESSIONS from SYNOPTIC CHARTS and cross sections of the atmosphere. The model includes the ARCTIC FRONT, POLAR FRONT, and between them a third front that develops at the boundary between maritime Arctic (mA) and maritime polar (mP) air or between cold and warm mP air (mPk and mPw). The three fronts mark the boundaries of four air masses: mA, mP, continental tropical (cT), and maritime tropical (mT). The fronts are approximately parallel and run from northwest to southeast.

threshold limit value (TLV)   The greatest concentration of a specified airborne pollutant to which workers may be legally exposed day after day. The TLV is calculated to produce no adverse effect on persons experiencing that level of exposure.

threshold velocity   The minimum speed at which wind raises dust, soil, or sand particles from the ground. This speed varies according to the size of the particles and the amount of moisture, because water adhering to mineral grains holds them together.

throughfall   The proportion of the total precipitation falling over a forest that reaches the ground. This consists of the sum of the amount that is not intercepted by plants and reaches the ground surface directly, the amount that drips from leaves, and the amount of STEM FLOW. It is equal to the total amount of precipitation minus the amount lost by EVAPORATION from plant surfaces.

thunder The sound that is caused by the discharge of energy during a LIGHTNING flash. As the flash moves along its LIGHTNING CHANNEL it raises the temperature of the ionized air inside the channel by up to 54,000° F (30,000° C) in less than one second. This causes the air to expand violently, increasing the pressure inside the channel to as much as 100 times its normal value. This expansion is so rapid as to be explosive, and it emits shock waves that immediately become the sound waves we hear as thunder. The sound waves travel at about 670 mph (1,080 km h^{-1})—the speed of sound in air. Light from the flash travels at the speed of light, which is approximately 1 million times faster. Consequently, a distant observer sees the lightning flash before hearing the clap of thunder associated with it. By counting the time that elapses between seeing the lightning and hearing the thunder it is possible to calculate the approximate distance to the storm: every five seconds represents a distance of about one mile (three seconds per kilometer).

As they travel, the sound waves are damped by the air. Those with a short wavelength, carrying sounds of a high pitch, disappear before those with a long wavelength, carrying the low pitches. The greater the distance from the lightning, therefore, the deeper the note of the thunder. Very distant thunder is heard as a deep rumble.

The sound often continues for several seconds. This is because of the length of the lightning flash that causes it. A lightning flash can be more than 1 mile (1.6 km) long, and because of its forked shape some parts of it are closer to the observer than others. Sound from the nearest part of the flash reaches the observer first and sound from the more distant parts arrives later, extending the duration of the sound.

Thunder can seldom be heard from more than about 6 miles (10 km) away, because by then all the sound waves have been either absorbed by the air or refracted upward by the effect of the decrease in air temperature with height. Lightning can be seen over a much greater distance, especially at night. Silent flashes of sheet lightning are most often seen on warm summer nights and are sometimes called *heat lightning*.

The cause of thunder puzzled people for centuries. Germanic peoples believed it was the sound of the god Thor, either beating his huge anvil or throwing his hammer at other gods. Our word *thunder* is from the Old Norse *thórr*. According to a poem called *De*

Rerum Natura (On the nature of things) written in about 55 B.C. by the Latin poet Lucretius (his dates of birth and death are not known, but his full name was *Titus Lucretius Carus*) thunder is the sound of great clouds crashing together. Native Americans thought it was the sound of huge thunderbirds flapping their wings.

thunderbolt A rock, piece of metal, or dart that storm gods such as Zeus (Jupiter), Yahweh (Jehovah), and Thor were believed to hurl. As they traveled through the air, thunderbolts could be seen as LIGHTNING, and they were believed to cause the damage that occurs when lightning strikes an object or person. When lightning strikes sand the sudden discharge of energy is often sufficient to melt the grains, producing an irregular mass of glass. A piece of such glass was assumed to be a thunderbolt. In Ancient Greece and Rome the area around a lightning strike was fenced off and considered sacred and persons killed by lightning were buried where they died, rather than in the usual burial ground. It was not until 1752 that the true, electrical nature of lightning was proved by BENJAMIN FRANKLIN and, independently, by the French scientist Thomas-François d'Alibard (1703–99) and the thunderbolt was shown to be entirely mythical.

thunderhead A CUMULONIMBUS cloud that extends vertically to the TROPOPAUSE, where it spreads downwind to produce an anvil-shaped INCUS of CIRRIFORM cloud. The incus is the "head" of a cloud that is likely to produce a THUNDERSTORM.

thunderstorm A violent storm that causes heavy precipitation, strong gusts of wind, and THUNDER and LIGHTNING. Every day more than 16 million thunderstorms occur in the world as a whole and about 2,000 are taking place at any moment. About 100,000 occur in the United States every year.

Development of a storm begins with the growth of CUMULUS clouds. Warm air is rising by CONVECTION inside the clouds. As it rises it cools in an ADIABATIC process and some of its WATER VAPOR condenses. This releases LATENT HEAT of CONDENSATION, warming the surrounding air, and causing it to continue rising. The cloud builds rapidly, with updrafts traveling at up to 100 mph (160 km h^{-1}), and as its upper part passes the FREEZING LEVEL the BERGERON–FINDEISEN MECHANISM

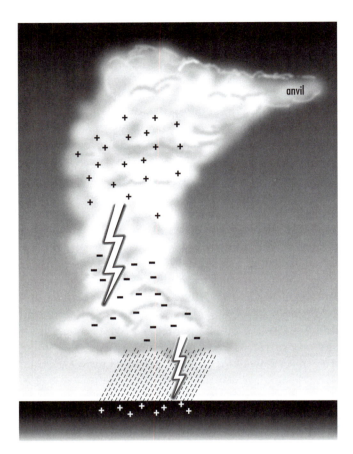

Thunderstorm. Electric charge has become separated inside the cumulonimbus cloud. Positive charge accumulates in the upper part of the cloud, negative charge in the lower part, and the negative charge induces a positive charge on the ground surface beneath the cloud. Lightning is the spark that discharges the charge distribution. Near the tropopause, the top of the cloud is swept into an anvil shape by the wind.

causes precipitation to begin, carried in the downdraft below the cloud base. The precipitation may fall as HAIL; the size of the HAILSTONES indicates the vertical extent of the cloud and the force of its upcurrents. There is also rain or snow, depending on the temperature in the lower part of the cloud and in the air between the cloud base and the surface.

By this stage the cloud has become a CUMULONIMBUS and it extends vertically to a height of 50,000 feet (15 km) or more. Upper-level winds draw out the ICE CRYSTALS at its top into an anvil shape. The storm is then in its most active stage.

The storm derives its energy from the latent heat of condensation and so it requires a constant inflow of

warm, moist air to sustain it. Once precipitation commences, the falling ice crystals and hailstones drag with them small envelopes of chilled air. These accumulate and form downdrafts, carrying precipitation and cold air out of the base of the cloud. As the precipitation intensifies, so do the downdrafts until they dominate the cloud and suppress the updrafts. The storm is then deprived of energy and the cloud of moist air. The precipitation becomes lighter and the cloud begins to dissipate.

It is while it is in its most active stage that the storm produces lightning. This occurs because electrical charge becomes separated inside the cloud. Positive charge accumulates in the upper regions and negative charge in the lower regions. The negative charge near the base of the cloud induces a positive charge on the ground beneath the cloud. Lightning consists of electrical discharges between the separated charges. This neutralizes them, but only temporarily in an active storm. It takes no more than about 20 seconds for the charges to separate again.

Scientists are not certain what it is that causes charge separation, but it is known that the small particles or droplets acquire positive charge and the large ones acquire negative charge. It is gravity that separates them, therefore, as the heavy ones sink to the bottom and the light ones are carried to the top. When a droplet freezes, positive IONS migrate toward the colder regions, leaving the less mobile negative ions in the warmer regions. Droplets freeze from the outside in, so a shell of positive charge surrounds a core of negative charge. As the core starts to freeze it expands, shattering the shell into minute fragments. The fragments carry their positive charge upward and the core carries its negative charge downward.

It may also happen that small ice splinters collide with hailstones the outsides of which have been warmed by the release of latent heat as SUPERCOOLED water freezes onto them. At each collision positive ions move toward the colder end of the splinter. This increases the negative charge both at the warmer end of the splinter and on the hailstone. The charge changes by only a very small amount at each collision, but because of the very large number of collisions there is a cumulative effect. The splinters, with their positive charge, are carried upward, and the hailstones, with their negative charge, drift downward.

thunderstorm day   A day on which a THUNDER-STORM is observed at a WEATHER STATION.

Tibetan high   An ANTICYCLONE that develops in summer over the Tibetan Plateau. In early summer the ground warms strongly. Air rises by CONVECTION, producing a shallow layer of low pressure near the ground and high pressure at heights above the 500-millibar level. The ANTICYCLONIC flow is from the east on the southern side of the anticyclone. This contributes to the breakdown of the westerly JET STREAM. The change from a midlatitude westerly flow to an easterly flow and the disappearance of the jet stream are linked to the onset of the MONSOON over southern Asia and the MAI-U rains over China.

Tico   A HURRICANE that struck the coast of Mazatlán, Mexico, on October 10, 1983. It killed 105 fishermen, whose boats were lost at sea.

tidal range   The difference in height between the high and low TIDES in a particular place along a coast. This varies according to the phase of the Moon. Tidal range is greatest during spring tides and least during neap tides. Mean tidal ranges take account of these cyclical variations to provide a general value. Tidal range varies according to the configuration of coastlines. Tides propagate as waves with a PERIOD similar to that of the forces generating the tides. When these sea waves arrive at coastlines or enter bays and estuaries they may be reflected. They may then form a standing wave, or seiche, with a period determined by the length and depth of the basin that contains it. If this period coincides with that of the tides, the AMPLITUDE of the tide can be increased greatly. This is the cause of the huge tidal range in the Bay of Fundy, in eastern Canada. The bay is about 168 miles (270 km) long and its depth averages 230 feet (70 m). These dimensions mean the standing wave produced by the tide has a period of 12 hours, with the result that the tidal range at spring tides can exceed 50 feet (15 m). Elsewhere, tidal ranges are smaller. The mean tidal range at Boston, Massachusetts, is about 9 feet (2.7 m), for example. A tidal range of less than 6.5 ft (2 m) is called *microtidal*, one of 6.5–13 ft (2–4m) is *mesotidal*, and one of more than 20 ft (6 m) is *macrotidal*.

tidal wind   A wind that is produced by the ATMOSPHERIC TIDES. The term is also applied to a wind that is produced in some tidal inlets by the displacement of air when the TIDE rises strongly. The wind is very light and can be felt only when the air is otherwise calm.

tides   The regular movements of surface waters, the atmosphere, and the solid Earth that are caused by the gravitational attraction of the Moon and, to a lesser extent, of the Sun. This attraction produces two bulges at the surface. The bulges move around the Earth in step with the orbit of the Moon, producing two tidal cycles in every 24-hour period.

Tidal forces affect the liquid core of the Earth and the resulting "Earth tides" can be measured by the bulges they produce in surface rocks. These are small, however, and the greatest tidal movement nowhere exceeds about 3 ft (1 m). ATMOSPHERIC TIDES are synchronized to the daily solar cycle, rather than the lunar cycle. It is in the oceans that the tides are most clearly seen.

All the parts of the Earth, including the atmosphere and oceans, are drawn toward the center of the Earth by the gravitational force. Because the Earth is rotating, the gravitational force balances the inertial

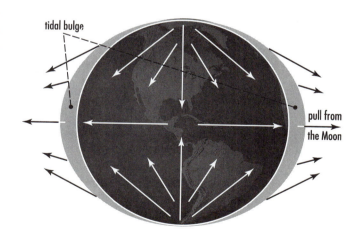

Tides. Lunar gravity reduces the magnitude of terrestrial gravity. Because the Earth is rotating, this increases the tendency of every part of the Earth to continue moving in a straight line and fly away from the Earth. The result is two bulges, produced where terrestrial and lunar gravity pull in opposite directions. One bulge is directly beneath the Moon and the other is on the opposite side of the Earth.

tendency of a moving body to continue to move in a straight line (*see* CENTRIPETAL ACCELERATION).

Both the Moon and the Sun also exert a gravitational force. Their effect is much smaller than that of the Earth's own gravitational force, because gravity is subject to the INVERSE SQUARE LAW: that is, its magnitude decreases rapidly with increasing distance. Lunar and solar gravity act in the opposite direction to terrestrial gravity and therefore reduce its influence. This process alters the balance between the terrestrial gravitational force and the inertial tendency of each part of the Earth, producing two bulges. One bulge lies on the surface of the Earth that is directly beneath the Moon. The lunar attraction acts in a straight line, reducing the magnitude of terrestrial gravity everywhere along that line. Consequently, the other bulge lies on the side of the Earth directly opposite the Moon. As the Moon orbits the Earth the two bulges follow it.

The Moon takes 24 hours and 50 minutes to orbit the Earth, so high and low tides occur at intervals of 12 hours 25 minutes. This is why the times of the tides change from day to day. The DECLINATION of the Moon changes during the month and is sometimes as much as 28.5°. This means that the lunar gravitational attraction acts at an angle to the equator, and therefore the tidal bulge is also at an angle to the equator. At any location on the surface of the Earth, therefore, one tide rises higher than the other tide and the two tides have the same AMPLITUDE only when the Moon is directly above the equator.

The gravitational force exerted by the Sun is about 47 percent of the force exerted by the Moon. This is because, despite being much more massive than the Moon, the Sun is much more distant. That is why the lunar influence on tides predominates.

The influence of the Sun is felt at times of spring and neap tides. Spring tides occur when the Earth, Sun, and Moon are aligned, so the gravitational influence of the Sun is added to that of the Moon. Spring tides rise to a higher level than average and ebb to a lower level. The height of spring tides varies according to the accuracy of the Earth–Moon–Sun alignment: the more closely the three bodies are aligned, the higher the spring tides are. Neap tides occur when lines drawn from the Sun and Moon meet in a right angle at the center of the Earth. The solar and lunar gravitational forces then act partly against each other, reducing the overall tidal effect. Neap tides are smaller than the average tides. As with the spring tides, their height varies according to the accuracy of the misalignment.

Although the two tidal bulges circle the Earth as waves, their magnitude and timing vary from place to place because of friction between the waves and the ocean floor and because of the shape and orientation of coastlines. The amplitude of the waves also varies with the volume of water through which they move. Ocean tides are much larger than the tides in seas that are almost completely enclosed by land, such as the Mediterranean and Baltic Seas. The difference between the mean height of high and low water at a particular place is known as the *TIDAL RANGE*.

Oceanic tidal movements play an important part in the transport of heat from the equatorial to polar regions. Tidal energy is dissipated in the deep oceans, and its dissipation causes some mixing of ocean waters. Some scientists now suspect that this mixing, combined with the wind-driven ocean currents, is more important than the THERMOHALINE CIRCULATION in the oceanic transport of heat.

Tides are also thought to be important in climate change. The DECLINATION of the Moon changes over a cycle that maximizes the tides every 1,800 years, and the amplitude of the cycle also varies over a cycle with a period of 5,000 years. Strong tides increase the amount of mixing in the oceans. Mixing with cold, deep water cools the surface water. Weak tides have the opposite effect. These cyclical variations coincide with, and may cause, abrupt fluctuations in climate that occur on much smaller time scales than the change between ice ages and INTERGLACIALS.

Very gradually, the tides are weakening. This is because the Moon is receding from the Earth by about 1.6 inch (4 cm) every year.

Tiglian interglacial   An INTERGLACIAL period in Europe that preceded the DONAU GLACIAL. It occurred at the very start of the PLEISTOCENE EPOCH (some authorities place it in the late Pliocene) and it is believed to have been a time when average temperatures were falling.

Tip   A TYPHOON that struck Japan on October 19, 1979, with winds of up to 55 mph (88 km h^{-1}). At least 36 people died.

tipping-bucket gauge   *See* RAIN GAUGE.

tjaele (frost table) A layer of frozen ground, or ground containing frozen water, that lies at the base of the ACTIVE LAYER in a PERMAFROST region. The tjaele moves downward during the summer thaw in the active layer.

TLV *See* THRESHOLD LIMIT VALUE.

TM *See* THEMATIC MAPPER.

TOGA *See* TROPICAL OCEAN GLOBAL ATMOSPHERE.

TOMS *See* TOTAL OZONE MAPPING SPECTROMETER.

tongara A hazy southeasterly wind that blows through the Makasar Strait, Indonesia.

Topex/Poseidon A joint mission by the NATIONAL AERONAUTICS AND SPACE ADMINISTRATION (NASA) and the Centre National d'Études Spatiales (CNES) in France that uses an ALTIMETER mounted on a satellite to measure the height of the surface of the ocean with unprecedented accuracy. The satellite, Topex, was launched from Kourou, French Guiana, in August 1992 on a Poseidon rocket. Topex orbits at a height of 830 miles (1,336 km) on a track with an INCLINATION of 66° and an orbital period of 112 minutes. It carries two altimeters, one built by NASA and the other by CNES. Both instruments measure the height of the ocean at the same points at intervals of 10 days. The resulting data are distributed to scientists in nine countries, who use them with other research programs investigating the global climate. The measurements show changes caused by meanders in strong ocean currents, local vortices, and large eddies associated with BOUNDARY CURRENTS and subtropical GYRES.

(You can learn more about the Topex/Poseidon mission at www-ccar.colorado.edu/research/topex/html/topex.html and www.tsgc.utexas.edu/spacecraft/topex.intro.html.)

tor *See* TORR.

tornadic storm *See* TORNADO.

tornado (1) The most violent of all weather phenomena, a tornado develops inside a huge CUMULONIMBUS storm cloud, extends from its base, and produces winds that can reach 300 mph (480 km h⁻¹). A storm that is capable of generating tornadoes is called a *tornadic* storm. Tornadic storms are sometimes isolated, but they are more likely to occur along a SQUALL LINE.

The cloud extends from below 1,000 feet (300 m) all the way to the TROPOPAUSE, at a height of 50,000 feet (15.25 km) or more. Inside the cloud the air is extremely unstable (*see* STABILITY OF AIR). As rising air cools its water vapor starts to condense, releasing latent heat. This heat warms the air and makes it continue rising. At the top of the cloud, rising air cannot penetrate the stable air of the stratosphere and it spreads horizontally, forming an ANVIL. The more vigorous the vertical air currents in the cloud, the bigger the anvil. The anvil extends at an angle of about 45° to the direction the storm is moving. In the Northern Hemisphere the anvil extends to the left of the direction of motion, as seen from above; in the Southern Hemisphere it extends to the right.

Near the top of the cloud the air is very cold. It sinks, warming and collecting moisture as it does so. This convective movement forms a number of cells. Ordinarily, the downcurrents in one cell interfere with the upcurrents in the neighboring cell, eventually suppressing them. When this happens, the storm dies. If the anvil is big enough, however, the upcurrents are carried clear of the downcurrents and a SUPERCELL forms, greatly extending the lifetime of the cloud.

MAMMATUS often forms on the underside of the anvil of isolated tornadic storm clouds, but rarely on those along squall lines. At this stage air is being drawn into the base of the cloud along its leading edge, so that strong gusts of wind blow toward the cloud as it approaches. This is the GUST FRONT. Air rises to the top and spreads into the anvil. Behind the anvil, equally strong downcurrents, producing first HAIL and then torrential rain as the storm passes, emerge as winds of up to hurricane force (more than 75 mph [120 km h⁻¹]) from the rear of the cloud.

WIND SHEAR deflects the upcurrent at about the midheight of the cloud. This causes the rising air to rotate about its own axis (*see* VORTICITY), creating a VORTEX with very low pressure at its center. The rotating air constitutes a MESOCYCLONE and its rotation begins to extend downward through the cloud. As the mesocyclone grows downward its diameter decreases. The conservation of its ANGULAR MOMENTUM causes the wind speed to accelerate around the vortex.

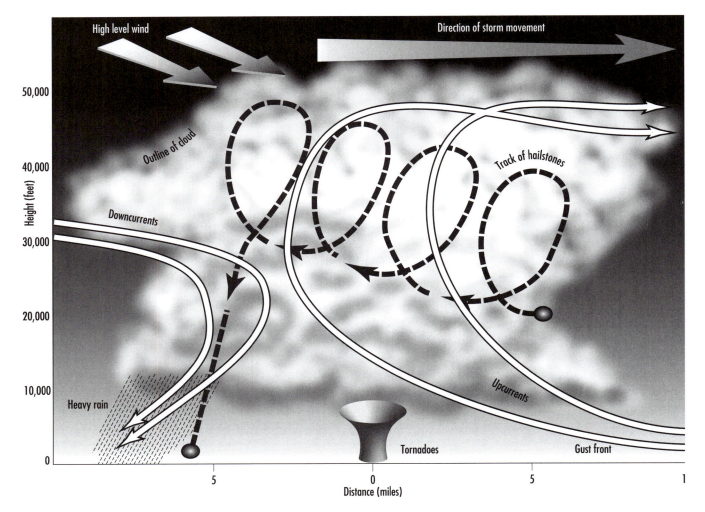

Tornado. A cross section through a storm cloud that is producing a tornado. Ahead of the cloud there is a gust front, with strong winds, and at the rear of the cloud, behind the tornado, there are hail and heavy rain. The storm is driven by fierce air currents.

Fragments at the base of the cloud start to rotate. Then part of the cloud base—in fact, the bottom of the mesocyclone—descends below the main cloud base, rotating slowly, to become the WALL CLOUD. A FUNNEL CLOUD, consisting of a rapidly spinning upcurrent, may emerge through the base of the wall cloud. If the funnel touches the ground it becomes a tornado.

When it touches the ground the tornado darkens because of the dust, debris, and other material swept into it and carried upward.

Tornadoes can happen anywhere and at any time of year, but they are most common in the Great Plains of the United States. There their frequency is greatest between May and September and they are most likely between 2 P.M. and 8 P.M. Tornadoes are classified by the damage they cause according to the FUJITA TORNADO INTENSITY SCALE. WATERSPOUTS, WHIRLWINDS, DUST DEVILS, and WATER DEVILS are related phenomena.

(For more information about tornadoes see Michael Allaby, *Dangerous Weather: Tornadoes* (New York: Facts On File, 1997)

tornado (2) A violent but brief thunderstorm that occurs in West Africa. It does not develop a MESOCYCLONE leading to a narrow column of rapidly rotating air that may extend below the storm cloud, and consequently it bears no relation to the "twister" type of storm. The West African tornado is associated with the southwesterly MONSOON and air from the HARMATTAN. It develops between March and May and in October and November near the coast, and between May and September inland. The storms lie along a SQUALL LINE, from 10 miles (16 km) sometimes to 200 miles (320 km) long, in Nigeria. They travel from east to west at about 30 mph (50 km h⁻¹) and produce huge dark clouds, frequent lightning, dust, and winds of up to 80 mph (130 km h⁻¹) inland but only half of that near the coast, and torrential rain once the squalls have passed. A tornado may last as little as 15 minutes and rarely longer than two hours.

Tornado Alley The area of the Great Plains, in the United States, in which tornadoes occur more frequently than they do anywhere else in the world and where the most violent tornado outbreaks are experienced. Tornado Alley is centered on Texas, Oklahoma, and Nebraska, but the area also covers Kansas, Iowa, Arkansas, Missouri, Alabama, and Mississippi, and tornadoes are also fairly frequent in northern Florida. In all these states there are an average of five tornadoes every year.

This region suffers more than any other because of its geographical features and the AIR MASSES that affect it. When it reaches the North American coast, air that has crossed the North Pacific is cool and moist. This maritime air rises to cross the Rocky Mountains, losing much of its moisture as it does so. On the eastern side of the mountains it descends gently, warming slightly, and advances slowly across the Great Plains behind a weak COLD FRONT that is aligned approximately southwest to northeast.

At the same time, continental tropical air forms over Mexico, New Mexico, and Texas and moves north. This air is warm and dry, and in spring and summer, when the land warms rapidly, it is heated further as it advances. It enters the plain as hot, dry air.

A third mass of air forms over the Gulf of Mexico. This air is warm and moist, and it moves in a northwesterly direction.

When the two air masses, from Mexico and the gulf, meet over northern Mexico and the southern United States the moist air from the gulf is held beneath the less dense continental air. The two move northwestward, the lower air warming still more as it crosses the land. Small clouds form, but although the air is being warmed from below it cannot rise very far by convection, because of the overlying layer of dry air.

On the western side of the Great Plains the two air masses meet the weak cold front advancing from the opposite direction. The cold air undercuts the warm air. There is now a "sandwich" of air, with dry air at the bottom and top and warm, moist air between them. The moist air is forced to rise. As it does so it expands then cools, and its water vapor starts to condense. At the same time its rate of cooling changes from the dry to the saturated LAPSE RATE. The air becomes very unstable (*see* STABILITY OF AIR) as it rises up the cold front. Eventually convection within it becomes so vigorous that upcurrents start breaking through the overlying dry air. When that happens, the clouds rise all the way to the TROPOPAUSE and often beyond it, into the lower stratosphere.

It is these clouds that produce some of the most violent thunderstorms known, and because they form along the cold front, which is still moving in a southeasterly direction, they tend to link together in SQUALL LINES. These are what produce the tornadoes that give Tornado Alley its name.

Tornado and Storm Research Organization A British organization, founded in 1974, that exists to gather data and undertake research into TORNADOES and other severe weather phenomena in Europe. It has representatives in Austria, France, Germany, Ireland, and Switzerland.

(More information about the organization can be found at http://www.torro.org.uk/index.html.)

tornadoes of the past *See* APPENDIX IV.

tornado warning A final warning issued by the NATIONAL WEATHER SERVICE to people living in an area where one or more TORNADOES have been reported. It means a tornado may occur at any time and everyone should immediately seek shelter.

tornado watch A preliminary warning issued by the NATIONAL WEATHER SERVICE to people living in an area where TORNADOES may occur in the next few hours. Tornadoes have not been reported, but the conditions are right for them.

torque (**couple, moment of a force**)   A twisting force that is equal to the product of a force and its distance from a point about which it is causing rotation. It is measured in newton meters (Nm) or pound-force feet (lbf ft). Torque is the force that causes air to rotate about a vertical axis.

torr   A unit of pressure that is equal to one millimeter of mercury. It is named after EVANGELISTA TORRICELLI and was used for a time in some European countries but is now little used. The tor, another unit named after Torricelli, and equal to one pascal, or one-hundredth of a millibar, was also proposed in 1913 but is rarely used.

Torricelli, Evangelista (1608–1647) Italian *Physicist and mathematician*   An Italian physicist and mathematician who discovered the principle of the barometer. Born at Faenza, he went to Rome in 1627 to study science and mathematics. In 1638, he read some of the works of GALILEO GALILEI. These inspired him to write a treatise developing some of Galileo's ideas on mechanics. Galileo invited him to Florence and, in 1641, the two met. Galileo was then old and blind. Torricelli became his assistant; Galileo died three months later. It was Galileo who suggested the problem that Torricelli solved.

The question was, why does water rise up a cylinder when a piston in the cylinder is raised, but only so far? The conventional explanation for water's rising in this way at all, which Galileo supported, was that raising the piston produces a vacuum in the cylinder and that nature abhors a vacuum, so the water has to rise. It can be made to rise only about 33 ft (10 m) above its ordinary level, however, and that was the puzzle. Galileo suspected that nature's abhorrence of a vacuum was limited, so vacuums can be tolerated under certain conditions. He asked Torricelli to investigate.

Torricelli rejected the idea of "vacuum abhorrence." He considered what might happen if the air possessed weight—in his day most scientists believed air had a property of "levity," making it tend to rise, so Torricelli's idea was fairly radical. If that were so, then the weight of air would press down on the surface of the water outside the cylinder. As the piston was raised there would be a space containing no air between the bottom of the piston and the surface of the water. The air would not press down inside the cylinder below the

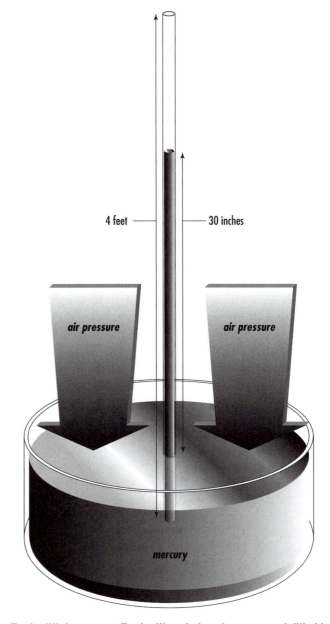

Torricelli's barometer. **Torricelli sealed a tube at one end, filled it with mercury, and inverted it in a bath of mercury. Enough mercury remained in the tube to produce a column 30 in tall. The weight of air pressing on the exposed surface in the bath forced mercury into the tube.**

piston, but it would still do so outside. Consequently, the pressure outside the cylinder would cause water to rise inside the cylinder as the piston was raised.

Suppose, though, that the pressure due to the weight of air was sufficient only to raise the water by

33 ft? In that case, 33 ft would be as far as the water could be made to rise, and withdrawing the piston further would have no effect.

In 1643, Torricelli tested this idea, choosing to use mercury, which is 13.6 times denser than water. He partly filled a dish with mercury and completely filled a glass tube 4 ft (1.2 m) long and open at one end. Placing his thumb over the open end, he inverted the tube and placed the open end below the surface of the mercury in the dish, holding the tube absolutely upright. When he removed his thumb, mercury flowed out of the tube and into the dish, but some remained in the tube. The level of mercury in the tube was about 30 in (760 mm) higher than the surface of the mercury in the dish.

Because the liquid had fallen from a full tube, rather than being drawn upward by the withdrawal of a piston, Torricelli had proved it was the pressure exerted by the weight of air on the mercury in the dish that supported the column of mercury in the tube. He had proved that air has weight, a finding with implications that were explored by BLAISE PASCAL.

Above the mercury, the tube was empty except for a small amount of mercury vapor. Torricelli was the first person to make a vacuum and a vacuum produced in this way is still known as a *Torricellian vacuum.*

Then he noticed that the height of the column of mercury varied slightly from day to day. He attributed this, correctly, to variations in the weight of the air. He had invented a device for measuring these small changes—the first barometer.

After Galileo's death, Torricelli was appointed mathematician to the grand duke and professor of mathematics in the Florentine Academy. He died in Florence.

Torrid Zone   A general name for the climatic region that lies between the tropic of Cancer (latitude 23.5° N) and the tropic of Capricorn (latitude 23.5° S). *See also* MATHEMATICAL CLIMATE.

Totable Tornado Observatory (TOTO)   A package of instruments that is designed to survive the conditions inside a TORNADO with winds up to 200 mph (322 km h⁻¹) and measure wind speed, atmospheric pressure, temperature, and electrical discharges. It was developed by Alfred L. Bedard and Carl Ramzy, scientists working at the NATIONAL OCEANIC AND ATMOSPHERIC ADMINISTRATION (NOAA) Environmental

Research Laboratory in Boulder, Colorado. It comprises a cylinder housed in a casing of half-inch aluminum set in a frame made from angle iron, with arms that hold the instruments extending from the casing. It is powered by batteries and records its measurements. Its acronym, *TOTO*, refers to Dorothy's dog in *The Wonderful Wizard of Oz.*

total degree-days   *See* DEGREE DAYS.

Total Ozone Mapping Spectrometer (TOMS)   The instrument that provided the first satellite evidence of the depletion of stratospheric OZONE over Antarctica. It was launched on October 24, 1978, on board the NIMBUS-7 satellite and transmitted daily maps of ozone distribution until 1993. A second TOMS, TOMS-METEOR, was launched in 1991 on the Russian METEOR satellite and a third, TOMS-ADEOS, in 1996 on a Japanese satellite. Although satellites carry other instruments that measure ozone concentrations, the data from TOMS are the most detailed.

TOMS consists of an instrument directed vertically downward that measures the intensity of radiation being reflected upward from the ground or ocean surface or from cloud tops. It samples radiation at six ULTRAVIOLET (UV) wavelengths between 312.5 nanometers (nm) and 380 nm in a repeating sequence. UV absorption varies according to the wavelength, so by comparing the amount reflected at each of the wavelengths it is possible to calculate the amount of UV being absorbed in the atmosphere. Since it is ozone that absorbs UV at UV-B wavelengths, the density of ozone in the column of air directly beneath the TOMS can be inferred from the amount of UV absorbed.

(Further information on TOMS, including data on the ozone layer, can be found at http://jwocky.gsfc. nasa/gov/reflect/reflect.html.)

TOTO   *See* TOTABLE TORNADO OBSERVATORY.

touriello   A southerly wind, of the FÖHN type, that descends from the Pyrenees and blows along the Ariège Valley, in southwestern France. The wind is especially violent in February and March, when it sometimes melts the snow, causing flooding and avalanches.

towering   The vertical stretching of a MIRAGE that occurs when the downward curvature of light due to

REFRACTION increases with altitude. To an observer, the distance between the top and bottom of the image is apparently increased. The FATA MORGANA is an example of towering.

Tower of the Winds (horologion)

What is believed to be the first device ever invented with the purpose of forecasting the weather was designed by the Greek astronomer Andronicus of Cyrrhus (who flourished around 100 B.C.E) and it was built in Athens at some time in the 1st century B.C.E. A substantial part of it is still standing.

The tower has eight sides. Figures representing the eight principal wind directions are carved at the top of each side. On top of the tower there was originally the bronze figure of Triton with a rod in his hand. The Triton (in some stories there are several Tritons) was a mythological being with the head, trunk, and arms of a human and the lower body of a fish. He was the son of Poseidon, the god of the sea, and Amphitrite, daughter of Oceanos and Tethys. The Triton figure turned in the wind, indicating the wind direction. This statue gave rise to the custom of placing WEATHER VANES, often in the form of a weathercock or other figure, on the tops of church steeples.

Each side also has a sundial. As the Sun crossed the sky in the course of the day, at least one dial would always be showing the time. The tower was also a public timepiece. Even in Athens, however, the Sun does not always shine. To help people tell the time on cloudy days the tower also contained a very elaborate clock driven by water (and called a *clepsydra*) that showed the hours on a dial. The tower also had a disk that rotated, showing the movements of the constellations and the Sun's yearly course through them. Andronicus also built another tower with sundials around its sides, on the Greek island of Tenos.

The principle behind the weather forecast was simple. Traditionally the Greeks had believed that their gods produced the weather. Make an appropriate offering to Zeus, father of the gods, and he would send good weather; please Poseidon and he would send a storm to destroy your enemies. ARISTOTLE, on the other hand, had maintained that the weather has entirely natural causes and that in principle it can be understood and even predicted. The 5th-century B.C.E. dramatist Aristophanes (c. 450–c. 388 B.C.E.) wrote a play about

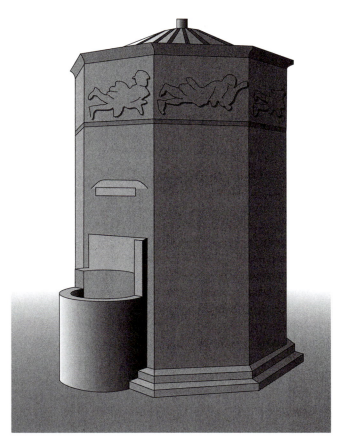

Tower of the Winds. **Most of the octagonal tower is still standing. It was built in the first century B.C.E. and helped Athenians predict the weather.**

the weather, called *Clouds,* that includes a debate between an educated philosopher and an unsophisticated person from the country over whether thunder is made by Zeus or whether it results from the collision of clouds.

It was the more scientific, Aristotelian attitude that underlay the tower. The direction of the wind hinted at the weather that would follow. Athenians could look at the tower, see the direction in which the Triton was pointing, and deduce from that what the weather would be over the next few hours or days. To help them, the gods of the winds, depicted on the faces of the tower, were also associated with particular types of weather. Boreas, the north wind, was a rude fellow who found it hard to breathe gently and was quite unable to sigh. Zephyrus, the west wind, was gentle. The sweetness of his breath brought forth flowers. Notus, the south wind, was wet, his forehead covered by dark clouds. Eurus, the east wind, would generate cold, dry weather.

While they thought about the kind of weather they might expect, the Athenians could also check the time.

TPC *See* TROPICAL PREDICTION CENTER.

Tracy A CYCLONE that struck Darwin, Australia, on December 25, 1974. It destroyed 90 percent of the city and killed more than 50 people.

trade cumulus (trade-wind cumulus) CUMULUS cloud that forms over the ocean in the Tropics. It develops in air that is trapped beneath the TRADE WIND INVERSION. This limits its vertical extent, producing clouds with flat tops at the height of the inversion, all of them much the same size and shape.

trade-wind cumulus *See* TRADE CUMULUS.

trade wind inversion A temperature INVERSION that is associated with the subsidence of air on the high-latitude side of the HADLEY CELL circulation. Air that has risen in the INTERTROPICAL CONVERGENCE ZONE loses its moisture through ADIABATIC cooling and CONDENSATION. It descends as dry air and warms at the DRY ADIABATIC LAPSE RATE. Near the surface, air is flowing toward the equator from higher latitudes. This forms a layer of air that is cooler than the subsiding air. The cool lower layer is thus trapped beneath the inversion. This limits the vertical development of cloud, restricting the capacity of the surface air to lose the moisture it accumulates in its passage over the ocean. The height of the inversion is at about 5,000–6,500 ft (1,500–2,000 m) at the outer edges of the Hadley cell and increases toward the equator. The increase in height is due partly to the fact that subsidence decreases closer to the equator, reducing the supply of warm air, and occurs partly because, as the lower air becomes increasingly moist and unstable, CUMULUS clouds frequently penetrate the inversion. The clouds evaporate in the dry air above the inversion, cooling it by the absorption of LATENT HEAT.

trade wind littoral climate In the STRAHLER CLIMATE CLASSIFICATION a climate in his group 1, which comprises climates controlled by equatorial and tropical AIR MASSES. Trade wind littoral climates are produced by maritime tropical air masses carried by the tropical easterly winds (the trades) from the western sides of oceanic subtropical regions of high pressure. They are found along narrow belts on eastern coasts in latitudes 10°–25° in both hemispheres. Temperatures remain fairly constant, and high, throughout the year. Rainfall is heavy, but with a strong seasonal variation. This type of climate is designated *Af–Am* in the KÖPPEN CLIMATE CLASSIFICATION.

trade winds The winds that blow toward the equator from each side, from the northeast in the Northern Hemisphere and from the southeast in the Southern Hemisphere. They are extremely dependable, especially on the eastern side of the Atlantic, Pacific, and Indian Oceans, blowing at an average speed of about 11 mph (18 km h^{-1}) in the Northern Hemisphere and 14 mph (22 km h^{-1}) in the Southern Hemisphere.

Air rises over the INTERTROPICAL CONVERGENCE ZONE (ITCZ). When it reaches the TROPOPAUSE the air moves away from the equator and it subsides in the TROPICS. As it descends into the SUBTROPICAL HIGHS, some of the air flows away from the equator, into higher latitudes, but most of it flows back toward the equator. It is joined by air flowing toward the equator from outside the Tropics. This circulation of air composes the HADLEY CELL and the air returning to the equator at a low level forms the trade winds.

As it moves from the subtropical highs to the EQUATORIAL TROUGH, its relative VORTICITY causes the air to start turning about vertical axes centered on the subtropical highs, clockwise in the Northern Hemisphere and counterclockwise in the Southern Hemisphere. This deflects the wind from a direction at right angles to the equator, producing the northeast and southeast trades.

Where the trades cross the equator, vorticity acts in an opposite way and the easterly winds can become westerlies. In summer, for example, when the ITCZ moves to about 5°–10° N, trade winds blowing from the southeast in the Indian Ocean south of the equator continue across the equator, where they become southwesterly winds that contribute to the onshore winds that produce the summer MONSOON over southern Asia.

There is some seasonal variation in the strength of the trade winds. They are strongest in winter and weakest in summer. This is because they originate in air flowing outward from the subtropical highs, which are most intense in winter.

The area affected by them also changes with the seasons. In March, the northeast trades are found between 3° N and 26° N over the Atlantic and between 5° N and 25° N over the Pacific. In September they occur between 11° N and 35° N over the Atlantic and between 10° N and 30° N over the Pacific. In March, the southeast trades occur between the equator and 25° S over the Atlantic and between 3° N and 28° S over the Pacific. In September they are found between 3° N and 25° S over the Atlantic and between 7° N and 20° S over the Pacific.

Weather systems in higher latitudes can also influence the strength of the trades. When these alterations in the distribution of pressure increase the PRESSURE GRADIENT away from the centers of the subtropical highs, the trade winds accelerate in what is called a *surge of the trades.* They are also affected by the SOUTHERN OSCILLATION. EL NIÑO events cause the trade winds to weaken, fade completely, or even reverse direction. LA NIÑA events cause the winds to strengthen.

Despite these variations, the trade winds are more constant, in both speed and direction, than any other wind system on Earth, and they are especially constant over the ocean. Their constancy was of great importance in the days when goods were traded in sailing ships and it was noted as soon as vessels were plying regularly through the Tropics.

It is their constancy that gives them their name. In early Saxon times, the word *trada* meant "footstep" or "track" and this came into English as *trade,* meaning "track" (*tread* has the same derivation). Any kind of established track or trail might be called a trade and so a wind that blows almost all of the time and almost always from the same direction was described as a trade wind. The pursuit of a particular occupation was also called a *trade* and in time the name came to be attached to the occupation itself.

Once the trade winds had been observed and their reliability verified scientists began trying to explain them. The first to do so, in 1686, was the British astronomer EDMUND HALLEY (1656–1742). In 1735 the British meteorologist GEORGE HADLEY (1685–1768) improved on this explanation, but it was not until 1856 that the trade winds were fully explained as part of the GENERAL CIRCULATION of the atmosphere by the American meteorologist WILLIAM FERREL (1817–91).

Trafalgar Square interglacial *See* IPSWICHIAN INTERGLACIAL.

tramontana A northerly or northeasterly wind that blows in the northwestern part of the Mediterranean region, carrying cool polar air.

transfrontier pollution The movement of pollutants that are carried in air or water across an international frontier. Transfrontier pollution cannot be regulated and reduced by the country that suffers it acting alone, but only by international agreement addressing the source of the pollutants. Several such agreements have been reached. For example, emissions of SULFUR DIOXIDE, especially from power-generating plants that burn coal and oil, have been reduced in order to reduce the damage to forests and freshwater that is caused by ACID RAIN. Acid rain damage was first reported from Sweden and was attributed to emissions from Britain, Germany, and other European countries.

translucidus A variety of clouds (*see* CLOUD CLASSIFICATION) that cover a large proportion of the sky but through which it is possible to discern the position of the Sun or Moon, so the cloud is translucent. The variety most often occurs with the cloud genera ALTOCUMULUS, ALTOSTRATUS, STRATOCUMULUS, and STRATUS.

The name of the variety is derived from the Latin verb *translucere,* which means "to shine" (*lucere*) "through" (*trans*).

transmissivity A measure of the transparency of the atmosphere to incoming solar radiation. It is the fraction of the solar radiation incident on the top of the atmosphere that reaches the surface in a direct beam (not as diffuse radiation). Transmissivity varies with the state of the atmosphere—HAZE and AIR POLLUTION reduce it—and with the distance the radiation must travel through the atmosphere, or PATH LENGTH.

transparency The capacity of a medium for permitting radiation to pass through it with no significant SCATTERING or absorption. The transparency of the atmosphere is usually measured by its TRANSMISSIVITY when the Sun is at its zenith (and the PATH LENGTH is 1).

transparent sky cover The proportion of the sky that is covered by cloud that does not completely hide whatever may be above it. Higher clouds or blue sky can be seen through the layer of cloud. The term is used in United States meteorological practice and the amount is usually expressed in tenths of the total sky.

transpiration Plants need water. They use it as a source of hydrogen for PHOTOSYNTHESIS, the mineral nutrients that enter through the roots are transported in solution to all parts of the plant, and it is water that fills plant cells and keeps them rigid. If a plant is deprived of water it wilts, and unless the supply of water is restored it may die.

Plants obtain their water from the soil. It travels through the stems along channels that form tissue called *xylem,* and it evaporates through the tiny pores, called STOMATA, in the leaves. A much smaller amount evaporates from pores called *lenticels* in the stem. The evaporation of water from its surfaces helps keep the plant cool in very hot weather. This loss of water from the plant is transpiration and the amount involved is large. In summer, a silver birch tree (*Betula pendula*), with about 250,000 leaves, may transpire 95 gallons (360 liters) of water a day and a full-grown oak tree (*Quercus* species) transpires up to 185 gallons (700 1) a day. A maple (*Acer* species) transpires about 53 gallons (200 liters) an hour. This is water that the plant takes from the ground and returns to the air. Not surprisingly, this process dries the ground and increases the HUMIDITY of the air.

If you cut a plant stem liquid seeps from it. This liquid, called *sap,* is mainly water and it is flowing upward. A small herb may lift water a few inches above the ground, but a Sierra redwood tree (*Sequoiadendron giganteum*) can grow to a height of 300 ft (90 m) and has roots that extend below ground to a depth of several feet. It raises water all the way from its roots to the leaves on its topmost branches.

Transpiration is driven from the leaves and not from the ground; the water is pulled from above, not pushed from below. Beneath the surface cells with their stomata leaves have a layer of tissue called *mesophyll.* The inside of mesophyll cells is coated in a film of water, and when the stomata are open some of this water evaporates. Hydrogen bonds between water molecules draw in more water to replace the water that has been lost. ADHESION and COHESION—which function in living tissues in just the same way as they do in the soil—then transmit this attraction all the way through the plant to the tips of its roots. There the attraction exerts a force that draws water toward the roots. The magnitude of this force varies from one plant species to another. In some it is very weak, but in others it can amount to more than 200 lb in^{-2} (1,380 kPa), which is about 13.8 times sea-level atmospheric pressure. The force can be so great that on a hot day, when the evaporation rate from the leaves is very high, water moves through the plant at up to 30 in (76 cm) a minute and the sides of the xylem vessels are pulled in, making a measurable difference in the diameter of the stem.

It is possible to measure the rate of transpiration from individual plants under laboratory conditions, but extremely difficult to do so reliably outdoors. For practical purposes it is impossible to distinguish the water vapor entering the air by evaporation and that entering by transpiration. The two are therefore considered together and the combined process is called *EVAPOTRANSPIRATION*.

(You can read more about transpiration in Michael Allaby, *Ecosystem: Temperate Forests* [New York: Facts On File, 1999], and in Michael Kent, *Advanced Biology* [Oxford: Oxford University Press, 2000].)

trapping One of the patterns a CHIMNEY PLUME may make as it moves away downwind. With increasing distance from the smokestack, the plume widens a little

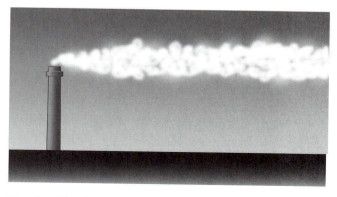

Trapping. **The plume of gases and particles widens and descends a little as it moves away, but there is little dispersion.**

and descends slightly, but there is little dispersion of the gases and particles. Trapping occurs when the DRY ADIABATIC LAPSE RATE is greater than the ENVIRONMENTAL LAPSE RATE in the layer of air extending from the surface to the height of the smokestack and there is a weak INVERSION in the air above the stack.

traveling wave A wave that moves through a medium, although the particles from which the medium is composed oscillate about a fixed point. A wave is a regular pattern of vertical or horizontal displacements. If the wave is traveling it is the pattern of displacements that progresses, but not the particles that are displaced. Sound waves (*see* SPEED OF SOUND) and most ATMOSPHERIC WAVES are traveling waves.

tree climate Any climate in which trees are able to grow. This includes all climates in which the mean summer temperature is at least 50° F (10° C) except desert and some savanna and steppe climates that are too arid to support trees.

tree line The climatic limit beyond which temperatures are too low to permit trees to grow. The tree line marks the boundary between tundra, with vegetation that includes small, stunted trees, and the bare rock, snow, and ice of the high Arctic and Antarctic. On mountains the height of the tree line varies with latitude and with the CONTINENTALITY or OCEANICITY of the climate. The variation with latitude occurs because the tree line is determined by temperature, the temperature on a mountainside is determined by the ENVIRONMENTAL LAPSE RATE (ELR), and so the temperature at any given elevation measured from a sea-level DATUM depends on the sea-level temperature. In the central Alps of Europe the tree line is at 6,500–7,000 ft (2,000–2,100 m), in the Rocky Mountains it is at about 12,500 ft (3,800 m), and in the mountains of New Guinea it is at 12,200–12,600 ft (3,700–3,800 m).

Trees do not usually grow where the mean summer temperature is lower than 50° F (10° C), so it is possible to calculate the approximate height of the tree line from the temperature at the foot of the mountain. Assume that the temperature at the TROPOPAUSE is –74° F (–59° C) and the height of the tropopause is 36,000 ft (11 km). Subtract the low-level temperature from –74° F (–59° C) and divide the result by 36 (or 11) to give the ELR per 1,000 ft (or kilometers). Then calculate the height at which the temperature falls below 50° F (10° C). For example, the mean summer temperature in Seattle, Washington, is 71° F (22° C), so if there were a mountain in Seattle the tree line on it would be at a little over 5,000 ft (1.6 km).

tree rings The concentric rings that can be seen in a cross section of the trunk or large branch of a tree. This pattern results from secondary growth.

Trees, like all plants, grow by extending the length of the main stem and branches. This is called *primary growth*. Most plants, including trees, also grow thicker stems and branches. This thickening is called *secondary growth* and it is secondary growth that produces the rings in woody plants.

The outside of a tree trunk is protected by bark. This is not simply a rough layer of dead tissue, however, but three layers, two of which are living. The rough outer layer consists of dead cells with a waxy coating. They are called *cork cells* and they form the outer skin to protect the living cells beneath from injury and to keep out water. Immediately beneath the cork cells there is a layer of living cells, called the *cork cambium*. These cells divide, but after a few weeks they die and become cork cells. A while later the outer skin splits and more cork cambium cells are produced to fill the gaps, and then they also die and become cork cells. This is how the trunk acquires its rough surface.

Beneath the cork cambium, the innermost layer of the bark consists of tubular arrangements of cells composing tissue called *phloem*. Sugars produced in the leaves by PHOTOSYNTHESIS are transported through the phloem to every part of the plant. If the bark of a tree is cut all the way around the trunk, sugars can no longer reach the roots because the phloem has been severed, and the tree dies. Cutting the bark in this way is called *ring-barking*.

There is another layer of cambium beneath the phloem. This is called the *vascular cambium*. It consists of cells that divide to produce more cambium cells, phloem cells, and xylem cells. Xylem cells also form a tubular system for transport, in this case to carry water upward from the roots to every part of the plant. Xylem and phloem tissues are composed of dead cells. Each time a cell of the vascular cambium divides into two, one of the new cells remains as a cambium cell and the other dies to become either a xylem cell or a phloem cell. Xylem cells accumulate on the inside of

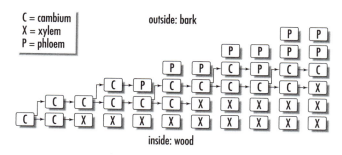

```
C = cambium
X = xylem
P = phloem
```
outside: bark

inside: wood

Tree rings. **The history of a single cell in the vascular cambium. Each time it divides, one of the resulting cells remains as a cambium cell and divides again, and the other dies to become either a phloem cell or a xylem cell.**

the vascular cambium. The accumulation of xylem cells shifts the cambium layer farther from the center of the trunk. This constitutes the secondary growth that increases the thickness of the trunk and branches.

On the inside of the xylem, old cells fill with metabolic waste products, including the lignin that toughens cell walls. The inner part of the trunk or branch, consisting of dead tissue, constitutes the heartwood and the outer part, containing the active xylem and vascular cambium, is the sapwood.

In those parts of the world with a seasonal climate, there is a period during which plant growth ceases. In temperate regions growth ceases in the winter and in other parts of the world it ceases during the dry season, which may be the summer. The cessation is of both primary and secondary growth.

When the dormant period ends the vascular cambium starts producing new xylem. The cells are large and have thin walls. They are a pale color. These cells form a cylinder surrounding the trunk or branch. Toward the end of the growing season the vascular cambium produces xylem made from smaller, darker cells with thicker walls. These form as another cylinder on the inside of the earlier xylem. When growth ceases for the year, the plant has produced two layers of xylem. In a cross section through the trunk or branch these are visible as a pale, circular band and a narrower, dark band. Each pair of circles, or rings, marks one year in the life of the plant. These are tree rings; by counting them it is possible to determine the age of the tree.

Tree rings also reveal information about the conditions for plant growth during the year when they

formed. If the weather was good the rings are broad, because growth was vigorous. If the weather was poor, the rings are narrow. If the weather was very bad, no growth at all may have occurred and in this case one year's rings are missing. This can be detected only by cross-dating the rings from that tree with a standard reference compiled from many trees. The dating of wood by means of tree rings is called DENDROCHRONOLOGY.

Using very long-lived trees, such as BRISTLECONE PINES, scientists have a record of tree rings that extends over more than 8,000 years. The use of tree rings to help in the reconstruction of past climates is DENDROCLIMATOLOGY.

(You can read more about how tree rings form, and about trees in general, in Michael Allaby, *Ecosystem: Temperate Forests* [New York: Facts On File, 1999].)

triple point *See* BOILING.

tropical air Warm air that originates either over oceans in the SUBTROPICAL HIGH-PRESSURE BELT, over continents at the edge of these high-pressure areas, or in the interior of continents in summer, when subsiding air produces high surface pressure (*see* SOURCE REGION) and tropical AIR MASSES. In summer, CONTINENTAL tropical (cT) air forms in the Northern Hemisphere over the Sahara, southern Europe, and Asia between about latitude 50° N and the Himalayas, and over the southwestern United States and Mexico. It carries hot, dry weather to the United States. The air is unstable (*see* STABILITY OF AIR), but so dry that it generates little cloud and it can produce DROUGHT. In the Southern Hemisphere cT air develops over southern Argentina, southern Africa, and much of the interior of Australia. In winter, cT air forms over a rather larger area of the southwestern United States and Mexico. In Eurasia it is pushed farther to the south, with its northern boundary at about latitude 40° N. It disappears from South America but covers a larger area in both southern Africa and Australia. MARITIME tropical (mT) air forms in winter over all the oceans between the equator and about latitude 40° N and S in both hemispheres. In summer it extends farther north, to about 55° N. Maritime tropical air is very warm; its temperature increases in an ADIABATIC process in the subsiding air. It is also very humid and stable.

Tropical Analysis and Forecast Branch *See* TROPICAL PREDICTION CENTER.

Tropical Atmosphere Ocean (TAO) A monitoring network that uses an array of moored buoys to gather data from the sea surface and sea-level atmosphere.

tropical cyclone An area of low surface pressure that generates fierce winds and rain to become the biggest and most violent type of atmospheric disturbance experienced on Earth. TORNADOES often have greater wind speeds, but they are very local. Tropical cyclones affect a much larger area than tornadoes and contain storm clouds that often produce tornadoes.

Tropical cyclone is their scientific name, but they have several common names. Those that occur in the North Atlantic are called *hurricanes*. Tropical cyclones never occur in the South Atlantic, because over the Atlantic the EQUATORIAL TROUGH does not move far enough south of the equator to provide the conditions they need. Those that occur in the North and South Pacific and the East China Sea and South China Sea are called *typhoons*. In the vicinity of Indonesia a tropical cyclone is known as a *bagyo* and near Australia it is called a *willy-willy*. Those that form over the Bay of Bengal are called *cyclones*.

Tropical cyclones develop in or close to the equatorial trough around an area where the pressure is slightly lower than that of the surrounding air. This is called

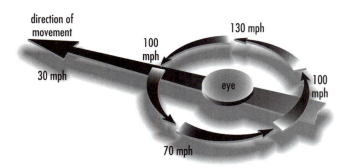

Tropical cyclone. **Wind speeds around the eye of a tropical cyclone vary as a result of the movement of the cyclone itself. On one side of the eye the speed of the cyclone must be added to the wind speed, and on the other side it must be subtracted from it. In this example, the cyclone is traveling at 30 mph (50 km h⁻¹) and the wind speed is 100 mph (160 km h⁻¹). On one side of the eye the actual wind speed is therefore 130 mph (210 km h⁻¹) and on the other it is 70 mph (110 km h⁻¹).**

a *TROPICAL DISTURBANCE* and in the Atlantic it is often associated with an EASTERLY WAVE.

Air flows toward the low-pressure area. As it does so, the CORIOLIS EFFECT swings it to the right, and as it spirals into the low-pressure center it is accelerated by the conservation of its ANGULAR MOMENTUM. The air then spirals upward. The magnitude of the Coriolis effect is zero at the equator. A tropical cyclone cannot develop closer than 5° north or south of the equator, because the Coriolis effect is too weak to set the air turning.

If the temperature of the sea surface is at least 80° F (27° C) over a large area, the rate of EVAPORATION is high and the rising air carries a large amount of water vapor. The rising air cools in an ADIABATIC process and as it does so its water vapor condenses, releasing LATENT HEAT. This warms the air, causing it to rise farther, and towering CUMULONIMBUS clouds form.

If the air rises vigorously enough it pierces the TRADE WIND INVERSION. If the rising air is then able to flow into a high-level ANTICYCLONE, present either as the remains of a weather system that has almost dissipated or as the result of the accumulation of rising air, both the upper-level high pressure and the low-level low pressure intensify. Air in the upper-level anticyclone flows in the opposite direction from the CYCLONIC circulation of the rising air. The rising air is swept away, allowing more air to rise and perpetuating the flow. Alternatively, rising air may encounter and be removed by the JET STREAM.

Three conditions must be met in order for a tropical cyclone to develop. The sea-surface temperature must be at least 80° F (27° C) over a large area. There must be an area of low pressure no closer to the equator than 5° north or south. There must be WIND SHEAR at high level to remove the rising air. The second and third of these conditions may be met at any time of year, but the surface water reaches a high enough temperature only after it has warmed through the summer. It is then as warm as it is possible for the sea to become, because at this temperature the rate of evaporation is such that the latent heat of vaporization absorbed from the sea prevents the temperature from rising further. Consequently, there is a season for tropical cyclones. It begins in late summer and ends in late autumn, when the equatorial trough moves toward the equator and the sea starts to cool.

When fully developed, a tropical cyclone consists of a central eye surrounded by an EYEWALL and then by several concentric bands of cloud. The air is cool inside the eye of a TROPICAL STORM. There are clouds, some of which give precipitation. As the storm intensifies into a tropical cyclone the air in the eye becomes warmer than the air surrounding the eye. Meteorologically, it is the warm air in the eye that distinguishes a tropical cyclone from a tropical storm, rather than the increase in wind speed. The sky clears and conditions are calm, with a wind of no more than 10 mph (16 km h^{-1}).

The strongest winds and precipitation are in the eyewall. This consists of cumulonimbus cloud towering sometimes to 59,000 ft (18 km). Beyond this bank of cloud, in which air is rising vigorously, there is a region of clear skies and subsiding air. This is surrounded by a further bank of CUMULIFORM cloud. The individual clouds become smaller with each band. The cyclone may extend to a diameter of up to 600 miles (965 km). Tropical cyclones are classified according to the surface pressure in the eye and the wind speeds in the eyewall on the SAFFIR/SIMPSON SCALE OF HURRICANE INTENSITY.

Low pressure in the eye causes the sea to rise. A fall in pressure of 1 millibar produces a sea-level rise of almost 1/2 in (1.2 cm). In the eye of a category 1 hurricane the sea level rises by about 14 in (35.6 cm) and in a category 5 hurricane by about 40 in (102 cm). The elevated sea level contributes to the severity of the STORM SURGE that is produced as a tropical cyclone crosses a coast.

Tropical cyclones move in a westerly direction, then turn away from the equator. In the case of hurricanes, this carries them into the Caribbean, and, depending on where they turn northward, they may cross the coast of the United States or miss it and remain over the sea. Those that remain over the sea for most of the time may travel all the way to Canada and even reach northwestern Europe, weakening all the time.

They travel initially at 10–15 mph (16–24 km h^{-1}), but as they move away from the equator they accelerate, sometimes to double that speed. Because the entire system is moving, the wind speed on one side of the eye is greater than that on the opposite side. If the storm is moving at 30 mph (50 km h^{-1}) and the wind speed in the eyewall is 100 mph (160 km h^{-1}), then on one side the two speeds combine to produce winds of 130 mph (210 km h^{-1}) and on the other side one must be sub-

tracted from the other, producing winds of 70 mph (110 km h^{-1}).

A tropical cyclone derives its energy from the condensation of water vapor, and therefore it retains its strength only for as long as it has an ample supply of warm water. As soon as it crosses land or seawater cooler than 80° F (27° C) the cyclone begins to weaken.

EL NIÑO episodes raise the tropical sea-surface temperature, thereby increasing the number and intensity of tropical cyclones. LA NIÑA events lower the water temperature, with an opposite effect.

The number of tropical cyclones in a season also varies in a cycle of 20–30 years according to some scientists and 25–50 years according to others. From the 1970s until 1994 tropical cyclones were fairly uncommon. Their frequency increased in 1995. Between 1950 and 1990 there were an average of 9.3 tropical storms, 5.8 hurricanes, and 2.2 major hurricanes a year in the Atlantic. Between 1995 and 1999, there were an average 13 tropical storms, 8.2 hurricanes, and 4.0 major hurricanes a year. Forecasters expect the increase to be maintained into the early decades of the 21st century. Overall, however, there has been no increase in either the frequency or severity of tropical cyclones since 1940 and between 1940 and 1990 the mean sustained wind speed decreased in hurricanes developing in the Atlantic Ocean and Caribbean Sea.

(You can learn more about tropical cyclones in Michael Allaby, *Dangerous Weather: Hurricanes* [New York: Facts On File, 1997].)

Tropical Cyclone Program   A branch of the INTEGRATED HYDROMETEOROLOGICAL SERVICES CORE of the United States NATIONAL WEATHER SERVICE that coordinates monitoring and research into TROPICAL CYCLONES. The program focuses mainly on increasing the accuracy of predicting tropical cyclone tracks and the intensity of cyclones, in order to provide better protection for communities that may be exposed.

(You can learn more about the program at www.nws.noaa.gov/om/tropical.htm.)

tropical depression   An area of low pressure that develops in the TROPICS through the CONVERGENCE of air at low level, which causes the air to rise, producing low pressure at the surface. Tropical depressions are not associated with FRONTAL SYSTEMS. Winds around the depression blow at less than 38 mph (61 km h^{-1}). If

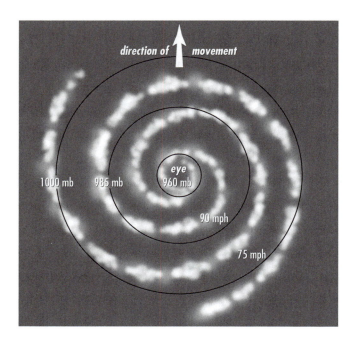

Tropical cyclone structure. **The structure of a tropical cyclone seen from above. The shaded region around the eye represents the eyewall, which is the area of densest cloud and strongest wind. Pressures and wind speeds at varying distances from the eye are indicated.**

the depression deepens and these speeds increase, the depression is reclassified as a TROPICAL STORM.

tropical desert The part of a SUBTROPICAL DESERT that lies within the TROPICS, in a latitude lower than 23.5° N or S. Tropical desert climates fall within type BWh of the KÖPPEN CLIMATE CLASSIFICATION, group 3 of the STRAHLER CLIMATE CLASSIFICATION, and type EA'd of the THORNTHWAITE CLIMATE CLASSIFICATION.

tropical desert and steppe climate In the STRAHLER CLIMATE CLASSIFICATION a climate in his group 1, which comprises climates controlled by equatorial and tropical AIR MASSES. This climate affects land areas in latitudes 15°–35° in both hemispheres and is associated with continental tropical air masses that develop in high-pressure cells in the upper TROPOSPHERE above land areas lying on the TROPICS of Cancer and Capricorn. The climate is hot, with a moderate range of temperature over the year, and is arid or semiarid. This climate is designated *BWh* (desert) and *BSh* (steppe) in the KÖPPEN CLIMATE CLASSIFICATION.

tropical disturbance An incipient TROPICAL STORM that is not associated with a FRONTAL SYSTEM. It is caused by the CONVERGENCE of air at a low level, which forces the air to rise and produces a region of low atmospheric pressure at the surface. A disturbance may produce nothing more than single CUMULUS clouds that survive for only a few hours. The clouds are often aligned to form CLOUD STREETS. More intense disturbance can produce much bigger clouds and SQUALL LINES. TROPICAL CYCLONES begin as tropical disturbances.

tropical monsoon climate A warm, tropical climate that has abundant rainfall and supports luxuriant rain forest vegetation but has a dry season in winter.

Tropical Ocean Global Atmosphere (TOGA) A program that ran from 1985 until 1994. It implemented an observational system for oceanic and atmospheric measurements with the aim of improving the understanding of ENSO events and, from that, their prediction. TOGA comprised satellite and surface measurements.

Tropical Prediction Center (TPC) The part of the NATIONAL WEATHER SERVICE that issues watches and warnings about dangerous weather conditions in the TROPICS. The TPC is based at the campus of Florida International University, in Miami. It has three branches. The NATIONAL HURRICANE CENTER (NHC) maintains a watch on TROPICAL CYCLONES. The Tropical Analysis and Forecast Branch (TAFB) concentrates on forecasting, especially for ships and aircraft, and interprets satellite data and provides satellite rainfall estimates. The TAFB also supports the NHC. The Technical Support Branch (TSB) provides assistance with the computer and communications systems.

(You can learn more about the TPC at www.nhc. noaa.gov/aboutintro.html.)

tropical rain forest *See* RAIN FOREST.

tropical rain forest climate *See* WET EQUATORIAL CLIMATE.

tropical rainy climate *See* TROPICAL WET–DRY CLIMATE.

tropical regime *See* RAINFALL REGIME.

tropical storm A TROPICAL DEPRESSION that has deepened until the winds around it are blowing at speeds of 38–74 mph (61–119 km h⁻¹). When the mean wind speed exceeds 74 mph (119 km h⁻¹) the storm is reclassified as a TROPICAL CYCLONE. For purposes of identification tropical storms are given names, which they retain if they subsequently strengthen to become cyclones. Between 80 and 100 tropical storms develop in most years. Up to about half of those that survive long enough to cross an ocean develop into full cyclones.

tropical storm warning A notification of the approach of a TROPICAL STORM to an inhabited area of the United States that is issued by the National Hurricane Center, in Miami, Florida, and broadcast on local radio, television, and WEATHER RADIO. A hurricane warning is broadcast when the storm is expected to arrive within 24 hours or less. Persons receiving a tropical storm warning should prepare for the imminent arrival of the hurricane and should leave a radio or television switched on and tuned to the local station. Updated information and safety instructions will be broadcast and should be obeyed promptly.

tropical storm watch A preliminary notification of the approach of a TROPICAL STORM to an inhabited area of the United States that is issued by the National Hurricane Center, in Miami, Florida, and broadcast on local radio, television, and WEATHER RADIO. Based on observations of the advancing storm and predictions of its future track, a tropical storm watch is issued one or two days before the expected arrival of the storm. It is given to people residing in a belt of coastline and its hinterland centered on the point where it is anticipated that the STORM CENTER will cross the coast. The affected area extends for a distance equal to about three times the radius of the hurricane to each side of this point. On receipt of a tropical storm watch people should prepare for the arrival of the storm.

tropical wave *See* EASTERLY WAVE.

tropical wet–dry climate In the STRAHLER CLIMATE CLASSIFICATION a climate in his group 1, which comprises climates controlled by equatorial and tropical

AIR MASSES. Tropical wet–dry climates occur in latitudes 5°–25° in both hemispheres. They are marked by a seasonal alternation between moist maritime tropical or maritime equatorial air and dry continental tropical air. This seasonal change produces a rainy season in summer and a dry season in winter. There are two types of tropical wet–dry climate: the tropical rainy climate, including the savanna climate, and the temperate rainy, or humid mesothermal, climate. These are designated *Aw* and *Cwa*, respectively, in the KÖPPEN CLIMATE CLASSIFICATION.

tropical year *See* ORBIT PERIOD.

Tropics The two lines of latitude at which the Sun is directly overhead at noon on one of the two SOL-

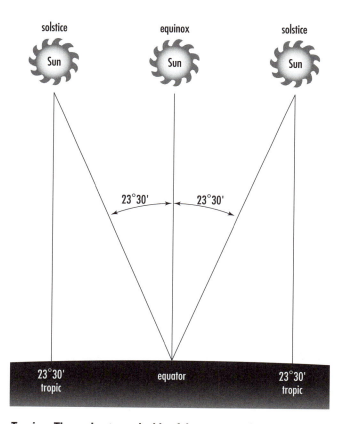

Tropics. The region to each side of the equator where the Sun is directly overhead at noon on at least one day every year. The Tropics are located at 23°30′ N and 23°30′ S. This is also the angle between a point on the equator and the Sun when the Sun is overhead at one or the other Tropic, and the angle by which the rotational axis of the Earth is tilted with respect to the plane of the ecliptic.

STICES. There is one tropic on each side of the equator, at latitudes 23°30' N and S. The northern tropic is known as the *tropic of Cancer* and the southern as the *tropic of Capricorn*. Tropical regions (the Tropics) are those that lie between the two tropics. With respect to the PLANE OF THE ECLIPTIC, the rotational axis of the Earth is tilted 23°30' from the vertical. Because of this, in the course of its yearly orbit about the Sun, first one hemisphere and then the other is tilted toward the Sun. To an observer at the equator, the noonday Sun is directly overhead at each EQUINOX, but at the SOLSTICES, in mid-December and mid-June, it is 23°30' to the south and north, respectively.

tropopause   The boundary that separates the TROPOSPHERE from the STRATOSPHERE. It is almost horizontal and clearly defined. Below the tropopause, the air temperature decreases with height. Above the tropopause, there is sometimes a temperature INVERSION, but there is always a layer in the lower stratosphere where the temperature remains constant with height—an ISOTHERMAL LAYER. Air can rise through the troposphere by convection, but it is unable to cross the tropopause, because its temperature and density are then equal to those of the surrounding air. Since air cannot cross the tropopause, neither can water vapor, AEROSOL particles, or other substances carried in the air. Very large CUMULONIMBUS clouds may generate vertical air currents that are vigorous enough to create TROPOPAUSE BREAKS that penetrate the tropopause, carrying moisture into the stratosphere, and violent volcanic eruptions can inject both gases and particles. Apart from these exceptions, however, water vapor and particles are held below the tropopause and stratospheric air is extremely dry.

The height of the tropopause varies according to the vigor of the convective air movements in the troposphere. These are most vigorous in the TROPICS and least vigorous in the Arctic and Antarctic. Consequently, the height of the tropopause averages about 10 miles (16 km) over the equator, 7 miles (11 km) in middle latitudes, and 5 miles (8 km) at each Pole. Pressure at the tropopause ranges from about 100 mb at the equator to about 300 mb at the Poles, reflecting the greater height of the equatorial tropopause.

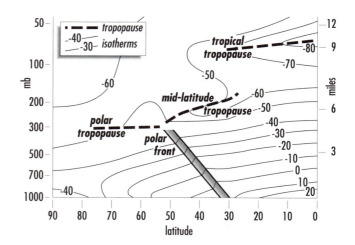

Tropopause. **A diagram that relates pressure along the left vertical axis, height along the right vertical axis, and latitude along the horizontal axis to temperature.**

Temperature decreases with height throughout the troposphere. Consequently, the deeper the troposphere the lower the temperature is at the top. This means the decrease in surface temperature from the equator to the Poles is reversed at the tropopause. At the equator, the temperature at the tropopause averages –85° F (–65° C), in middle latitudes it is about –67° F (–55° C), and over the Poles it is about –22°F (–30° C).

tropopause break   A discontinuity in the TROPOPAUSE through which tropospheric air and water vapor are able to cross into the STRATOSPHERE and stratospheric OZONE sometimes enters the TROPOSPHERE. Breaks occur where the north–south temperature gradient is most sharply defined, at about latitudes 30° and 50° N and S.

tropopause chart   A SYNOPTIC CHART that shows the vertical distribution of pressure through the TROPOSPHERE by height, ISOTHERMS, height of the TROPOPAUSE, and TROPOPAUSE BREAKS.

tropopause fold   A downward depression in the TROPOPAUSE that carries stratospheric air into the troposphere, and sometimes as far as the surface. The stratospheric air is very dry and enriched in OZONE.

tropophyte   A plant that is adapted to a climate with pronounced wet and dry seasons.

troposphere   The lower layer of the atmosphere that extends from the surface to the TROPOPAUSE. It is the layer in which vertical and horizontal currents cause the air to be turbulent and it is the layer that contains virtually all of the atmospheric water vapor. The turbulent transport of heat combined with the condensation of water vapor to form clouds and deliver precipitation make the troposphere the atmospheric layer in which almost all weather phenomena occur.

trough   A long, tonguelike protrusion of low pressure into an area of higher pressure. The waves in the POLAR FRONT JET STREAM associated with the INDEX CYCLE that extend toward the equator are also called *troughs.*

trowal   Canadian term for a TROUGH of warm air that is held high above the surface by an OCCLUSION. It often produces layered clouds similar to those associated with a WARM FRONT, and precipitation. As it passes and cold air replaces it, the sky clears. Trowals are often shown on Canadian weather maps.

true altitude   *See* CORRECTED ALTITUDE.

TSB   *See* TROPICAL PREDICTION CENTER.

tsunami   An ocean wave that is sometimes very large when it reaches the coast and that can cause great devastation. Unlike other waves, it is caused neither by wind nor by tidal movements—despite its old common name *tidal wave.* The modern name, *tsunami,* is Japanese for "harbor wave," which is a much more accurate description. Tsunamis are caused by major disturbances on the ocean floor. These include earthquakes, eruption of submarine volcanoes, and large submarine mudslides that sometimes occur when sediments become unstable and slide down the outer edges of a continental shelf.

A disturbance of this kind sends a shock wave through the entire depth of ocean around it. This can be seen in war movies, where an explosion at depth causes a momentary "shudder" at the surface before the water displaced by the explosion is thrown into the air. A tsunami, therefore, is a shock wave transmitted through the ocean. Its wavelength and period are very long (*see* WAVE CHARACTERISTICS). Typically, a tsunami has a wavelength of about 160 miles (200 km) and a wave period of 15–20 minutes. Across the open sea they travel at a speed given by $\sqrt{(gd)}$, where g is the acceleration due to gravity (32 ft per second per second, or 9.81 ms^{-2}) and d is the depth of water. Because the average depth of the oceans is about 13,000 ft (4,000 m), a tsunami wave travels through the open ocean at about 385 KNOTS (715 km h^{-1}). The wave is very small, however: its amplitude is rarely more than about 20 in (50 cm). It is so small, in fact, and travels so fast, that out at sea sailors on ships fail to notice it.

When the wave enters shallow coastal waters it slows. Its wavelength shortens and its height increases. The size and form of tsunamis vary greatly, depending on the type, magnitude, and location of the disturbance that causes them and the configuration of the coasts they reach. Some appear as a breaker or a series of breakers. Others do not break as surf does but resemble a rapidly rising TIDE, which continues to rise until it has traveled far beyond the ordinary tidal limit. Often there is a warning of the impending approach of a tsunami. Water that is rising and falling against the shore with the normal movement of the waves retreats much farther than usual and remains for a few moments at a very low level. Then, when it advances, its advance takes it very much higher than usual. Anyone observing this phenomenon is well advised to seek safety immediately, by moving as quickly as possible to the highest ground within reach.

(For more information about tsunamis, see Michael Allaby, *Dangerous Weather: Floods* [New York: Facts On File, 1998]; Frank I. González, "Tsunami!" *Scientific American,* 280, 5, pp. 44–55, May 1999; and Richard Monastersky, "Waves of Death," *Science Online,* October 3, 1998, at www.sciencenews.org/sn_arc98/10_3_98/Bob2.htm.)

tsunami warning   A notification that is broadcast to people living in a coastal area when a TSUNAMI is approaching. The warning is similar to a FLASH FLOOD WARNING. Persons receiving it should move inland immediately and should not attempt to return to their homes until the emergency services tell them it is safe to do so.

tuba A supplementary feature of clouds of the types that may give rise to TORNADOES or WATERSPOUTS. It consists of a tapering, funnel-shaped projection beneath the cloud base. *Tuba* is the Latin word that means "trumpet."

tundra climate In the CLIMATE CLASSIFICATION devised by MIKHAIL I. BUDYKO, a climate in which the RADIATIONAL INDEX OF DRYNESS has a value of less than 0.33. *See also* DRY SUBHUMID CLIMATE.

In the STRAHLER CLIMATE CLASSIFICATION a climate in his group 3, comprising climates controlled by polar and arctic AIR MASSES. A tundra climate occurs in latitudes north of 55° N and south of latitude 50° S along coasts exposed to FRONTAL ZONES where maritime polar and continental polar air masses interact with arctic air masses to produce CYCLONIC storms. The climate is moist and very cold, although the proximity of the ocean moderates temperatures, which are milder than those of an ICE CAP CLIMATE. There is no summer or warm season. This is also designated a POLAR CLIMATE. *See also* MARINE SUBARCTIC CLIMATE.

turbidity A reduction in the TRANSPARENCY of the atmosphere that is caused by HAZE or AIR POLLUTION.

turbopause The poorly defined upper boundary of the TURBOSPHERE, at a height of about 60 miles (100 km), above which the equilibrium of air is not maintained by CONVECTION. Above the turbopause air molecules move by DIFFUSION, which causes them to form layers according to their weights. Heavier gases lie beneath lighter gases. Above about 75 miles (120 km) PHOTODISSOCIATION by solar ULTRAVIOLET RADIATION separates oxygen molecules, O_2 + photon (UV) → O + O, and more than half of the oxygen is present as single atoms.

turbosphere The region of the atmosphere in which the equilibrium of the air is maintained mainly by CONVECTION. This means the air in the turbosphere is being constantly mixed by vertical convective motion. Air expands and becomes less dense as it rises and is compressed and becomes more dense as it sinks, so within the turbosphere the layering of air is based on its density, rather than the molecular weight of its constituent gases. The turbosphere extends from the surface to a height of about 60 miles (100 km), although

Turbulent flow. A flow of air, or other fluid, that is irregular, with many eddies.

its upper boundary, called the TURBOPAUSE, is poorly defined.

turbulence *See* TURBULENT FLOW.

turbulent flow (turbulence) The movement of a fluid in which elements of the fluid follow streamlines that cross each other, so the flow passing any particular point changes speed and direction in an irregular and unpredictable fashion. Except in the laminar boundary layer (*see* LAMINAR FLOW) air movement is almost always turbulent. Turbulent flow is the cause of the GUSTINESS and continually changing speed and direction of the wind felt at ground level, but turbulent flow acts vertically as well as horizontally. Where two belts of air are moving at different speeds, air at the boundary between them can be set rotating in a vertical plane. Turbulent flow perpetuates itself, because erratic movement in one place jostles the adjacent air and sets it moving, and FRICTION, convective movement, and pressure differences are continually introducing new disturbances to the flow.

twister A popular name for a TORNADO, which is a narrow column of rapidly rotating air that projects below the base of a storm cloud and follows an erratic path as it moves over the surface.

Tyndall, John (1820–1893) Irish *Physicist* John Tyndall was born at Leighlin Bridge, County Carlow, in southwestern Ireland, on August 2, 1820, the son of a police officer. He was educated at the school in the nearby town of Carlow, where he acquired a sound background in mathematics and English. He left school at 17 and two years later, in 1839, he joined the Ordnance Survey, the British government agency responsible for mapping the country, to train as a surveyor. In the 19th century the survey was part of the military establishment. Tyndall worked on the survey of Ireland (which was then part of Britain), and when that task was completed in 1842 he transferred to the English survey. He was dismissed in 1843, however, because he had complained about the efficiency of the Ordnance Survey and about its treatment of Irish people. He returned to Ireland but then obtained a position with a firm of surveyors in England and conducted surveys for the companies that were rapidly expanding the railroad network.

The railroad boom ended in 1847 and Tyndall found a post as a mathematics teacher at Queenwood College, Hampshire. There he became friendly with the science teacher, the chemist Edward Frankland (1825–99), and through Frankland's influence he developed an interest in science. The two men decided to further their education and in October 1848 they enrolled together at the University of Marburg, Germany. Tyndall was then 28 years old.

Tyndall studied physics, calculus, and chemistry at Marburg. His chemistry teacher was Robert Bunsen (1811–99), after whom the laboratory Bunsen burner is named. Bunsen helped and encouraged Tyndall, who began the course with only a limited knowledge of German. Despite this he qualified for a doctorate in 1850. He specialized in the study of magnetism and optics and continued his research for an additional year in Germany. Short of funds, Tyndall returned to his post at Queenwood College in 1851; there he spent the next two years, supplementing his salary by translating and reviewing scientific articles from foreign journals.

By this time Tyndall was beginning to earn the respect of other scientists. He was elected a fellow of the Royal Society in 1852 and in 1853 it was arranged that he should give a lecture at the Royal Institution in London. This was so successful that he was invited to give a second lecture, and then a whole course of lectures. A few months later he was appointed professor of natural philosophy at the Royal Institution. Michael Faraday (1791–1867) was the superintendent of the Royal Institution. The two men were close friends and

John Tyndall. **The Irish physicist who discovered that certain atmospheric gases are transparent to radiation and that others absorb it.** *(John Frederick Lewis Collection, Picture and Print Collection, The Free Library of Philadelphia)*

when Faraday died Tyndall succeeded him as superintendent.

Tyndall acquired a reputation as a brilliant and entertaining lecturer. In 1872 and 1873 he made a lecture tour of the United States. Tyndall paid the proceeds from those lectures into a trust for the advancement of American science. He was also a talented writer and journalist and did much to popularize science in Britain and the United States. From 1859 to 1868 he was also professor of physics at the Royal School of Mines, where he gave a series of popular lectures on science, working in collaboration with another close friend, Thomas Henry Huxley (1825–95). The success of these lectures prompted the British Association for the Advancement of Science to hold similar lectures at its annual meetings.

In addition to lecturing, teaching, and writing, Tyndall was conducting his own research in the laboratories in the basement of the Royal Institution. At first these were on diamagnetism. Then, starting in

January 1859, he studied radiant heat. He found that although oxygen, nitrogen, and hydrogen are completely transparent to radiant heat (INFRARED RADIATION), other gases, especially water vapor, carbon dioxide, and ozone, are relatively opaque. Tyndall said that without water vapor the surface of the Earth would be permanently frozen. Later he speculated about the way changing the concentrations of these gases might affect the climate. This was the first suggestion of what is now known as the GREENHOUSE EFFECT.

In 1869 Tyndall discovered what is now called the *Tyndall effect*. He was investigating the way light passes through liquids and found that light passes unimpeded through a solution or pure solvent, but that the light beam becomes visible in a colloidal solution (*see* COLLOID). This suggested that although the colloidal particles cannot be seen, they are big enough to scatter light. From this he reasoned that particles in the atmosphere should also scatter light and that air molecules should scatter blue light more than red light, causing the blue color of the sky. LORD RAYLEIGH confirmed this in 1871. Tyndall used the effect to measure the pollution of London air. It also led him to suspect that the air may contain microscopic living organisms, and he was able to show, in 1881, that bacterial spores are present in even the most carefully filtered air. This confirmed the rejection by Louis Pasteur (1822–95) of the idea that life is generated spontaneously and provided added support to the germ theory of disease.

Tyndall had many interests. He was a keen mountaineer. He climbed Mont Blanc (15,781 ft [4,813 m]) several times and was the first person to climb the Weisshorn (14,804 ft [4,515 m]). In 1849 he and Huxley began to take annual vacations in the Alps. This led them to publish a treatise, *On the Motion and Structure of Glaciers,* about glacial movement. He also invented many scientific instruments. He led an active civic life, acting as a government adviser and helping to investigate the causes of mining and other industrial accidents. He was especially concerned about safety at sea. Tyndall also devoted considerable energy to championing the cause of those he believed had been badly treated.

In the 1860s, for example, he lectured widely on the work of the German physicist Julius Robert Mayer (1814–78). Mayer was years ahead of all his contemporaries in calculating the mechanical equivalent of heat, proposing the conservation of energy (*see* THERMODYNAMICS, LAWS OF), and suggesting that solar energy is the ultimate source of all the energy on Earth and that solar energy is produced by the conversion of KINETIC ENERGY into radiant energy. Yet Mayer's work aroused little interest, and more famous scientists, including James Prescott Joule (1818–89), Hermann Ludwig Ferdinand von Helmholtz (1821–94), and Lord Kelvin (1824–1907), were given the credit for discoveries Mayer had made first. Mayer became so depressed at the lack of recognition and his complete failure to claim priority for his discoveries that in 1849 he attempted suicide, and in 1851 he was admitted to a mental institution that would be judged harsh and cruel by modern standards. Tyndall worked tirelessly to put right the wrongs Mayer had suffered and he finally succeeded in having Mayer's achievements recognized.

In 1876, when he was 56, Tyndall married Louisa Hamilton. They had no children but lived together devotedly until Tyndall's death.

Despite his energy, throughout his adult life Tyndall slept badly and was often unwell. By the 1880s his health was deteriorating. He retired from the Royal Institution in 1887 and went to live at Hindhead, Surrey. He gave up most of his scientific work but took an active part in political campaigns, including the campaign against the Home Rule Bill, which would have made Ireland independent of Britain. In 1891, for the first time in 30 years, he was too ill to take his annual climbing vacation in the Alps. His insomnia became steadily worse and he experimented with a variety of drugs to treat it. He died on December 4, 1893, after his wife had accidentally given him an overdose of the sedative chloral hydrate.

(You can learn more about John Tyndall at www.tyndall.uea.ac.uk/tyndall.htm; www.irsa.ie/Resources/Heritage.tyndall.html; and earthobservatory.nasa.gov/Library/Giants/Tyndall.)

typhoon A TROPICAL CYCLONE that forms over the Pacific. Typhoons develop in both the North and South Pacific. They travel from east to west in both hemispheres, intensifying from TROPICAL STORMS to storms of full hurricane strength (*see* SAFFIR/SIMPSON HURRICANE SCALE) when they reach the western side of the ocean. VORTICITY then causes their tracks to turn away from the equator, toward the north in the Northern Hemisphere and toward the south in the Southern Hemisphere. Northern Hemisphere typhoons often reach the Philippines and South Asia, those in the Southern Hemisphere may reach Australia, and typhoons in either hemisphere may reach Indonesia, which straddles the equator. Typhoons are known as *bagyo* in the Philippines and Indonesia and sometimes called *willy-willies* in Australia. Of all the tropical cyclones that occur, 90 percent are either CYCLONES or typhoons, and they are often extremely severe.

U

UARS *See* UPPER ATMOSPHERE RESEARCH SATELLITE.

ubac Sloping ground that faces away from the direction of the equator and therefore is shaded. In the Northern Hemisphere an ubac is a north-facing slope. The opposite slope is called an *ADRET*.

Ulloa's ring *See* SUPERNUMERARY BOW.

ultraviolet index (UVI) A guide to the intensity of ULTRAVIOLET RADIATION, reported as an index value that is related to the duration of exposure that causes sunburn in the most susceptible people, who are those with pale skin. The index was developed by the U.S. Environmental Protection Agency, and since June 1994 predicted UVI values for certain cities have been issued regularly by the NATIONAL WEATHER SERVICE. The index runs from 0 to 15 and the reported values are usually accompanied by recommended precautionary measures, which include the appropriate sun protection factor (SPF) for sunscreens.

ultraviolet radiation (UV) Electromagnetic radiation at wavelengths between about 4 nanometers (nm) and 400 nm. This wave band is approximate and there is overlap at each end, with X rays at the short-wave end and visible violet light at the long-wave end. UV radiation at 400–300 nm is sometimes called *near UV*, UV at 300–200 nm is *far UV*, and UV at less than 200 nm is *extreme UV* or *vacuum UV*. The UV wave band is also divided into UV-A (315–380 nm), UV-B or soft UV (280–315 nm), and UV-C or hard UV (shorter than 280 nm).

UV Index

UV category	UVI value	Time to burn (minutes)	Precautions
Minimal	0–2	30–60	Wear a hat
Low	3–4	15–20	Wear a hat; use sunscreen SPF 15+
Moderate	5–6	10–12	Wear a hat; use sunscreen SPF 15+; keep in shade
High	7–9	7–8.5	Wear a hat; use sunscreen SPF 15+; keep in shade; stay indoors between 10 A.M. and 4 P.M.
Very high	10–15	4–6	Stay indoors as much as possible; outdoors wear a hat and use sunscreen SPF 15+

Uma A CYCLONE that struck Vanuatu on February 7, 1987. It killed at least 45 people.

umbra *See* PENUMBRA.

Umkehr effect A reversal (German *Umkehr*) in the relative intensities of light from directly overhead at two WAVELENGTHS as the Sun's ZENITH ANGLE varies. One of the two wavelengths is more strongly absorbed by OZONE than is the other, and so a series of measurements of the change in their relative intensities can be used to determine the vertical distribution of ozone.

unc *See* UNCINUS.

uncinus (unc) A species of CIRRUS cloud (*see* CLOUD CLASSIFICATION) that consists of long filaments that end in hooks, in the shape of commas, or in tufts, but not with a protruding upper part. Uncinus forms when the cloud is being swept out by strong winds in the upper TROPOSPHERE. *Uncinus* is a Latin word that means "hook."

unconfined aquifer *See* AQUIFER.

undercast A complete cloud cover (ten-tenths or 8 OKTAS) as it appears to, and is reported by, the pilot of an aircraft flying above the cloud.

undulatus A variety of cloud (*see* CLOUD CLASSIFICATION) in which sheets, layers, or patches of cloud undulate like waves. It is most often seen with the cloud genera CIRROCUMULUS, CIRROSTRATUS, ALTOCUMULUS, ALTOSTRATUS, STRATOCUMULUS, and STRATUS.

The name of the variety is derived from the Latin *unda,* which means "wave."

UNEP *See* entries listed under UNITED NATIONS ENVIRONMENT PROGRAM.

United Nations Environment Program *See* EARTHWATCH PROGRAM, GLOBAL ENVIRONMENTAL MONITORING SYSTEM, GLOBAL ENVIRONMENT FACILITY, GLOBAL RESOURCE INFORMATION DATABASE, INTERGOVERNMENTAL PANEL ON CLIMATE CHANGE, MONTREAL PROTOCOL ON SUBSTANCES THAT DEPLETE THE OZONE LAYER, VIENNA CONVENTION ON THE PROTECTION OF THE OZONE LAYER.

United States Weather Bureau The federal agency that was instituted to gather meteorological data and to prepare and issue weather forecasts and warnings of severe weather. The inspiration to form the bureau was initially that of Thomas Jefferson (1743–1826), who had an abiding interest in meteorology and kept weather records over a number of years. In 1849 JOSEPH HENRY inaugurated the collection of meteorological data at a central point, and on February 9, 1870, President Ulysses S. Grant signed a joint resolution of Congress that authorized the secretary of war to establish a weather service within the army. The Army Signal Corps operated the service until July 1, 1891, when the Weather Bureau was created as a civilian service within the Department of Agriculture. The bureau was transferred to the Department of Commerce on June 30, 1940; became part of the Environmental Science Services Administration on July 13, 1965; and in 1967 was renamed the NATIONAL WEATHER SERVICE.

In 1898 the bureau started regular kite observations, which continued until 1933, and began regular balloon soundings in 1909. From 1925, data were also collected by aircraft, and in 1926 the Air Commerce Act made the bureau responsible for providing weather services for aviation. Experiments with radio communication began in 1901, and in 1939 the bureau introduced the first telephone weather service, in New York City.

When the WORLD METEOROLOGICAL ORGANIZATION was formed in 1951, the chief of the Weather Bureau, Francis W. Reichelderfer, was elected its first president.

(You can learn more about the history of the Weather Bureau and National Weather Service at www.nws.mbay.net/history.html and www.crh.noaa. gov/lmk/history1.htm).

universal gas constant *See* GAS CONSTANT.

Universal Time (UT) A name for GREENWICH MEAN TIME that was introduced in 1928 on the recommendation of the International Astronomical Union to prevent confusion: Universal Time is counted from midnight, whereas at the time of its introduction Greenwich Mean Time was counted from noon.

unrestricted visibility Horizontal VISIBILITY that is not obstructed or obscured for at least 7 miles (11 km).

unsettled   An adjective that describes weather conditions that are fine but may change at any time in the near future with the development of cloud and possibly PRECIPITATION.

upper air   The upper part of the TROPOSPHERE. There is no definite height at which the upper air commences, but the lowest UPPER-AIR CHARTS typically depict conditions at a level where the AIR PRESSURE is 925 mb (27 in mercury, 13.4 lb in⁻²).

upper-air chart (**upper-level chart**)   A CONSTANT-PRESSURE CHART that depicts the condition of the atmosphere at a pressure level in the upper TROPO-SPHERE. Upper-air charts are usually prepared for the pressure levels at 925, 850, 400, 300, 250, 200, 150, and 100 mb. Charts are sometimes prepared for higher levels, but conditions there have little immediate effect on the weather experienced at the surface. The charts are prepared by using a STATION MODEL similar to those used for surface charts, but omitting information about cloud cover, precipitation, visibility, and present weather. The contour lines on a constant-pressure chart link points that are the same height above sea level.

upper atmosphere   The part of the atmosphere that lies above the TROPOPAUSE.

upper-atmosphere dynamics   The movement of air in the UPPER ATMOSPHERE, but especially at heights of more than 300 miles (500 km), where the motion consists predominantly of ATMOSPHERIC TIDES, ATMO-SPHERIC WAVES, TURBULENT FLOW, and sound waves.

Upper Atmosphere Research Satellite (**UARS**)   A National Aeronautics and Space Administration (NASA) satellite that was launched on September 12, 1991. It carries six instruments that are designed to measure the temperature, chemical composition, and winds in and above the stratosphere. These are the CRYOGENIC LIMB ARRAY ETALON SPECTROMETER, IMPROVED STRATOSPHERIC AND MESOSPHERIC SOUNDER, MICROWAVE LIMB SOUNDER, HALOGEN OCCULTATION EXPERIMENT, HIGH-RESOLUTION DOPPLER INTERFEROM-ETER, and WIND IMAGING INTERFEROMETER. Together, they generate the most complete set of observational data for the upper atmosphere that have ever been produced.

upper front   A weather front that is present in the UPPER AIR but that does not extend to the surface.

upper-level chart   *See* UPPER-AIR CHART.

upper mixing layer   A region in the upper MESO-SPHERE, at a height of between about 50 and 80 km (30 and 50 miles), through which there is a rapid decrease of temperature with height. This is associated with considerable turbulence that mixes the constituents of the atmosphere.

uprush (**vertical jet**)   The very strong upcurrent that is found in a rapidly developing CUMULUS cloud. A cloud with an uprush is likely to grow into a CUMU-LONIMBUS and produce a THUNDERSTORM.

upslope fog   *See* HILL FOG.

upslope snow   *See* SNOWSTORM.

Upton Warren interstadial   An INTERSTADE that occurred during the last ice age (known as the WISCON-SINIAN in North America, the DEVENSIAN in Britain, and the WEICHSELIAN in continental Europe). The interstade lasted from about 43,000 years ago until 42,000 years ago. POLLEN ANALYSIS indicates that the climate was similar to that of southern Sweden today. There were few trees and even pine (*Pinus* species) and birch (*Betula* species) trees were scarce. There could certainly have been no woodlands. The remains of insects suggest summer temperatures reached at least 15° C (59° F). This is warm enough for trees, and their absence is believed to be due to the short duration of the interstade, which did not allow sufficient time for them to spread northward and become established. These conditions probably extended over much of Europe. They were first identified at Upton Warren, a place in Worcestershire, England.

upwelling   The rise of water all the way from near the ocean floor to the surface that occurs in particular regions of the ocean. Deep water is cold; by carrying it to the surface, upwellings affect the temperature of the air in contact with the surface.

Upwellings are caused by the EKMAN SPIRAL. Ocean currents are driven by the wind, and, like the wind, they are subject to the CORIOLIS EFFECT (CorF). The balance between CorF and friction between the wind and ocean surface produces two component forces of equal strength, one acting in the direction of the wind and the other acting at right angles to the wind direction. This results in the current's flowing at an angle of 45° to the wind direction—to the right of the wind in the Northern Hemisphere and to the left in the Southern Hemisphere. Below the surface the influence of the wind decreases and so the CorF becomes dominant. This has the effect of deflecting the current, and the overall result is that water in the surface boundary layer, down to about 25 m (82 ft), is slowly transported in a direction at right angles to the wind. Deeper water then rises to the surface to replace it. It is a slow movement, often at less than 1 m (3.3 ft) per day. The surface boundary layer that is subject to the Ekman spiral is known as the *Ekman layer.*

Where winds blow parallel to a north–south coastline and drive a current that flows near the coast, the Ekman spiral pushes the water away from the coast and deep water rises to the surface. This is known as *coastal upwelling* and it happens along the coasts forming the eastern margins of oceans, such as the western coast of North, Central, and South America. There, the ocean currents flow toward the equator—the California Current in the North Pacific and the Peru (or Humboldt) Current in the South Pacific. These carry cold water, but upwelling increases their cold influence. The difference is very marked. In August, the temperature of the Atlantic off the coast of North Carolina is often 21° C (70° F), and it can be warmer. The Pacific, off the coast of California in the same latitude, is about 15° C (59° F).

On each side of the equator the TRADE WINDS drive currents from east to west in both hemispheres. The CorF acts on the currents, pushing them away from the equator. The result is that the Equatorial Currents are displaced a little way from the equator and there is a region of upwelling between them; the rising water divides at the surface and flows north and south. This is known as *equatorial upwelling.*

Open ocean upwelling also occurs. Atmospheric CYCLONES produce winds that drive currents. Like other currents, these are deflected by the CorF and water rises to the surface layer. As a cyclone crosses the ocean it leaves behind it a wake of relatively cool water. The deeper the cyclone the more pronounced this wake is and it is very evident to the rear of a TROPICAL CYCLONE.

Ocean water is constantly turning over, but the process is extremely slow. The deep water that wells up to the surface was at the surface centuries earlier (*see* ATLANTIC CONVEYOR and NORTH ATLANTIC DEEP WATER). It sank in high latitudes and was very cold. Since the solubility of oxygen in water is inversely proportional to the temperature, the sinking water was rich in dissolved oxygen. It carried its oxygen to the ocean depths, where it sustained marine organisms. Inorganic nutrients, particularly nitrate (NO_3) and phosphate (PO_4), sink from the upper layers of the ocean and accumulate on the ocean floor. As it moves across the deep ocean floor, the deep water becomes enriched with nutrients, and when it wells up to the surface it takes them with it. This action greatly enriches the ordinarily nutrient-poor surface waters and supports large marine populations. If the prevailing winds should fail, or change direction, the wind-driven current may also cease to flow, and without its movement the upwellings also cease. This is what happens during an EL NIÑO. Its social as well as its meteorological effects can be serious. Fisheries that are sustained by the upwelling nutrients fail, and populations that are economically dependent on them suffer badly.

upwind In the direction from which the wind is blowing. WIND DIRECTION always refers to the direction from which the wind blows, not the direction in which it is blowing, so if the wind is a westerly, for example, the upwind direction is to the west. The opposite of upwind is downwind, meaning in the direction the wind is blowing.

upwind effect The increase in precipitation that occurs over the land or sea that lies on the UPWIND side of a range of hills or mountains. It occurs because air moving toward the OROGRAPHIC barrier rises before it reaches the solid surface. Consequently, air is cooling and its RELATIVE HUMIDITY is increasing for some distance upwind of the barrier.

urban boundary layer The layer of air over a city that extends from the top of the URBAN CANOPY to the uppermost limit of the region in which the climatic properties of the air are modified by the surface below. The urban surface is usually rougher and warmer and is often drier than that of the surrounding countryside. The roughness reduces WIND SPEED and generates EDDIES, and the resulting TURBULENT FLOW mixes the air. The slowing of the wind causes air to accumulate and expand upward, and the upward expansion is increased by convection due to the HEAT ISLAND effect. Vertical expansion of the air produces an URBAN DOME and during the day raises the upper margin of the urban boundary layer until it approximately coincides with the top of the PLANETARY BOUNDARY LAYER. At night the urban boundary layer contracts to about one-fifth of its maximum daytime depth. This is because air in the planetary boundary layer is stable (*see* STABILITY OF AIR) at night, thus restricting vertical movement.

urban canopy layer   The layer of air that lies below the level of the rooftops in a city. The climate in this layer is strongly modified by the many MICROCLIMATES produced by the streets and buildings. Winds blowing in from the surrounding countryside are slowed and deflected by the buildings and other obstructions they encounter, but they are also funneled (*see* FUNNELING) along URBAN CANYONS. The burning of FOSSIL FUELS releases WATER VAPOR into the air, and some of the water piped into the city from outside evaporates. Together these modify the HUMIDITY of the air in the canopy layer. Its temperature is also modified by heat released from vehicle and other engines, industrial processes, and the air-conditioning and heating of buildings. The magnitude of these effects changes in the course of the day, but it does so as a reflection of human activity rather than being wholly due to the daily cycle of sunshine and darkness.

urban canyon   A city street that is lined on both sides by tall buildings, so that physically it resembles a canyon. Wind tends to be funneled along the street (*see* FUNNELING), making it windier, especially if the street is aligned with the PREVAILING WIND. The canyon also affects the way solar radiation is received and absorbed. Depending on the orientation of the canyon, the faces of buildings on one side may receive different amounts of radiation or similar amounts but at different times of day. If the street runs north to south, for example, buildings on the western side face the Sun in the morning and sunshine reaches buildings on the eastern side in the afternoon. If the street runs east and west, both sides receive the same amount of sunshine, but protrusions from the buildings cast deep shadows.

urban climate   Weather conditions in a large city are markedly different from those in rural areas adjacent to the city. The difference between them occurs through-

out the year, and so city dwellers experience a climate that is generally warmer, wetter, and less windy than the climate in which country dwellers live. The urban climate is a genuine phenomenon.

Various factors combine to make cities warmer than rural areas. This is called the HEAT ISLAND EFFECT. In winter the city is an average 1°–2° C (1.8°–3.6° F) warmer than the countryside, and over the year it is 0.5°–1.5° C (0.9°–2.7° F) warmer. Because the city is warmer, there are 10 percent fewer HEATING DEGREE-DAYS there than in rural areas.

The city air is also dustier, however. There are 10 times more small particles in the air than in country air. High buildings also shade much of the ground surface. Together these two factors reduce the amount of sunshine at street level by 15–30 percent. They also reduce the intensity of ULTRAVIOLET RADIATION by 5 percent in summer and 30 percent in winter.

Warmer temperatures mean the RELATIVE HUMIDITY is about 6 percent lower in the city. There are also less EVAPORATION, because surface water from precipitation is removed rapidly through storm drains, and less TRANSPIRATION, because there are fewer plants. The amount of precipitation over the city is 5–15 percent higher than in rural areas, however. This difference is probably due to the larger num-

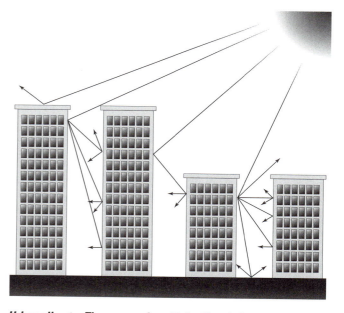

Urban climate. **The process by which tall and short buildings reflect solar radiation. Little of the radiation reaches the ground. Almost all of it is absorbed by the fabric of the buildings.**

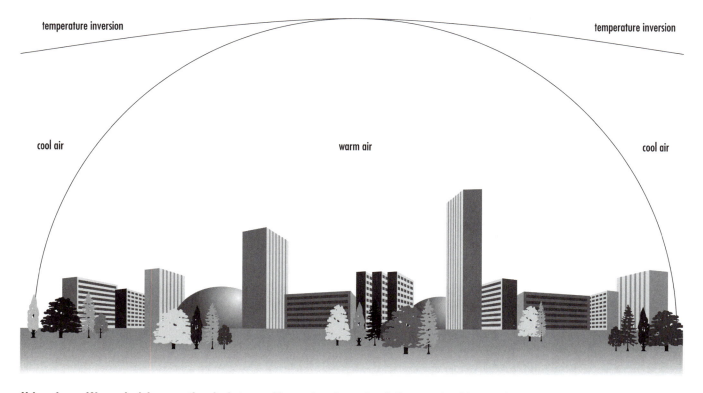

temperature inversion   temperature inversion

cool air   warm air   cool air

Urban dome. **Warm air rising over the city is trapped beneath an inversion. It flows to the sides, cools, and returns to the ground, forming a domed shape.**

ber of particles that are available as CLOUD CONDEN-SATION NUCLEI, although scientists are not yet sure. The city is 5–10 percent cloudier than the countryside. The warmer air increases INSTABILITY and the formation of CUMULIFORM clouds. Some of these develop into CUMULONIMBUS, and there are 5 percent more THUNDERSTORMS in winter and 29 percent more in summer.

Wind speeds are 25 percent lower than in the countryside because of the effect of buildings and there are 5–20 percent more days when the air is calm. At night there is often a COUNTRY breeze. A temperature INVERSION often forms above the city, especially in winter. This traps gaseous pollutants and particles, increasing the frequency of FOG. Fog occurs in winter twice as frequently in the city and 30 percent more often in summer.

urban dome   The effect of the temperature INVERSION that is associated with an urban HEAT ISLAND. Warm air rises over the city, encounters the inversion,

and spreads to the sides. As it moves it radiates away some of its own heat. This cools it, increases its density, and so it subsides over the countryside just beyond the city boundary. From there it flows back into the city, toward the low-pressure region at the center. There are thus CONVERGENCE in the inner part of the city and DIVERGENCE above the city, and the warm air beneath the inversion has an approximately domed shape. The urban dome is most pronounced on calm nights when the sky is clear, because that is when the heat island is most strongly developed.

U.S. Weather Bureau   *See* UNITED STATES WEATHER BUREAU.

UT   *See* UNIVERSAL TIME.

UV   *See* ULTRAVIOLET RADIATION.

UVI   *See* ULTRAVIOLET INDEX.

V

V *See* VOLT.

vacillation The irregular change that takes place in the ROSSBY WAVES surrounding each hemisphere as their AMPLITUDE oscillates between a maximum and a minimum. The series of steps by which the oscillation occurs constitute the INDEX CYCLE.

Val A TYPHOON, rated as category 5 on the SAFFIR/SIMPSON SCALE, that struck Western Samoa from December 6 to 10, 1991. It generated winds of up to 150 mph (241 km h⁻¹) and killed 12 people and rendered 4,000 homeless.

valley breeze An ANABATIC WIND that blows during the day in some mountain areas. It most commonly occurs when conditions are calm and the sky is clear and is most frequent in summer. As the ground surface warms in the sunshine, air in contact with it is also warmed. This air expands and becomes less dense. Cooler, denser air that is farther from the surface subsides beneath the warm air, pushing the warm air up the mountainsides as a warm breeze. This sometimes causes the development of CUMULIFORM cloud above the mountain, leading to showers or thunderstorms. Mountainsides where mountain and valley breezes occur regularly are often preferred for growing fruit, because the constant air movement prevents the static conditions in which frost can form.

valley fog A type of RADIATION FOG that forms in the bottom of valleys. Valleys are especially prone to fogs of this kind, because at night, as the hillsides cool by radiating the warmth they absorb during the day, air chilled by contact with the cold ground sinks to the valley floor. Valley fog often takes a long time to clear in the morning, because it reflects much of the sunshine (*see* ALBEDO), and although the fog droplets are warmed by the Sun they also lose heat as INFRARED RADIATION. The fog blanket slows the rate at which the ground warms and so the low-lying air remains saturated.

Valley breeze. **A warm wind that blows up the side of a mountain during the day, when air is being warmed by contact with the ground surface and rises up the mountainside.**

valley glacier A long, narrow GLACIER that is confined between rocks on either side. There are two types. Alpine glaciers are fed by cirque glaciers. A cirque glacier is a mass of ice that lies in a hollow high in the mountains. Snow accumulates in the hollow, compressing the ice at the base, which spills over the edge. Alpine glaciers are common in the European Alps and in the coastal mountains of Alaska. Outlet glaciers are fed by an ice sheet or ice cap. The Vatnajökull ice cap in Iceland feeds several outlet glaciers, and the GREENLAND ICE SHEET feeds several. The biggest is the Humboldt Glacier. It is 62 miles (100 km) wide and flows north to enter the sea as a wall of ice more than 300 ft (118 m) high. It was an ICEBERG that calved from the Jakobshavn (or Illulissat) Glacier, which sank the *Titanic* on the night of April 14–15, 1912.

(You can read more about Greenland (called *Grønland* in Danish and *Kalaallit Nunaat* in Inuit) in Michael Allaby, *Ecosystem: Deserts* [New York: Facts On File, 2001].)

valley inversion *See* FROST HOLLOW.

Van Allen radiation belts Two toroidal (doughnut-shaped) regions of the MAGNETOSPHERE that were discovered in 1958 by the American physicist James Albert Van Allen (born 1914). The belts consist of charged particles, trapped from the SOLAR WIND and part of the PLASMA that composes the MAGNETOSPHERE. The centers of the belts are about 1,865 miles (3,000 km) and 9,940 miles (16,000 km) above the Earth.

vapor Any gas, although the word is often used to mean WATER VAPOR. A gas, or vapor, is a substance that fills any container in which it is confined, regardless of its quantity. The relationships among its temperature, its volume, and the pressure it exerts on the walls of its container are described by the GAS LAWS. Its molecules move freely and bounce when they collide with each other or with a solid surface. When a vapor is cooled its molecules lose energy and move more slowly. If it is cooled sufficiently the molecules join in temporary groups and occupy a volume that reflects the number of molecules present, at which point the vapor has condensed and become a liquid.

vapor pressure The PARTIAL PRESSURE exerted on a surface by water vapor present in the air. Over an

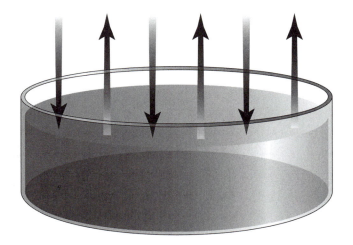

Vapor pressure. Water molecules are constantly escaping from an exposed water surface and being absorbed into it. This process is indicated by the broad arrows. Water molecules present in the air exert pressure on surfaces. This is the vapor pressure.

open surface of water or ice water molecules are constantly escaping into the air by EVAPORATION or SUBLIMATION. These molecules supplement those already present in the air, thus increasing the vapor pressure. That pressure also drives water molecules to merge with the exposed surface. Consequently there is a two-way motion of molecules leaving and entering the air. If the rates at which molecules are leaving and entering the air are equal, so there is no net gain or loss of water vapor, the water vapor is said to be *saturated*, although it is usually the air that is said to be *saturated*.

vapor-pressure deficit *See* SATURATION DEFICIT.

vapor trail *See* CONTRAIL.

vardar *See* VARDARAC.

vardarac (vardar) A cold, northwesterly wind resembling the MISTRAL that blows in the fall along the valley of the Vardar River, in Macedonia, from Skopje to the Aegean Sea.

variable ceiling A CEILING that changes in height rapidly and repeatedly while its height is being measured. The height of the ceiling must then be given as the average of the measured values. Variable ceilings are reported only if their height is lower than 3,000 ft (915 m).

variable gas   An atmospheric constituent gas the amount of which varies from place to place or time to time as a proportion of the whole. WATER VAPOR, CARBON DIOXIDE, and OZONE are the most important variable gases. The proportion of water vapor ranges from almost 0 percent to about 4 percent by volume. The amount of carbon dioxide varies during the day and also seasonally, in inverse proportion to the rate of plant PHOTOSYNTHESIS. It also increases as a result of the burning of FOSSIL FUELS. The amount of ozone present in the air also varies. The gas enters the lower atmosphere as a pollutant and is formed in the STRATOSPHERE by the action of ULTRAVIOLET RADIATION, so the concentration increases to a maximum of about 10 parts per million by volume in the OZONE LAYER. Although particles suspended in the air do not constitute a gas, they behave as one, and their concentration varies from place to place, especially between air over continents and over the ocean. Consequently, AEROSOLS are often treated as a variable gas. The atmosphere also contains CONSTANT GASES.

variable visibility   The condition in which VISIBILITY increases and decreases rapidly while it is being measured. The visibility must then be given as the average of the measured values. Variable visibility is reported only if it is less than 3 miles (4.8 km).

varves   A sequence of light and dark bands that is visible in vertical sections taken from the sediments on the beds of some glacial lakes. The study of varves, called *varve analysis,* helps scientists to measure how long the lake existed and to determine the rate at which climate changed in the past. One varve comprises one light band and one dark band. Within each varve, the pale layer is thick and has a coarse texture and the dark layer is thin and fine-grained.

In spring and summer water that has melted from a nearby GLACIER flows rapidly into the lake. It carries small pebbles, sand grains, silt particles, and particles of clay and deposits all of them in the lake water. The bigger particles quickly settle to the bottom, forming a thick, pale layer of pebbles, sand, and some bigger silt particles. In winter the edge of the glacier and surface of the lake both freeze and the supply of meltwater ceases.

All of the large particles now lie on the lakebed, but much smaller particles, of silt and especially of clay, sink much more slowly. They continue to settle through the winter, forming the thin, fine-grained, dark layer.

Since one layer is formed in summer and the other in winter, it is possible to measure the age of the lake by counting the varves in the same way that TREE RINGS can be counted to determine the age of a tree. The technique for doing this was introduced in 1878 by the Swedish geologist Gerhard Jacob de Geer (1858–1943). By counting the varves in Scandinavian lakes de Geer concluded that southern Sweden was still covered by ice 13,500 years ago. As temperatures rose the glaciers retreated to the north and then, 8,700 years ago, they separated into two small ice caps.

Varves also form in milder climates that allow aquatic algae (single-celled plants) to grow for part of the year. These varves comprise one layer that is rich in organic matter and one that contains little organic matter.

In summer there is a bloom of algae. The remains of dead algae settle to the lakebed to form the organic-rich layer. At the end of the summer the blooms die down, and so the winter layer contains much less organic matter. This is how the Green River Formation developed over an area of 48,250 square miles (125,000 km²) in southwestern Wyoming, northwestern Colorado, and northeastern Utah during the Eocene epoch (56.5–35.4 million years ago).

Where the surface of the lake freezes in winter the process is a little different. Sunlight can penetrate the ice and the algae continue to grow through the winter, albeit slowly. Their remains settle on the bottom, forming a thin, organic-rich layer. In spring, when the ice melts, water rushes into the lake carrying pebbles, silt, and sand. This mixes with the dead algae as it settles to the lakebed, and so although the algae are growing more vigorously, the summer layer contains a smaller proportion of organic matter.

The thickness of varves provides an indication of the warmth of the weather during each year.

vaudaire (vauderon)   A strong, southerly FÖHN WIND that blows across Lake Geneva, Switzerland.

vauderon   *See* VAUDAIRE.

vector quantity   A physical quantity that acts in a direction so that its description must include both the magnitude of the amount and its direction of action. This is contrasted with a SCALAR QUANTITY. Velocity is a vector quantity. Wind velocity is the speed of the wind and the direction from which it is blowing. The speed of the wind can also be stated, omitting the direction.

veering   A change in the wind direction that moves in a clockwise direction, for example, from the southwest to the northwest. If the wind direction is given as the number of degrees from north, a veering wind increases the number.

vegetation index mapping   The use of satellite images to identify areas of the Earth that are experiencing deforestation, DROUGHT, or desert encroachment. The greenness of plant cover can be measured from images transmitted by satellites in polar orbit and the

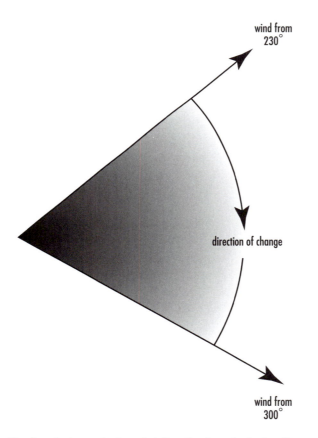

Veering. **A change in the wind direction in a clockwise direction, in this case from 230° to 300°.**

health of vegetation inferred by comparing the color to a series that have been compiled into an index.

VEI   *See* VOLCANIC EXPLOSIVITY INDEX.

veil of cloud   A layer of cloud that is very thin. Objects can be seen through it.

velocity   The speed at which a body is traveling in a specified direction. Speed is a scalar quantity: that is, only its amount is relevant and not the direction in which it acts. Temperature is also a scalar quantity. Velocity is a VECTOR QUANTITY: that is, its direction of action must be specified. It is correct, for example, to report the wind speed as, say, 25 mph (40 km h⁻¹), but it would be incorrect to describe this as the wind velocity. The wind velocity might be 25 mph from 240°, often abbreviated as 25/240; 240° is the compass direction from which the wind is blowing.

velum   An ACCESSORY FEATURE of clouds that consists of a layer of cloud extending horizontally for a considerable distance above other clouds and sometimes connecting the tops of CUMULUS clouds. *Velum* is a Latin word that means "curtain," "veil," or "covering."

vendaval   A strong, sometimes gale-force, southwesterly wind that blows along the coast of Spain, carrying heavy rain and high seas. In the Strait of Gibraltar and along the coast to its east, the vendaval usually originates from between the southwest and northwest.

vent du Midi   A southerly wind that blows through the central region of the Massif Central and Cevennes in southern France, carrying warm, moist air. The name means "wind of the Midi" (the Midi is the south of France).

ventifact   A desert pebble that has been worn away by wind-blown sand in such a way that it has clearly defined faces. Provided the pebble is not moved, the positions of the faces indicate the direction of the PREVAILING WIND.

ventilation   The removal of pollutants from the air by the action of the wind, which introduces clean air. If air is trapped beneath an INVERSION and pollutants that are being emitted into it are mixed thoroughly into the air beneath the inversion, then when a wind blows unpolluted air into the trapped air, the concentration of pollutants decreases from the boundary where the wind enters, at a rate proportional to the wind speed. The extent to which the wind removes pollutants is known as the *ventilation factor* and it is the product of the wind speed and the depth of the polluted air.

ventilation factor *See* VENTILATION.

Venturi effect The acceleration of the wind that occurs when the flow of air is constricted, for example, by tall buildings or hills. Air enters and leaves the constricted area at the same rate, but in order to do so it must travel faster at the constriction. Constriction has the effect of moving STREAMLINES closer together. The effect was first discovered by the Italian physicist G. B. Venturi (1746–1822), for whom it is named.

Venus atmosphere Venus is of approximately similar size to Earth. Its mean diameter is 7,521 miles (12,104 km); that of Earth is 7,918 miles (12,742 km). It is closer than Earth to the Sun, orbiting at 0.72 astronomical unit (AU) (1 astronomical unit is the average distance between the Earth and Sun). Venus orbits at an average distance of 66,932,000 miles (107,712,000 km) from the Sun and Earth orbits at 92,961,000 miles (149,600,000 km). Gravity at the surface of Venus is 8.87 m s^{-2} and at the surface of Earth it is 9.8 m s^{-2}.

Earth and Venus are very similar in these respects and are sometimes described as twin planets. The atmosphere and climate of Venus are very different from those of Earth, however.

The atmosphere of Venus consists of 96 percent CARBON DIOXIDE and 3.5 percent NITROGEN with traces of about 150 parts per million (ppm) of SULFUR DIOXIDE, 70 ppm of ARGON, 20 ppm of water vapor, 17 ppm of CARBON MONOXIDE, 12 ppm of HELIUM, and 7 ppm of NEON. The atmospheric pressure at the surface is 92 bars, compared with 1 bar on Earth.

This dense atmosphere, consisting almost entirely of carbon dioxide, produces a strong GREENHOUSE EFFECT, and combined with its closer proximity to the Sun this gives Venus an extremely hot, dry climate. The global average surface temperature is 737 K (867° F [464° C]). On Earth, lead melts at 621.5° F (327.5° C); the melting point would be higher under the greater atmospheric pressure on Venus, but probably lead would flow as a liquid, and this gives an indication of just how hot the climate is. It is hot enough to make rocks glow. There is little variation in temperature between the equator and poles or between day and night, because of the strong and efficient transport of heat by the atmospheric circulation in both hemispheres. Surface winds blow at about 2 mph (3 km h^{-1}) or less, but there are high-level winds blowing parallel to the equator at about 224 mph (360 km h^{-1}). Such winds could not blow on Earth because they would be balanced by the CORIOLIS EFFECT (CorF). The magnitude of the CorF is proportional to the rotational speed of the planet, and Venus turns much more slowly than Earth, so 1 day on Venus is equal to 243 days on Earth. Venus also rotates in the opposite direction (the rotation is retrograde), so the Sun rises in the west and sets in the east.

Viewed from outside, the surface of Venus is almost entirely hidden by cloud. At one time the clouds were believed to be of water. Today they are known to consist of sulfuric acid droplets. Spacecraft have visited Venus on several occasions and the surface has been comprehensively mapped, but the high pressure and temperature and strongly acidic atmosphere mean instruments last for only a very short time on the surface.

(You can learn more about Venus and about questions to which scientists would like to find answers from ruby.kordic.re.kr/~vr/CyberAstronomy/Venus/HTML/index.html; Ellen Stofan, "A Strategy for Venus Exploration: 2000–2010," at www.planetary.brown.edu/tepswg/venus_strategy.html; and

Venturi effect. **The constriction, which might be the walls of buildings or the sides of hills, draws streamlines closer together, accelerating the air through the constricted region.**

"New Climate Modeling of Venus May Hold Clues to Earth's Future" at www.eurekalert.org/releases/uconvm021899.html.)

Vera Two TYPHOONS, the first of which struck Honshu, Japan, in September 1959. It killed nearly 4,500 people, destroyed about 40,000 homes, and left 1.5 million people homeless. The second Typhoon Vera struck Zhejiang Province, China, on September 16, 1989. It killed 162 people and injured 692.

veranillo The name given along the western coasts of tropical Central and South America to a short dry season, a "little VERANO," that interrupts the long wet season. It lasts for just a few weeks and produces hot, dry weather.

verano The name given in parts of tropical Central and South America to the long dry season. It lasts from November until April.

vertebratus A variety of cirrus cloud (*see* CLOUD CLASSIFICATION) in which the elements are arranged in a pattern reminiscent of the skeleton of a fish, with the long ribs clearly displayed. *Vertebratus* is a Latin word that means "jointed."

vertical anemometer An instrument that measures the vertical component of wind speed. It consists of a pressure ANEMOMETER with its plate mounted horizontally rather than vertically, or a propeller mounted on a vertical axis. Vertical air motion high above the ground and in clouds is measured by specialized instruments carried on aircraft. The atmosphere forms horizontal layers, and because of this horizontal air movements are almost always more than 10 times greater than vertical movements and sometimes as much as 100 times greater. Consequently, measuring only horizontal wind speed is usually sufficient. Nevertheless, knowledge of vertical air movements is important for understanding atmospheric phenomena such as cloud formation and local turbulence.

vertical differential chart A diagram that shows values for a particular atmospheric feature, such as temperature or pressure, at two different heights. A THICKNESS CHART is a vertical differential chart.

vertical jet *See* UPRUSH.

Vesuvian eruption *See* VOLCANO.

VFR weather Weather conditions in which the surface VISIBILITY and CLOUD BASE allow aircraft to take off, land, and cruise under visual flight rules: that is, pilots require no assistance from ground-based guidance systems and there is no restriction on the airfields they can use. Visual flight rules apply when the surface visibility is at least 3 miles (5 km) and the cloud base is no lower than 1,000 feet (300 m). Under poorer weather conditions pilots must operate under IFR TERMINAL MINIMA.

Victor A TYPHOON that struck Guangdong and Fujian Provinces, China, in August 1997. It killed 49 people and destroyed 10,000 homes.

Vienna Convention on the Protection of the Ozone Layer An international agreement that was reached in 1985 under the auspices of the United Nations Environment Program (UNEP). Under the convention governments undertook to protect the OZONE LAYER and cooperate in the scientific research that was needed to understand the atmospheric processes involved in the formation and destruction of OZONE.

Violet A TYPHOON that crossed Japan on September 22, 1996, with winds of 78 mph (125 km h^{-1}). It killed at least seven people, most of them in the Tokyo area, and caused about 200 landslides in Honshu, where it destroyed more than 80 homes and flooded more than 3,000. By the following day, as it moved out into the Pacific, Violet had weakened to a TROPICAL STORM.

virazon A southwesterly sea breeze (*see* LAND AND SEA BREEZES) that blows onto the western coasts of Chile and Peru, where the Andes Mountains descend steeply toward the sea. It occurs with great regularity, commencing at about 10 A.M. and reaching its greatest force at about 3 P.M. In summer it is often so strong that boats are unable to put to sea. A westerly sea breeze affecting the Atlantic coasts of Spain and Portugal is also called *virazon*.

virga (fallstreaks) A wispy veil that is seen beneath the base of a cloud but that does not reach to the ground. It is PRECIPITATION falling from the cloud into relatively dry air, where it evaporates. Air currents may carry water vapor from the evaporation of virga aloft to a height where it condenses once more. Virga may consist of water droplets or ice crystals, depending on the type of cloud from which it falls.

Viroqua A city in Wisconsin that was struck by one of the most severe tornadoes on record on the afternoon of June 28, 1865. On the basis of the damage it caused, the tornado has been judged F4 on the FUJITA TORNADO INTENSITY SCALE. The tornado produced multiple vortices that sometimes merged into a single vortex, and it moved at an estimated 60 mph (96 km h⁻¹) along a path that was 300 yards (275 m) wide and 30 miles (48 km) long. The tornado began to the southwest of Viroqua, passed through the southern part of the town, passed to the south of Rockton, and then dissipated. It lifted a schoolhouse, with its teacher and 24 students inside, from the ground, then dropped it, killing the teacher and 8 students. A total of 22 people were killed and at least 100 injured.

(You can read more about the Viroqua tornado at www.wx-fx.com/viroqua.html.)

virtual temperature The temperature dry air would have if it were at the same density and pressure as moist air. The virtual temperature of moist air is a little higher than its actual temperature. By correcting for the effect due to the density of water vapor, the use of the virtual temperature makes it possible to apply the EQUATION OF STATE to moist air by using a single GAS CONSTANT, rather than calculating constants separately for dry and moist air.

viscosity The resistance a fluid presents to shear forces, and therefore to flow. It is caused by the random intermingling of molecules at the boundary between two fluid bodies, one of which is moving in relation to the other. Such molecular mingling also transfers energy from one fluid body to the other. This is important in the transport of energy by liquids, but EDDY VISCOSITY is by far the most important mechanism for energy transport in air.

visibility The distance from which an observer is able to distinguish an object such as a tree or building with the naked eye. Expressed another way, it is the transparency of the air to visible light. Visibility is measured at a weather station with reference to a number of familiar objects at known distances from an observation point. It is necessary to make a number of measurements in different directions to determine the all-around visibility. The PREVAILING VISIBILITY is the visibility that is reported on a STATION MODEL.

An object is visible if the eye can detect a contrast between it and the sky. For most people this requires the object to be at least about 5 percent darker than the sky. The minimum detectable contrast is symbolized by the Greek letter epsilon (ϵ).

Visibility is reduced by the presence of water droplets and solid particles in the air between the object and the observer and, to a very much smaller extent, by the air itself. Objects are visible because light is reflected from them to the observer. Between the object and the observer some of the reflected light is scattered (see SCATTERING) and some is absorbed, so only a proportion of the reflected light reaches the observer. This light mixes with AIRLIGHT, from other directions, which "dilutes" the light from the object, making it harder to see.

The loss of reflected light between the object and the observer is known as *extinction* and the fraction of the light lost per unit of distance under given conditions is known as the *extinction coefficient*, abbreviated as b_{ext}. The extinction coefficient varies according to the size and density of droplets or particles.

If the extinction coefficient is known, the visual range (r_v) is given by

$$r_v = \log_e \epsilon/b_{ext}$$

visible radiation Electromagnetic radiation to which the human eye is sensitive. This is shortwave radiation at WAVELENGTHS between 0.4 µm and 0.7 µm. At 0.4- to 0.44-µm wavelength the radiation is perceived as violet light, 0.44–0.49 µm gives blue light, 0.49–0.53 µm gives green light, 0.53–0.58 µm gives yellow light, 0.58–0.64 µm gives orange light, and 0.64–0.7 µm gives red light. These colors mix to give white light. At wavelengths shorter than 0.4 µm there is ULTRAVIOLET RADIATION and at wavelengths higher than 0.7 µm there is INFRARED RADIATION. The atmosphere is completely transparent to radiation at wavelengths between 0.35 µm and 0.8 µm. This means radiation passes through the atmosphere without being absorbed, but it is scattered by air molecules (*see* SCATTERING).

visual range *See* VISIBILITY.

volatile An adjective that describes any substance that vaporizes readily at temperatures ordinarily found near the surface.

volcanic explosivity index (VEI) A classification of volcanic eruptions that includes the estimated amount of material each category injects into the atmosphere. This is relevant to climate, because the more violent the explosion the more likely it is that some of the fine particles it throws into the air will penetrate the STRATO-SPHERE, where they may remain for months or even years. Particles in the stratosphere reflect incoming sunlight and may therefore have a climatic cooling effect experienced over a wide area. The VEI also includes the frequency with which eruptions of each type occur, on the basis of historical records. The full VEI is given in the table.

volcano Volcanic eruptions inject ash, dust, rocks, and a variety of gases into the air. They play an important part in the cycles of elements by returning chemical elements to the atmosphere from below the Earth's crust. Carbon, sulfur, and other elements that spend part of their time in the air eventually find their way into the sea, and a proportion form insoluble compounds that accumulate in the sediment on the seabed.

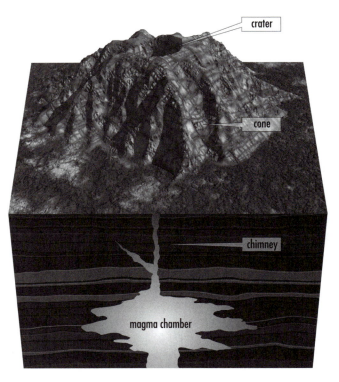

Volcanoes. A cross section through a typical volcano of the Strombolian type.

Compression beneath the weight of overlying sediment and heat from the lower crust slowly turn this sediment into sedimentary rock. Seafloor spreading carries the layers of sedimentary rock lying on top of the rocks of the oceanic crust to a subduction zone, where one of

Index	Type	Plume height feet	Volume type cu.feet	Eruption	Frequency	Example
0	nonexplosive	<350	3,500+	Hawaiian	daily	Kilauea
1	gentle	350–3,500	35,000+	Hawaiian/Strombolian	daily	Stromboli
2	explosive	3,500–16,000	35 million+	Strombolian/Vulcanian	weekly	Galeras 1992
3	severe	10,000–50,000	35 million+	Vulcanian	yearly	Ruiz 1985
4	cataclysmic	33,000–82,000	350 million+	Vulcanian/Plinian	decades	Galunggung 1982
5	paroxysmal	>82,000	3.5 billion	Plinian	centuries	Saint Helens 1981
6	colossal	>82,000	35 billion+	Plinian/Ultra-Plinian	centuries	Krakatau 1883
7	supercolossal	>82,000	350 billion+	Ultra-Plinian	millenia	Tambora 1815
8	megacolossal	>82,000	3,500 billion+	Ultra-Plinian	tens of millenia	Yellowstone 2 Ma

(Ma means millions of years ago.)

the Earth's crustal plates is sinking beneath another (*see* PLATE TECTONICS). This returns the sediment to the Earth's mantle, beneath the crust, and with it the elements that were once part of the air. Eventually the mantle rock containing those elements may return to the surface in a volcanic eruption and the elements may be hurled high into the air.

A volcano begins as a space among the rocks of the crust into which mantle rock rises. Once it is this close to the surface the hot, semimolten rock is called *magma* and the space where it accumulates is a *magma chamber*. It continues to accumulate until its pressure forces it upward. Finding its way through weaknesses in the overlying rock it makes one or more chimneys that eventually reach the surface. What happens then depends on the composition of the magma. It may seep over the ground surface, be thrown into the air, or explode violently. Magma that pours out at the surface is called *lava*.

There are different types of volcanic eruption. A Hawaiian eruption produces fountains of fire and very fluid lava that flows down the side of the volcano. Peléean eruptions, named after Pelé, the Hawaiian goddess of volcanoes, are violent and explosive. Plinian eruptions are also explosive and eject large amounts of material into the air. They are named after Pliny the Elder, who died in A.D. 79 when Vesuvius erupted in this way. Strombolian eruptions, named after the Italian volcano Stromboli, throw out thick lava that falls back to build a steep-sided cone. Surtseyan eruptions are named after the island of Surtsey, near Iceland, that was formed by one in 1963. They are very violent. They happen when water pours into the vent leading to the magma chamber, causing a huge explosion. Vesuvian eruptions are explosive but occur after long periods of dormancy. Vulcanian eruptions happen when the pressure from trapped gases blows away the overlying crust of solidified lava.

The materials ejected by volcanoes can affect the climate by reflecting incoming sunlight and lowering temperature. The volcanoes that have had a climatic effect in modern times include EL CHICHÓN, MOUNT AGUNG, MOUNT ASO-SAN, MOUNT KATMAI, MOUNT PINATUBO, and MOUNT TAMBORA.

volt (V) The derived SYSTÈME INTERNATIONAL D'U-NITÉS (SI) UNIT of electromotive force, electric potential, or potential difference, which is defined as the potential difference between two points on an electric conductor that is carrying a constant current of 1 AMPERE, and the power being dissipated between the two points is 1 WATT. The unit was adopted internationally in 1881 and is named in honor of the Italian physicist and inventor of the battery Alessandro Giuseppe Anastasio, Count Volta (1745–1827).

Voluntary Observing Ship (VOS) A merchant ship that is equipped to act as a weather station. These ships are recruited by PORT METEOROLOGICAL OFFICERS, who also supervise the provision and installation of the necessary instruments. The VOS scheme began in 1853, at a conference in Brussels that was convened by MATTHEW F. MAURY and attended by 10 maritime countries. In 1984–85 there were about 7,700 VOS in the world as a whole. Numbers have declined since then (probably because of the decline in the merchant fleets of several major industrial nations), and in 1994 the VOS fleet comprised about 7,200 vessels from 49 countries. The United States has the largest number of VOS, more than 1,600.

(You can learn more about the VOS scheme at www.vos.noaa.gov/wmo.html and www.srh.noaa.gov/bro/vos.htm.)

Volz photometer An instrument for measuring the intensity of direct sunlight that was invented by Frederick E. Volz. It makes measurements that are defined by filters, which isolate particular WAVE BANDS. The measurements are not very precise, but readings can be taken by relatively unskilled workers and the sky does not have to be completely cloudless. The measurements indicate the amount of sunlight that is being scattered by haze and atmospheric particles. The instrument is therefore used to monitor AIR POLLUTION.

von Kármán constant A value that was discovered by the Hungarian-born American aerodynamicist Theodore von Kármán (1881–1963). The constant is used in calculations of WIND PROFILES, and its value is approximately 0.4. The von Kármán constant holds under all circumstances, but the difficulties involved in wind experiments make it impossible to measure it precisely.

Vonnegut, Bernard (1914–1997) American *Physicist* Bernard Vonnegut was born at Indianapolis, Indiana, on August 29, 1914. He was educated at the Massachusetts Institute of Technology (MIT) and graduated in 1936. He obtained his Ph.D. from MIT in 1939 for research into the conditions that produce ICING on aircraft. From 1939 until 1941 he worked for the Hartford Empire Company, and from 1941 until 1945 he was a research associate at MIT.

In 1945 Vonnegut moved to the laboratories of the General Electric Corporation in Schenectady, New York, where he continued his research into icing in collaboration with VINCENT SCHAEFER. After the discovery that dry ice (solid carbon dioxide) was effective at CLOUD SEEDING, Vonnegut turned his attention to the search for other materials that might perform the same task. Deciding that crystals of silver iodide were the right size and shape to act as FREEZING NUCLEI he experimented with them and was proved correct. Silver iodide largely replaced dry ice as a seeding medium. Unlike dry ice, it can be stored indefinitely at room temperature and it does not have to be released from an airplane flying above the target cloud. Silver iodide can be released at ground level and is carried into the cloud by vertical air currents. If dry ice were released in this way it would vaporize before it could cause ice DEPOSITION.

Vonnegut moved to the Arthur D. Little Corporation in 1952, and in 1967 he was appointed Distinguished Research Professor of the State University of New York, a position he held until his death at Albany, New York, on April 25, 1997.

vortex   A spiraling movement in a fluid that affects only a local area. The fluid at the center of a vortex is usually stationary or slow-moving, but this calm center may be surrounded by gas or liquid that is moving very rapidly. A TORNADO is a vortex and water flowing from a bathtub usually forms one. On a much larger scale, the POLAR VORTEX that forms each winter over Antarctica and in some winters over the Arctic encloses a large mass of very cold air. The tendency of a moving fluid to form a vortex is VORTICITY.

vorticity   The tendency of a mass of fluid that is moving relative to the surface of the Earth to turn about a vertical axis. It is caused partly by the fact that the mass of fluid is moving in relation to the fluid adjacent to it. This generates forces arising from the difference in velocities of the two fluids, so that the faster

tends to curve around the slower. Near the ground, friction causes moving air to turn about a horizontal axis that is at right angles to the wind direction. This generates EDDIES but has no large-scale effect on weather. In the more important type of vorticity air turns about a vertical axis, so its movement is horizontal. It is sometimes called *vertical vorticity* and it arises when air converges or diverges.

CONVERGENCE and DIVERGENCE generate vorticity in respect to a fixed frame of reference, in this case the surface of the Earth. This is known as *relative vorticity* and in meteorology it is always shown by the symbol ζ (the lowercase Greek letter zeta).

Vorticity is also caused by the rotation of the Earth. This is necessarily so because the fixed reference frame of the Earth's surface is also rotating about a vertical axis with an angular velocity that is proportional to the angular velocity of the Earth and also to the latitude of the rotational axis of the mass of fluid. This aspect is known as *planetary vorticity* and its magnitude is always equal to that of the CORIOLIS EFFECT, known as the *Coriolis parameter*. Both are designated by the symbol f.

Seen from a position directly above either the North or South Pole, the Earth rotates in a counterclockwise direction. By meteorological convention, vorticity in a similar direction is said to be positive and vorticity in the opposite, clockwise direction is said to be negative.

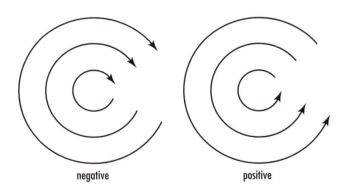

negative   positive

Vorticity. **The tendency of a moving fluid to rotate about an axis. Vorticity is conventionally described as negative if the direction of flow is clockwise and positive if the flow is counterclockwise. In the case of air or water movements across the surface of the Earth, the axis is vertical and the direction of motion is as seen from above.**

The sum of the planetary vorticity and the relative vorticity $(\zeta + f)$ is known as the *absolute vorticity*. Because of the conservation of ANGULAR MOMENTUM, the absolute vorticity remains constant. If minor components are omitted, such as those arising from the forces of buoyancy, friction, and torque, its value is given by the vorticity equation

$$d/dt\,(\zeta + f) = -(\zeta + f)D$$

where d/dt is the rate of change and D is the rate of convergence (or divergence in the case of negative convergence, $-D$).

Because planetary vorticity is equal to the Coriolis parameter and absolute vorticity is constant $(d/dt\,(\zeta + f) = 0)$ a change in the latitude of moving air causes a change in f that must be compensated by a change in ζ. If air in the Northern Hemisphere is diverted northward, for example, f increases, ζ decreases to compensate, and the vorticity becomes more negative, or anticyclonic. This turns the air in a southerly direction, decreasing f and increasing ζ, and so vorticity becomes more positive (cyclonic), turning the air northward again. This is how waves with a long wavelength develop in air that is flowing zonally (parallel to the equator). ROSSBY WAVES and LEE WAVES are examples of this effect.

VOS *See* VOLUNTARY OBSERVING SHIP.

Vostok, Lake A lake that lies beneath the Russian VOSTOK STATION in Antarctica. The presence of water beneath the ICE SHEET was recognized in the 1970s, and the size of Lake Vostok was revealed by satellite RADAR ALTIMETRY in 1996. The largest of about 70 subglacial lakes, Lake Vostok measures approximately 139 × 30 miles (224 × 48 km) in area and is about 1,588 ft (484 m) deep—about the same volume as Lake Ontario. Drilling of the Vostok ICE CORE was stopped about 330 ft (100 m) above the surface of the lake while ways were devised to sample its waters without contaminating them. It is possible the lake is populated by microorganisms and molecules of biological origin in an environment that has been isolated from the rest of the world for hundreds of thousands or even millions of years.

(For more information about Lake Vostok see www.jpl.nasa.gov and www.nerc-bas.ac.uk/public/press/vostok.html.)

Vostok Station A Russian research station in Antarctica, located at 78°27'51" S 106°51'57" E, and at an elevation of 11,401 ft (3,475 m). The name *Vostok* means "east," and it was opened on December 16, 1957. It is sited at the geomagnetic South Pole and at the center of the East Antarctic ICE SHEET. It is also at the southern COLD POLE. Scientists from many other countries work there. The primary project at Vostok has been the drilling of an ICE CORE. Drilling started in 1980 and in 1985 reached a depth of 7,225 ft (2,202 m), beyond which it was impossible to continue with that core. A second core was started in 1984. In 1989 it became a joint Russian–French–U.S. project and in 1990 reached a final depth of 8,353 ft (2,546 m). A third core, started in 1990, reached 8,202 ft (2,500 m) in 1992 and 11,887 ft (3,623 m) in 1998. Ice from this depth is about 420,000 years old.

(For more information about Vostok, see deschutes.gso.uri.edu/~cara/Field_Work/Vostok/Vostok.html.)

Vulcanian eruption *See* VOLCANO.

W *See* WATT.

Waalian interglacial (Donau–Günz interglacial) An INTERGLACIAL period in northern Europe that began about 1 million years ago and ended about 800,000 years ago. It followed the DONAU GLACIAL and preceded the GÜNZ GLACIAL.

wake A region of TURBULENT FLOW that lies DOWNWIND of a surface obstruction or behind a body that is moving through a fluid. The fluid passes over the surface of the object and becomes detached from the surface on the LEE side. The region where it becomes detached is called the *cavity*. The wake forms on the

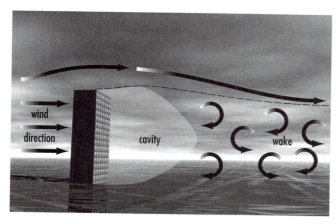

Wake. Immediately downwind of a building there is a region called the *cavity* and beyond that is the wake, a region of turbulence, with erratic, gusty wind.

downwind side of the cavity. As the wind detaches, EDDIES form within the flow. These produce GUSTS of wind and rapidly changing wind directions in the wake of a building or other obstruction. If there is no further obstruction the wake extends for a distance equal to about 10 times the height of the obstruction. The turbulent flow in the wake of a boat is clearly visible. The wake behind an aircraft is invisible, but it represents a serious hazard to any aircraft that enters it.

wake low A small area of low AIR PRESSURE that is found to the rear of a fully developed THUNDERSTORM. The cold downdrafts at the leading edge of the advancing thunderstorm produce a local area of high pressure. Wake lows form behind the main cloud mass and are associated with clearing skies and the end of the PRECIPITATION. Wake lows are common behind SQUALL LINE storms.

Waldsterben The name, which means "forest death," that was given to the damage in German forests that was attributed to ACID RAIN. At first it was attributed to SULFUR DIOXIDE transported by rain. This proved not to be the cause, however. Lichens that are highly intolerant of sulfur grew abundantly in the damaged forests. Gradually the term *Waldsterben* fell from use as the condition of forest trees came to be understood better.

(For more information about Waldsterben see Michael Allaby, *Ecosystem: Temperate Forests* [New York: Facts On File, 1999]. There is a more technical and detailed description in *Air Pollution's Toll on Forests and Crops,* edited by James J. MacKenzie and Mohamed T. El-Ashry [New Haven: Yale University Press, 1989].)

Walker cell *See* WALKER CIRCULATION.

Walker circulation A movement of tropical air that was proposed in 1923 by Sir Gilbert Thomas Walker (1868–1958) and that has since been found to be correct. The Walker circulation is a slight but continuous latitudinal movement that is superimposed on the HADLEY CELLS. It occurs between the equator and about latitude 30° in both hemispheres as a series of cells, the *Walker cells.* Air rises over the tropical western Pacific Ocean and over the eastern Indian Ocean, both near Indonesia. CONDENSATION in the rising air produces towering clouds and heavy rain. At high levels the rising air currents separate into two streams, flowing east and west. The high-level streams from neighboring cells converge and subside over the eastern Pacific, near the South American coast, and over the western Indian Ocean, near the coast of Africa. They diverge at low level (*see* DIVERGENCE.)

This circulation produces regions of low surface pressure over land in tropical South America, Africa, and Indonesia, where air is converging and rising. These are separated by areas of high pressure over the oceans, where air is subsiding and diverging. This distribution of surface pressure produces a prevailing easterly (east-to-west) flow near the surface, which strengthens the easterly TRADE WINDS of the Hadley cells.

The Walker circulation produces a very wet climate in Indonesia and a dry climate over western South America. Every few years the pattern changes and the Walker circulation over the Pacific weakens or reverses. This change is associated with the SOUTHERN OSCILLATION.

Sir Gilbert Walker was a British meteorologist and professor of meteorology at Imperial College, London. He had been appointed head of the Indian Meteorological Service, and in 1904 he was asked to look for a pattern in the occurrence of the Indian MONSOONS. Both a failure of the monsoon and excessively heavy monsoon rains could destroy the crops and cause famine. The monsoon had caused famines in 1877 and again in 1899; it was the 1899 famine that prompted

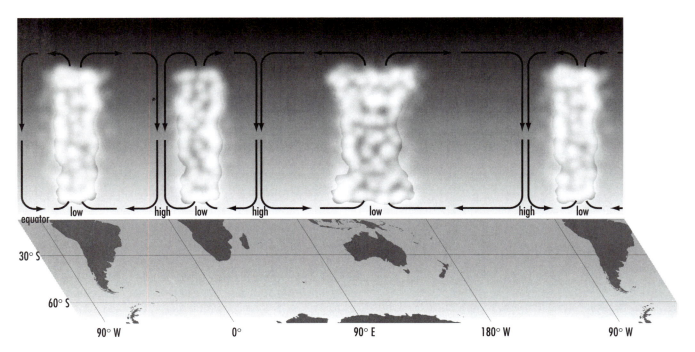

***Walker circulation.* A pattern of latitudinal air movements over the Tropics that produce an easterly flow of air near the surface of the Pacific Ocean.**

the British authorities (who then ruled India) to see whether the variations in monsoons could be predicted in time to take steps to minimize the suffering they caused.

Walker approached the task by studying climate records from all over the world. In those days weather was believed to be a fairly local phenomenon, but Walker noticed that events in one place were sometimes accompanied by different events a long way away. When there was low pressure over Tahiti, pressure was often high over Darwin, Australia. When the Indian monsoon failed, the winter in Canada was mild. Relationships such as these, between events that occur great distances apart, are now known as TELECONNECTIONS. It was Walker who first discovered them.

It was from these studies that Walker discovered a relationship between oscillations in the pressure over the eastern and western Pacific Ocean and the Indian monsoon and rainfall in Africa. He called this period change in pressure distribution the *Southern Oscillation*, but he did not link it to the EL NIÑO effect. That connection was not made until 1960, by JACOB BJERKNES.

(You can learn more about the Walker circulation and its link to the Southern Oscillation at library. thinkquest.org/20901/overview_2.htm and at www.cotf. edu/ ete/modules/elnino/cratmosphere.html.)

Walker, Sir Gilbert   *See* WALKER CIRCULATION.

wall cloud   The extension to a CUMULONIMBUS cloud that appears when a MESOCYCLONE has developed inside the cloud and it may be expanding downward to become a TORNADO. The wall cloud descends below the cloud base. It rotates in a CYCLONIC direction and no PRECIPITATION falls from it. Although the wall cloud appears to be an extension of the main cloud, in fact it marks the region where warm, moist air is being drawn into the main updraft of the SUPERCELL. The warm air cools by an ADIABATIC mechanism as it rises, causing some of its moisture to condense, and strong CONVERGENCE makes it rotate.

Wally   A CYCLONE that struck Fiji in April 1980. It caused floods and landslides in which at least 13 people died and thousands were left homeless.

warm anticyclone   *See* WARM HIGH.

Wall cloud. **The wall cloud hangs beneath the main storm cloud, rotating cyclonically. When a wall cloud appears, a tornado is imminent.**

warm cloud   A cloud in which the temperature is above freezing throughout.

warm-core anticyclone   *See* WARM HIGH.

warm-core cyclone   *See* WARM LOW.

warm-core high   *See* WARM HIGH.

warm-core low   *See* WARM LOW.

warm cyclone   *See* WARM LOW.

warm front   The weather front that marks the boundary between cold air and advancing warm air. Fronts are named *warm* or *cold* in respect to the air behind them and the designation is relative. Air behind a warm front is warmer than the air ahead of the front, but *warm* implies no particular temperature. A warm front slopes at an angle of 0.5–1°, the FRONTAL ZONE is 60–120 miles (100–200 km) wide, and the front advances at an average speed of 15 mph (24 km h⁻¹). A warm front is shown on a weather map as either a red

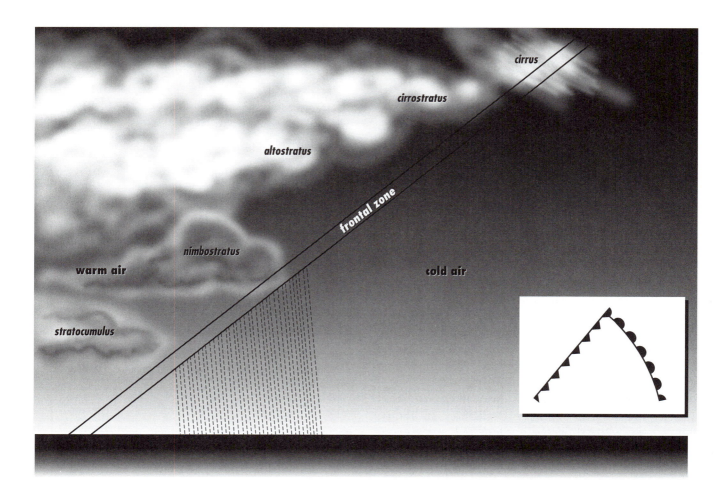

Warm front. **A front with the warmer air behind it. Warm fronts are associated with stratiform cloud and are shown on weather maps as lines with semicircles along the leading edge.**

line or a black line with a row of black semicircles along its leading edge.

The weather associated with a warm front varies, depending on whether it is an ANAFRONT or a KATAFRONT. In an anafront, which is the more active type, STRATIFORM cloud occurs all the way to the top of the front, at the TROPOPAUSE. CIRRUS appears overhead about 600 miles (1,000 km) ahead of the point at which the front touches the ground. Precipitation falls throughout a belt about 250 miles (400 km) wide.

warm glacier *See* TEMPERATE GLACIER.

warm high (**warm anticyclone, warm-core anticyclone, warm-core high**) An ANTICYCLONE that is warmer at its center than it is near the edges.

warm low (**warm-core cyclone, warm cyclone, warm-core low**) A CYCLONE that is warmer at its center than it is near the edges.

warm pocket *See* PRECIPITATION INVERSION.

warm rain Rain that falls from a WARM CLOUD. The RAINDROPS have not been frozen during their formation, and consequently they are at a higher temperature than those that fall from COLD CLOUDS or MIXED CLOUDS.

warm seclusion A pool of warm air that lies above the center of a CYCLONE. A warm seclusion develops in the course of CYCLOGENESIS over the ocean when the BACK-BENT WARM FRONT completely encircles the warm air behind the COLD FRONT.

warm sector The region between the WARM FRONT and COLD FRONT of a FRONTAL SYSTEM that contains the wedge of warm air inside the FRONTAL WAVE. STRATIFORM cloud often forms in the warm air behind the warm front and in winter ADVECTION FOG may develop where warm moist air crosses a cold surface. Elsewhere cloud in the warm sector is often broken, although NIMBOSTRATUS and CUMULONIMBUS may develop along a cold ANAFRONT. The tropical air on the low-latitude side of the POLAR FRONT is drawn into higher latitudes by the frontal wave. This constitutes the air of the warm sector, and in winter it produces a marked rise in temperature. Showers increase as the cold front approaches, and with the passage of the cold front the temperature falls again, often sharply.

warm water sphere The part of the oceans where the temperature is higher than 46°F (8°C).

warm wave A sudden rise in temperature that occurs in middle latitudes, usually in summer. It is caused by the arrival of air from a lower latitude on the eastern side of a CYCLONE or on the western side of an ANTICYCLONE in the Northern Hemisphere. In either case the warm wave often heralds wet weather from an approaching DEPRESSION.

Wasatch A strong, easterly, JET-EFFECT WIND that blows across the plains of Utah from the mouths of canyons in the Wasatch Mountains.

washout The removal from the air of solid particles by collision with falling raindrops. The raindrops engulf the particles and carry them to the ground. Particles are also removed from the air by FALLOUT, IMPACTION, and RAINOUT.

water Hydrogen oxide (H_2O) is able to exist in all three PHASES (gas, liquid, and solid) at temperatures that are common at the surface of the Earth. A pond in winter, when its surface is partly frozen, contains liquid water and solid ice, and the air immediately above the surface contains water vapor, which is an invisible gas. In this example the gas, liquid, and solid are at different temperatures, but at 32.018° F (273.16 K [0.01° C]) and at a pressure of 6.112 mb (0.089 lb in^{-2} [0.18 in mercury]) pure water exists in all three phases simul-

taneously. This is called the *triple point* of water (*see* BOILING).

Water is the only common substance that can exist in all three phases under the conditions found at the surface of the Earth. Ammonia (NH_3) is also a common substance, but at ordinary SEA-LEVEL PRESSURE it freezes at –107.9° F (–77.7° C) and boils at –29.83° F (–34.35° C). Hydrogen chloride (HCl) or hydrochloric acid freezes at –173.6° F (–114.22° C) and boils at –121.27° F (–85.15° C). HCl is stored as a liquid in glass bottles by dissolving it in water, so the acid in the bottle in fact is a solution.

When water freezes its molecules form an open pattern. This causes the water to expand as it freezes and it also causes ice to be less dense, and therefore to weigh less, than liquid water and to float on top of it. This is another unusual property. Most substances are denser as solids than they are as liquids. If ice were denser than water, ponds, lakes, and the sea would freeze from the bottom up, rather than from the top down, and life would be impossible for many of the plants and animals that inhabit the bottom sediments.

Water also has a much larger HEAT CAPACITY than any other common substance. The very high heat capacity of the oceans means they absorb large amounts of heat with very little change in temperature and then release their stored heat very slowly. This has a powerful moderating effect on air temperatures. Without the heat capacity of the oceans summer temperatures would be a great deal higher and winter temperatures a great deal lower.

The water molecule is polar (*see* POLAR MOLECULE). This property makes it an excellent solvent, and the HYDROGEN BONDS that link liquid water molecules also contribute to the capacity of water to form solutions from a wide variety of substances. This works in three ways. Some compounds, such as ethanol (C_2H_5OH), form hydrogen bonds with water molecules, allowing the ethanol and water molecules to mix freely with each other. Other polar compounds held together by COVALENT BONDS become ionized (*see* IONIZATION) in water. Hydrochloric acid (HCl), for example, separates into H$^+$ and Cl$^-$ ions. The H$^+$ then joins a water molecule, changing it into a hydronium ion (H_3O^+), and the Cl$^-$ ions attach themselves to the H$^+$ ends of water molecules. Compounds linked by IONIC BONDS, such as common salt (NaCl), are pulled apart by the strong forces of attraction exerted by the positive and negative ends of water molecules. The Na+ and Cl⁻ ions then attach themselves to water molecules and move with them.

Water is almost a universal solvent—a solvent in which anything dissolves. Because of this it never occurs naturally in its pure form, but always as a solution. Its capacity as a solvent also means living organisms can use it to transport nutrients and metabolic waste products, and the chemical reactions by which organisms grow and maintain their tissues take place in solution. Without water, life on Earth would be impossible.

water balance (**moisture balance**) The difference between the amount of water that reaches an area as PRECIPITATION and the amount that is lost by EVAPO-TRANSPIRATION and RUNOFF. If the amount of precipitation is greater than the sum of evapotranspiration and runoff there is water available for plant growth and for support of animals and people. The water balance is therefore of great importance agriculturally, and it is also a strong influence on the type of natural vegetation an area supports. It is calculated for a specified area and for a specified time (usually one year) by

$$p = E + f + \Delta r$$

where p is precipitation; E is evapotranspiration; f is filtration, which is the amount of water absorbed and retained by the soil; and Δr is net runoff.

water cloud A cloud that is composed entirely of water droplets. It does not extend above the FREEZING LEVEL and therefore contains no ICE CRYSTALS.

water cycle *See* HYDROLOGICAL CYCLE.

water deficit The difference between the amount of water that is needed to sustain the healthy growth of plants (usually crop plants) and the amount that is supplied by precipitation and is available to plants, where the amount available is smaller than the amount required.

water devil A phenomenon that resembles an aquatic DUST DEVIL, but that is smaller, less violent, and shorter in duration. A water devil can result from vigorous

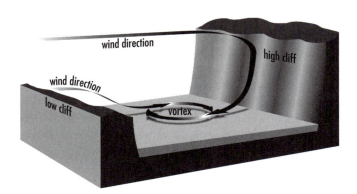

Water devil. **Wind blows over the low cliffs, blows across the lake, and rebounds from the face of the high cliffs, so two streams of air, moving in opposite directions, meet over the lake and produce a vortex.**

CONVECTION over a water surface that is warmed unevenly. A column of air rises vigorously above a "hot spot" and surrounding air is drawn in at the base of the column. CONVERGENCE may then cause the air to rotate, forming a spiraling funnel of rising air. Pressure is low at the center of the vortex. Air approaching it expands rapidly, then cools in an ADIABATIC process, and its water vapor condenses. This makes the funnel visible and, like a WATERSPOUT, it has a ring of spray around its base.

There is a second way a water devil can develop. Where a lake is bounded by low cliffs on one side and high cliffs on the opposite side, suitable wind conditions can produce eddies that generate a vortex over the water.

If the wind blows over the low cliffs, over the lake, and into the high cliffs, air striking the high cliffs is deflected down the face of the high cliffs and back across the lake. When that happens, somewhere over the lake air moving away from the high cliffs meets air moving toward the high cliffs. The resulting WIND SHEAR may set the air rotating. VORTICITY and the conservation of ANGULAR MOMENTUM may then be sufficient to sustain a vortex, with air spiraling upward around it. Water devils of this type are freak occurrences. They last no more than a few minutes and are seldom large, although they have been known to rise to a height of about 10 ft (3 m) and to have the strength to lift a small rowboat clear of the water and then drop it.

water equivalent   The depth of a layer of snow after it has been converted to an equivalent fall of rain. Snow traps small pockets of air between its grains and flakes. This makes snow bulky, but to an extent that varies according to the type of snow. When snow is expected, weather forecasts predict the depth of fall people can expect. This is valuable information, but it cannot be used to compare the snowfall in different places and at different times. For comparisons, the depth of snow must be converted into an equivalent depth of water. This is done by pressing an open-ended cylinder vertically through the layer of snow, sliding a plate beneath the lower end to seal it, then removing the cylinder and melting the snow. The depth of liquid water is then recorded as the amount of precipitation. Liquid water cannot be compressed and so it provides a standard measure. As an approximate guide, the water equivalent is one-tenth the depth of snow.

water infiltration   *See* PERCOLATION.

water sky   The appearance of a cloudy sky over water, when the Sun is high and the ALBEDO of the water is low. Little light is reflected to illuminate the base of the cloud, which is consequently dark in color.

waterspout   A TORNADO, or column of spiraling air that occurs over water. It resembles a tornado and is larger than a WATER DEVIL. Some waterspouts are true tornadoes. They form in a CUMULONIMBUS cloud that contains a MESOCYCLONE and may originate over water or form over land and then drift over water. A waterspout of this type is as powerful as any other tornado and if it moves from water to land it is just as dangerous.

Tornado funnels are usually dark because of the dust and debris that are drawn into them and then spiral upward. Waterspouts are white and if a tornado crosses from land to water its color quickly changes, because while it remains above the surface of the sea or a lake there is no dust and debris to darken it. The funnel consists entirely of air and water.

Around its base, air accelerating into the vortex whips up spray that forms a white cloud called a *spray ring*. Some of this water is carried up into the funnel, but most of the water in the funnel forms there, by the CONDENSATION of water vapor.

Air above the water surface is moist. As it is accelerated toward the center of the vortex it moves down a very steep PRESSURE GRADIENT to a region where the atmospheric pressure is much lower than it is outside the vortex. The change in pressure causes the incoming air to expand rapidly. As it expands it also cools, because of the conversion of heat to the KINETIC ENERGY required for expansion. Its temperature falls below the DEW POINT and water vapor condenses. Condensation releases LATENT HEAT, adding to the instability of the rising air.

Waterspouts can also form in the absence of a mesocyclone. They are most likely to do so over shallow water when the weather is very hot, especially in sheltered places such as bays. Because most occur close to the shore on fine afternoons in summer they are often visible to people relaxing on the beach.

They are caused by CONVECTION. Summer sunshine warms the water. Where the water is deep, it is only the surface layer that is warmed and there is a limit to the temperature it can reach, because the warm surface water mixes with the colder water beneath it. Over the open sea or a large lake, winds ripple the surface, increasing the mixing. Sheltered, shallow water, however, can be warmed all the way to the bottom and its temperature can rise much further.

Air is warmed by contact with the water surface and rises by convection, carrying a large amount of water vapor with it. As it rises, the air cools in an ADIABATIC process and its water vapor starts to condense, releasing latent heat that increases its INSTABILITY. CUMULUS CONGESTUS cloud starts to form. This does not extend to a height where the temperature is below freezing and the cloud may produce no precipitation, but if the air is rising rapidly enough the air being drawn in below the cloud may start to rotate about a vertical axis. The resulting vortex may then extend below the cloud as a spiraling column of air and water.

Waterspouts of this type are weaker than tornadoes. Most have a funnel about 150 ft (45 m) in diameter, although some can reach 300 ft (90 m), and they produce wind speeds of about 50 mph (80 km h^{-1}). They occur along tropical coastlines, in the Gulf of Mexico, and in the Mediterranean, and are most frequent in the southern Florida Keys, where up to 100 form every month during the summer.

water surplus   The difference between the amount of water that is needed to sustain the healthy growth of plants (usually crop plants) and the amount that is supplied by precipitation and is available to plants, where

wavelength (m)	name	frequency (Hz)
10^{-11}–10^{-14}	gamma rays	300–30 EHz
10^{-9}–10^{-11}	X rays	3 EHz–300 PHz
10^{-7}–10^{-9}	ultraviolet	30–3 PHz
4–7×10^{-7}	visible light	1 PHz–300 THz
10^{-3}–10^{-6}	infrared	300 THz–300 GHz
10^{-1}–10^{-3}	microwave	300 GHz–300 MHz
1–10^9	radio	300 MHz–3 Hz

(EHz (exahertz) = 10^{18} hertz; PHz (petahertz) = 10^{15} hertz; THz (terahertz) = 10^{12} hertz; GHz (gigahertz) = 10^9 hertz; MHz (megahertz) = 10^6 hertz.)

the amount available is greater than the amount required.

water table *See* AQUIFER.

water vapor The gaseous phase of WATER in which the molecules of H_2O are no longer attached to one another by HYDROGEN BONDS but can move freely and independently. Energy must be applied in order to break the hydrogen bonds and change liquid water into a gas. This energy, called the *LATENT HEAT* of vaporization, is 2,501 J g^{-1} (600 cal g^{-1}) at 32° F (0° C). A similar amount of latent heat is released when water vapor condenses into a liquid (*see* CONDENSATION).

When ice is exposed to very dry air, some of its molecules enter the air, turning directly from the solid to the gaseous PHASE without passing through a liquid phase. This change is called *SUBLIMATION*. The change in the opposite direction, from gas to solid, is called *DEPOSITION*. Sublimation requires the absorption of an amount of latent heat that is equal to the sum of the latent heat of melting and the latent heat of vaporization. At 32° F (0° C) this is 2,835 J g^{-1} (680 cal g^{-1}). Deposition releases exactly the same amount of latent heat.

Water vapor absorbs INFRARED RADIATION at WAVELENGTHS of 5.3–7.7 µm and beyond 20 µm. It is the principal GREENHOUSE GAS in terms of the amount of BLACKBODY RADIATION it absorbs. However, water vapor is not usually considered a greenhouse gas because its atmospheric concentration is widely variable and impossible to control.

water-vapor absorption *See* WATER VAPOR.

watt (W) The derived SYSTÈME INTERNATIONAL D'UNITÉS (SI) UNIT of power, which is equal to 1 JOULE per SECOND. 1 kW = 1.341 horsepower; 1 horsepower = 745.7 W. The unit was adopted in 1889 and is named in honor of the Scottish engineer and instrument maker James Watt (1736–1819).

wave band A range of WAVELENGTHS within which all electromagnetic radiation is similar in character. The electromagnetic spectrum contains seven wave bands, listed in the table.

wave characteristics A wave traveling through water (or air) possesses certain features that can be used to describe it. Its wavelength is the distance from one wave crest (or trough) to the next. The wave amplitude is the distance from the crest or trough to a point midway between the crest and trough—the distance by which the wave moves up and down. The height of the wave is the vertical distance between crests and troughs and is equal to twice the amplitude. The steepness of the wave is given by the hypotenuse of a right-angled triangle, the other two sides of which are the wavelength and the wave height. The number of crests (or troughs) that pass a fixed point in a unit of time is the wave frequency, and the wave period is the time that elapses between one crest (or trough) and the next passing a fixed point.

Most sea waves are generated by the action of the wind, and the height of a wind wave is determined by the strength of the wind and the distance across which it blows (known as the *FETCH*). The speed at which waves travel is proportional to their period ($c = \lambda/T$, where c is speed, λ is wavelength, and T is period).

Waves travel in groups. Those with the longest wavelength are at the front of the group and those with the shortest wavelength are at the rear. It follows (by $c = \lambda/T$) that the waves at the rear of the group are traveling fastest. They overtake the waves ahead of them, advancing through the group, but as they do so they lose height, and when they reach the front of the group they disappear. The group as a whole advances at half the speed of the individual waves that constitute it.

Waves travel through water or air, but they do not carry the water or air with them. A molecule of water or air, or a cork bobbing about on the water, describes a small circular motion. That is why a boat is rocked by waves, but not transported by them.

Molecules are affected by wave motion to a depth equal to half the wavelength and at that depth the amount of molecular movement is negligible. Land slopes into the sea rather than meeting it abruptly. Consequently, as a sea wave approaches a coast the sea becomes shallower. When the depth of water is less than half the length of the waves, the motion of particles near the bottom becomes flattened and the wave slows. This cannot alter the wave period, however,

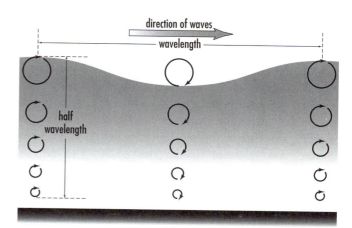

Wave characteristics. Waves cause the water to move in a circular motion. The waves move through the water, but the water itself does not advance.

because waves continue to arrive at the same frequency. Instead, the wavelength decreases—the same number of wave crests pass in a given period of time, but the horizontal distance between them is reduced. The seafloor does not slope everywhere by the same amount, however, and the reduction in wavelength due to reducing water depth has the effect of aligning the waves with the submarine contours. It is why waves usually approach a shore at right angles.

Waves transmit energy imparted to them by the wind, tidal forces, or, in the case of TSUNAMIS, disturbances of the ocean floor. As they enter shallow water and slow, they continue to convey energy at the same rate. This causes the height of the waves to increase, and as their height increases so does their steepness. The circular motion of molecules at the crest is accelerated and when it exceeds that of the wave itself, the wave becomes unstable. Its crest curls forward and spills. It is then a breaker.

Objects floating at the surface move in the same way as the water molecules. In deep water they move only vertically, but close inshore, where the waves are breaking and molecules at the wave crests are traveling faster than the waves themselves, they are carried toward the shore. That is what makes surfing possible.

wave cloud *See* LENTICULAR CLOUD.

wave cyclone *See* WAVE DEPRESSION.

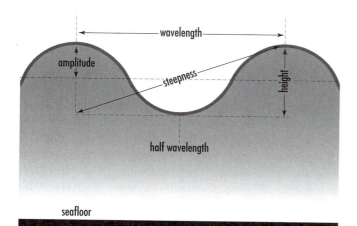

Wave characteristics. Waves are described in terms of wavelength—the distance between one crest or trough and the next; height—the vertical distance between crests and troughs; amplitude—half the height; and steepness—the angle between the horizontal and a line drawn from a point level with a trough and directly beneath a crest to the top of the adjacent crest. Wave motion ceases at a depth beneath the troughs equal to half the wavelength.

wave depression (wave cyclone) A DEPRESSION that forms around the point where a COLD FRONT and WARM FRONT meet and warm air is beginning to rise over, or be undercut by, the cold air. The FRONTAL SYSTEM is shaped like a wave on a SYNOPTIC CHART, with a center of low pressure and CYCLONIC circulation at the peak of the wave. Wave depressions are a feature of weather systems in middle latitudes. They move from west to east, carried by the PREVAILING WESTERLIES, and several often occur one after the other. These constitute a *depression family* and each member of the family is linked to a wave in the JET STREAM above it (*see* INDEX CYCLE). Depression families produce prolonged periods with gray skies and wet weather, interrupted only briefly by the RIDGES between one depression and the next.

wave equation A partial differential equation that represents the velocity (v) and vertical displacement produced by a wave (ψ) as a function of space and time (t), where space is described by three coordinates x, y, and z. The equation can then be written as

$$v^2\psi = \delta^2\psi/\delta\, x^2 + \delta^2\,\psi/\delta\, y^2 + \delta^2\,\psi/\delta\, z^2 = (1/v^2)\delta^2\psi/\delta\, t^2$$

wave front The line joining all the points that are at the same PHASE along the path of an advancing wave.

wave hole A hole that forms in a layer of cloud on the LEE side of a mountain. It is caused by a stream of air that descends from the mountain peak, warming in an ADIABATIC process as it does so and causing the EVAPORATION of CLOUD DROPLETS in its path.

wavelength The horizontal distance between the crest (or trough) of one wave and the crest (or trough) of the next. The wavelength (λ) of a system of waves is related to their FREQUENCY (f) by

$$\lambda = c/f$$

where c is the speed at which the waves advance.

wave theory of cyclones The theory that explains CYCLOGENESIS as the formation of waves along the interface (or fronts) between two fluids.

Wb *See weber.*

WCP *See* WORLD CLIMATE PROGRAM.

weakening In SYNOPTIC METEOROLOGY, a decrease in the PRESSURE GRADIENT that takes place over hours or days, causing a reduction in wind speed.

weak front A WARM FRONT that is overriding a mass of cold air, but behind which the RELATIVE HUMIDITY of the air is low. As the warm air rises and cools little or no cloud forms and consequently the front produces no appreciable change in the weather.

weather The state of the atmosphere as it is in a particular place at a particular time or over a fairly brief period, with special emphasis on short-term changes. A description of the weather includes references to the current or expected TEMPERATURE, AIR PRESSURE, HUMIDITY, VISIBILITY, CLOUD AMOUNT and type (*see* CLOUD CLASSIFICATION), and PRECIPITATION. Weather is contrasted with CLIMATE, which is a much broader concept.

weathercock *See* WIND VANE.

weather facsimile *See* WEFAX.

weatherglass A BAROMETER, in particular a household one that is mounted on the wall.

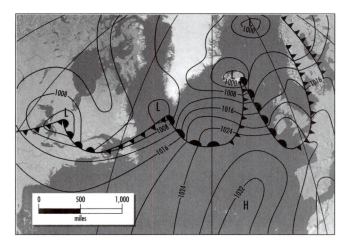

Wave depression. **A family of wave depressions seen as they are crossing the North Atlantic.**

weather house A household ornament that predicts the weather. It consists of a house with two doors and the figure of a person—usually a man and a woman or a young woman and an old woman—behind each. Only one figure at a time can emerge through its door. When one of the figures emerges it is supposed to mean rain is imminent and when the other figure emerges the weather is supposed to remain fine. In fact, the device is a type of HYGROMETER. The figures are made to move by the contraction or stretching of one or two lengths of hair in response to the HUMIDITY.

weathering All the processes by which solid rock is broken into ever smaller fragments and finally into particles ranging in size from those of clay, which are less than 0.00004 in (4μm) across, to sand grains up to 0.08 in (2 mm) across. Not all weathering is due to the physical processes associated with weather. Chemical solutions that originate deep below the surface and rise through fissures in crustal rocks react with particular minerals in the rock. This makes some minerals soluble, so they are removed in solution. PRECIPITATION delivers water to the surface. Precipitation is naturally acid, because of the carbon dioxide and other gases that have dissolved in it. As the water filters downward through the soil it dissolves some minerals, leaving rocks pitted. If air enters the cracks, other minerals are oxidized. These changes, known as *chemical weathering,* weaken and fragment rock. They also tend to smooth it, because sharp corners and protrusions present large surface areas on which the chemical reactions can take place, so they are attacked more rapidly than flat rock faces.

Rock at the surface is directly exposed to physical weathering processes. In middle and high latitudes, water penetrates small fissures and then freezes in winter. The water expands with great force when it freezes, widening the fissures and breaking off small flakes of rock. In spring the ice melts and the rock fragments are washed from the rock by the melting water or blown away by the wind. In warmer, drier climates, the rain reacts with minerals to form chemical salts that crystallize when the rain ceases and the rock dries, expanding as they do so. Crystallization, especially of common salt (sodium chloride, NaCl), causes rocks to crack. A violent storm may cause rocks that have been weath-

ered from below to fall down hillsides. As they fall the rocks accelerate and detach other rocks, to produce a rock fall.

Weathered rock is subject to EROSION. On sloping ground, rocks and rock particles that have been broken from the main rock mass may suddenly slide downhill. Landslides occur when heavy rain turns soil to mud, which lubricates the ground beneath the rocks. Heavy rain can also shift large masses of soil, which descend as mudslides.

Rock falls, landslides, and mudslides are known collectively as *mass wasting.* If they occur in populated areas they can cause appalling devastation.

weather lore People have always tried to predict the weather. They needed to know when to sow and harvest their crops, when to shelter their animals from an impending storm, and when it was safe to set sail on fishing expeditions or sea journeys. Until the invention of TELEGRAPHY there was no way weather observations made at points scattered over a large area could be communicated to a central point quickly enough to produce a SYNOPTIC picture of weather conditions that might make accurate forecasting feasible. Instead, people had to rely on experience and local knowledge to interpret the signs they could see around them. These signs and their meanings became incorporated in sayings and short verses that made them easier to remember.

Predictions were also associated with gods and in Christian cultures with saints. This link derives from the time when the weather was believed to result from the direct intervention of supernatural beings. Storms, hail, gales, and warm sunshine were all produced at the whim of these beings. Many of these old associations survive.

Lore, used in this sense (the word has other meanings), is a body of tradition or knowledge on a particular subject. Weather lore is the accumulated traditions and observations with which our ancestors attempted to interpret weather signs and forecast the weather. Possibly the earliest written collection of sayings about the weather was made by the Greek philosopher Theophrastus (371 or 370–288 or 287 B.C.E.), a student of ARISTOTLE. Evidently he talked well, because it was Aristotle who gave him the name *Theophrastus,* which means "divine speech." His real name was Tyrtamus, and after Aristotle retired he took over the

Lyceum (school) Aristotle had founded. Theophrastus is best known as the founder of the science of botany, but he also wrote *On Weather Signs* and *On Winds,* two short books that contain natural signs indicating rain, wind, storms, and fair weather. The Greek poet Aratus (c. 315–c. 245 B.C.E.) also collected some weather sayings. About half of his only surviving complete work, *Phaenomena,* is devoted to them. The collections made by Theophrastus and Aratus were passed down from generation to generation, translated into Latin, and repeated by Roman authors such as Virgil (70–19 B.C.E.), and absorbed into many European cultures.

Religious festivals, mostly held to celebrate particular saints, mark the progress of the Christian year and the weather on those days is often believed to set the pattern for the period that follows. Days that are traditionally held to be significant in this way are called CONTROL DAYS; they include CANDLEMAS, EASTER DAY, and CHRISTMAS DAY. The saints whose days predict the weather to come include SAINTS BARTHOLOMEW, HILARY, LUKE, MARTIN, MARY, MICHAEL AND GALLUS, PAUL, SIMON AND JUDE, SWITHIN, and VITUS. The DOG DAYS are inherited from Roman belief, and GROUNDHOG DAY is also one when the weather is foretold.

Other beliefs are based on direct observation. RED SKY in the evening indicates a fine day tomorrow, for example, as do DEW IN THE NIGHT, RAIN BEFORE SEVEN, and a GRAY MIST AT DAWN.

The appearance and behavior of familiar plants and animals are also held to foretell the weather. COWS lie down when rain approaches, but that is only one of the ways they can be used as forecasters.

Some of the traditional beliefs are accurate, but most are not. This may be because the original ideas behind them have become corrupted over the generations. There is not the slightest doubt that people whose lives sometimes depend on the weather, such as sailors and shepherds, are able to read signs of approaching wind, storms, and fine weather.

weather map   A map that shows the distribution of pressure, winds, and PRECIPITATION over an area of the Earth's surface at a particular time. Weather maps that are shown on television and printed in newspapers are based on the more detailed SYNOPTIC CHARTS used by meteorologists.

weather minimum   The poorest weather conditions in which aircraft are permitted to fly under visual (VFR) or instrument (IFR) flight rules.

weather observation   A record of weather conditions based on measurements made in a standardized fashion and written down according to a strict formula. This allows observations made by many people in many places to be compiled into an overall picture of weather over a large area at a certain time. In most countries, thousands of volunteers make regular observations and communicate them to a central point. In the United States, these volunteers constitute the CoOp Network. Its members are supervised by the NATIONAL WEATHER SERVICE, which forwards the data they submit to the National Climatic Data Center. Some of the volunteers have been collecting data for more than 70 years.

weather radar   Radar that is used to study processes inside clouds, especially the density of water droplets, where the water is most concentrated, and the level at which rising water droplets freeze and falling ice melts. This information is used to determine the likelihood of precipitation and the intensity of storms. It is possible because water droplets strongly reflect electromagnetic radiation with a wavelength of 5–10 cm (2–4 in). This was discovered early in the 1940s, but it was not until the 1960s that meteorologists first began using radar extensively. Today almost all severe storms are monitored by radar in the United States and many other countries.

Weather Radio   A network of more than 480 stations that broadcast continuous weather information 24 hours a day over the whole of the United States, U.S. coastal waters, Puerto Rico, the U.S. Virgin Islands, and the U.S. Pacific Territories. As well as ordinary information about weather conditions, the network broadcasts warnings and watches of hazards. It also provides warnings of other types of hazard, such as volcanic activity, earthquakes, and chemical and oil spills. A special radio receiver is required to pick up the signal. It is a public service provided by the NATIONAL OCEANIC AND ATMOSPHERIC ADMINISTRATION and broadcast at 162.400 MHz, 162.425 MHz, 162.450 MHz, 162.475 MHz, 162.500 MHz, 162.525 MHz, and 162.550 MHz.

(There is more information about the service, including details of how to obtain a receiver, at www.nws.noaa.gov/nwr.)

weather satellite A satellite in Earth orbit that carries instruments that produce images of the Earth from which meteorological and climatological information can be obtained. Satellite images are now of vital importance in weather forecasting and in monitoring climatic change. The first satellite dedicated to weather observation was the TELEVISION INFRARED OBSERVATIONAL SATELLITE (TIROS). There are now many, among them providing a complete coverage of the surface of the Earth through 24 hours every day of the year.

(There is more information about weather satellites in Robert Haynes, "Sentinels in the Sky: Weather Satellites" at http://ecco.bsee.swin.edu.au/chronos/metsat/ weather.html and at http://octopus.gma.org/surfing/ satellites/sat_weat.html.)

weather ship A WEATHER STATION that is mounted on a ship dedicated for the purpose. The ship is anchored permanently in one location (except when it needs to return to port for repairs or maintenance). Weather ships are sited away from shipping lanes, in sea areas that are not monitored by VOLUNTARY OBSERVING SHIPS. Many are in remote parts of the North Atlantic and North Pacific Oceans and the seas off Scandinavia.

weather shore The shore from which the wind is blowing, as seen by a ship at sea.

weather side The side of a ship that faces into the wind or weather.

weather station A place that is equipped with the instruments needed to make standardized measurements and observations of weather conditions and the technical and communications facilities to transmit weather reports to a central point. The station may be manned, but many modern weather stations are fully automated.

weber (Wb) The derived SYSTÈME INTERNATIONAL D'UNITÉS (SI) UNIT of magnetic flux, which is the amount of magnetism in a magnetic field calculated from the strength and extent of the field. The weber is equal to the magnetic flux that produces an electromotive force of 1 VOLT in a conducting coil of one turn, as the flux is reduced to zero at a uniform rate in one second. The unit was adopted internationally in 1948 and is named in honor of the German physicist Wilhelm Eduard Weber (1804–91).

wedge A RIDGE of high pressure in which the ISOBARS make a V-shaped point, like a wedge.

Wefax An abbreviation for *weather facsimile,* which is a system for transmitting by radio such material as graphic reproductions of weather maps, summaries of temperatures, and cloud analyses. Most Wefax transmissions are from GOES craft. Schools and individual enthusiasts can receive Wefax data provided they have suitable equipment.

(There is more information on Wefax and the equipment needed to receive data from it at http://www.aa6g.org/weather/goes.html and at http://octopus.gma.org/surfing/satellites/sat_weat/html.)

Wegener, Alfred Lothar (1880–1930) German *Meteorologist and geologist* Alfred Wegener is best known for having formulated the theory of CONTINENTAL DRIFT that developed later into the theory of PLATE TECTONICS. He was primarily a meteorologist, however, and studied the formation of RAINDROPS and the circulation of air over polar regions.

Wegener was born in Berlin on November 1, 1880. His father was a minister and director of an orphanage. He was educated at the Universities of Heidelberg, Innsbruck, and Berlin. In 1905 he received a Ph.D. in planetary astronomy from the University of Berlin and immediately switched his interest to meteorology, taking a job at the Royal Prussian Aeronautical Observatory, near Berlin. He used kites and balloons to study the upper atmosphere and also flew hot air balloons. In 1906 Alfred and his brother Kurt remained airborne for more than 52 hours, breaking the world endurance record.

In 1906 he joined a Danish two-year expedition to Greenland as the official meteorologist. Wegener studied the polar air, using kites and tethered balloons, and on his return to Germany in 1909 he became a lecturer in meteorology and astronomy at the University of Marburg. He collected his lectures into a book that was published in 1911, *Thermodynamik der Atmosphäre* (Thermodynamics of the atmosphere). This became a standard textbook throughout Germany. In it, Wegener pointed out that where ICE CRYSTALS and supercooled droplets (*see* SUPERCOOLING) are present together, the crystals grow at the expense of the droplets, because the equilibrium VAPOR PRESSURE is lower over the crystals. He suggested that this might lead to the formation of crystals large enough to sink through the cloud and melt at lower levels to become raindrops. Wegener never had an opportunity to test this idea in real clouds. TOR BERGERON and Walter Findeisen finally tested it in the 1930s. Bergeron acknowledged his debt to Wegener, and although this type of raindrop formation is usually known as the *BERGERON–FINDEISEN MECHANISM*, it is sometimes called the *Wegener–Bergeron–Findeisen mechanism.*

Alfred Wegener. **The German meteorologist who formulated the theory of continental drift that later became the theory of plate tectonics.** *(Copyright Alfred Wegener Institute for Polar and Marine Research)*

Wegener was also the first person to explain two kinds of arc that are occasionally seen in the Arctic opposite the Sun. They are caused by ice crystals that form in the very cold air. They are now called *Wegener arcs.*

Since 1910, Wegener had been intrigued by the apparent fit of the continental coastlines on each side of the Atlantic Ocean. Drawing together various strands of evidence to support the idea, in 1912 he published a short book, *Die Entstehung der Kontinente und Ozeane* (The Origin of the continents and oceans), in which he proposed what he called *continental displacement:* the idea that the continents were once joined and have moved slowly to their present positions.

It was also in 1912 that Wegener married Else Köppen, the daughter of WLADIMIR KÖPPEN, the most eminent climatologist in Germany. Wegener and Köppen collaborated in a book about the history of climate, *The Climates of the Geological Past.* Wegener then returned to Greenland. The four-man 1912–13 expedition crossed the ice cap and was the first to spend the winter on the ice.

On the outbreak of war in 1914 Wegener was drafted into the army, but he was wounded almost at once. He spent a long time recuperating, during which he elaborated on his theory of continental drift, publishing an expanded version of *The Origin of the Continents and Oceans* in 1915 (it was not translated into English until 1924). The book received a hostile reception from German scientists, and Wladimir Köppen strongly disapproved of his son-in-law's digression from meteorology into geophysics. Wegener spent the remainder of the war employed in the military meteorological service.

After the war he returned to Marburg. In 1924 he accepted a post created especially for him and became the professor of meteorology and geophysics at the University of Graz, in Austria.

In 1930 he returned to Greenland as the leader of a team of 21 scientists and technicians planning to study the climate over the ice cap. They intended to establish three bases, all at 71° N, one on each coast and one in the center, but they were delayed by bad weather. On July 15 a party left to establish the central base, called *Eismitte,* 250 miles (402 km) inland. The weather then prevented necessary supplies from reaching them, including the radio transmitter and hut in which they were to live. On September 21, Wegener, accompanied by 14 others, set off with 15 sleds to carry supplies to Eismitte. The appalling conditions forced all but Wegener, Fritz Lowe, and Rasmus Villumsen to give up and return. These three finally reached Eismitte on October 30. Lowe was exhausted and badly frostbitten. They stayed long enough to celebrate Wegener's 50th birthday on November 1, then Wegener and Villumsen began their return, leaving Lowe to recover. They never reached the base camp. At first it was assumed they had decided to overwinter at Eismitte, but when they had still not appeared in April a party went in search of them. They found Wegener's body on May 12, 1931. He appeared to have suffered a heart attack. Villumsen had carefully buried the body. They marked the site with a mausoleum made from ice blocks and later added a large iron cross. Despite a long search, Villumsen was never found.

There is now an Alfred Wegener Institute for Polar and Marine Research at Bremerhaven, Germany.

(You can learn more about Alfred Wegener at www.pangaea.org/wegener.htm, www.pbs.org/wgbh/aso/databank/entries/bowege.html, www.ucmp.berkeley.edu/history/wegener.html, www.dkonline.com/science/private/earthquest/contents/hall.html, and pubs.usgs.gov/publications/text/wegener.html.)

Wegener arc   *See* WEGENER, ALFRED LOTHAR.

Wegener–Bergeron–Findeisen mechanism   *See* BERGERON–FINDEISEN MECHANISM.

Weichselian glacial   *See* DEVENSIAN GLACIAL.

weighing gauge   *See* RAIN GAUGE.

weight of the atmosphere   The magnitude of the gravitational force that attracts the mass of the atmosphere and thereby produces AIR PRESSURE. The entire atmosphere weighs about 5.83×10^{15} tons (5.3×10^{15} t).

weir effect   The movement of air that is carried inland by sea breezes (*see* LAND AND SEA BREEZES) in western South America from northern Peru to central Colombia. Cool, moist air flows inland for about 37 miles (60 km), rising up the foothills of the Andes and

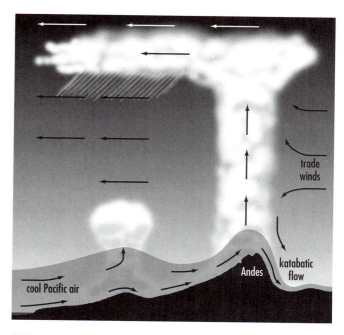

Weir effect. Cool air from the Pacific rises up the sides of the Andes, then spills over the top like water over a weir and descends into the north–south valleys as a katabatic wind. Orographic lifting sometimes triggers conditional instability, producing thunderstorms.

losing some of its moisture as it does so (*see* OROGRAPHIC LIFTING). Then it rises up the main part of the range and spills over the top like water over a weir, pushed by the air behind it, and descends as a KATABATIC WIND into the valleys that run longitudinally. The air forms a deep pool behind the mountains, and its forced rise can sometimes trigger CONDITIONAL INSTABILITY with severe THUNDERSTORMS.

West Australian Current   An ocean current that flows northward from the WEST WIND DRIFT, parallel to the western coast of Australia. The current flows strongly and steadily in summer but weakens in winter. Its water is cold, at 37°–45° F (3°–7° C), and its salinity, of 34.5 parts per thousand (‰), is below the average for seawater.

west coastal regime   *See* RAINFALL REGIME.

west coast desert climate   In the STRAHLER CLIMATE CLASSIFICATION a climate in his group 1, comprising climates controlled by equatorial and tropical AIR

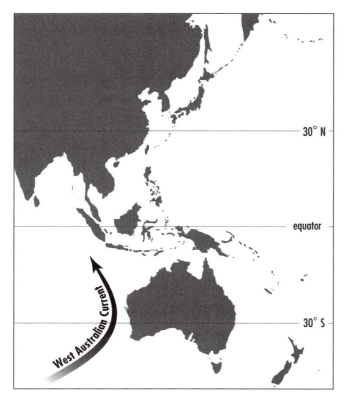

West Australian Current. **A cold ocean current that flows from the Southern Ocean, passing close to the western coast of Australia.**

MASSES. West coast desert climates occur in latitudes 15°–30° in both hemispheres along narrow belts on western coasts that border the oceanic subtropical high-pressure cells. Stable, dry maritime tropical air is subsiding in these cells. This produces a dry, fairly cool climate with a small range of temperature through the year and frequent FOG. In the KÖPPEN CLIMATE CLASSIFICATION this climate is designated *BWk* if it is cool and *BWh* if it is warm.

westerlies   *See* PREVAILING WESTERLIES.

westerly type   The kind of weather that is characteristic of middle latitudes and is generated and transported by the PREVAILING WESTERLIES. Weather conditions are variable, with successions of CYCLONES and ANTICYCLONES.

westerly wave   A wavelike disturbance that is embedded within the PREVAILING WESTERLIES of middle latitudes.

western intensification   The strengthening of ocean currents as they flow along the western margins of the oceans. The currents become narrower and faster as a result of the combined action of internal friction, WIND STRESS, and VORTICITY. They move northward in the Northern Hemisphere and southward in the Southern Hemisphere. Currents along the eastern margins of the oceans are broader and slower.

West Greenland Current   An ocean current that flows to the west of Greenland and that is an extension of the western branch of the NORTH ATLANTIC DRIFT. It carries relatively warm water northward parallel to the western coast of Greenland and into the Davis Strait. The current then divides, part of it continuing into Baffin Bay and part joining the LABRADOR CURRENT.

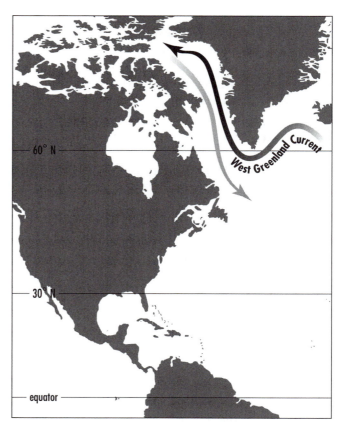

West Greenland Current. **A warm ocean current that flows parallel to the western coast of Greenland.**

West Wind Drift (Antarctic Circumpolar Current) An ocean current that flows from west to east around the coast of Antarctica. It is driven by the PREVAILING WINDS, which are from the west, and it is the only ocean current that flows all the way around the world. It carries water that is cold, with a temperature of 30°–40° F (–1°–5° C), and has a low salinity, of less than 34.7 parts per thousand (‰). This current is not to be confused with the ANTARCTIC POLAR CURRENT.

wet adiabat *See* SATURATED ADIABAT.

wet-bulb depression The difference between the temperature that is registered by a dry-bulb thermometer and that registered by a WET-BULB THERMOMETER adjacent to it. The difference is due to the LATENT HEAT of vaporization drawn from the bulb of the wet-bulb thermometer, and so a comparison between the dry-bulb temperature and the wet-bulb depression can be used to determine the RELATIVE HUMIDITY and the DEW POINT TEMPERATURE.

wet-bulb potential temperature The WET-BULB TEMPERATURE that a PARCEL OF saturated AIR would have if it were taken by an ADIABATIC mechanism to the 1,000-mb level (that is, to sea level).

wet-bulb temperature The temperature that is registered by a WET-BULB THERMOMETER. In saturated air (relative humidity 100 percent) the wet-bulb temperature is equal to the dry-bulb temperature, indicating that no evaporation is taking place. At any relative humidity below SATURATION the wet-bulb temperature is lower than the dry-bulb temperature.

wet-bulb thermometer A thermometer that is fitted with a layer of wetted cloth around its bulb. The cloth, usually muslin, extends below the bulb, where it is immersed in a reservoir of distilled water, so it acts as a wick. Water is drawn into the wick by CAPILLARITY and evaporates from it, thus maintaining a constant amount of moisture around the bulb provided the reservoir does not run dry and there is a free circulation of air around the cloth. LATENT HEAT of vaporization is taken from the thermometer bulb. This depresses the temperature registered by the thermometer. A wet-bulb thermometer is used in conjunction with a dry-bulb thermometer and the two together constitute a PSYCHROMETER.

wet equatorial climate In the STRAHLER CLIMATE CLASSIFICATION a climate of group 1, which comprises climates controlled by equatorial and tropical AIR MASSES. Wet equatorial climates are associated with warm, moist tropical maritime air and equatorial air. These climates affect regions between latitudes 10° N and 10° S and parts of Asia in latitudes 10°–20° N that have a MONSOON climate. Temperatures remain fairly constant throughout the year and rainfall is heavy, produced mainly by convectional storms. The wet equatorial climates include tropical rain forest climates, of which there are two types distinguished by whether or not they experience the monsoons. These are designated *Af* and *Am* in the KÖPPEN CLIMATE CLASSIFICATION.

wet scrubber A device that is used to remove pollutants from a stream of waste gas. It consists of an ABSORPTION TOWER in which the gas stream is put into contact with a liquid that absorbs the pollutants. Different liquids are used, depending on the pollutants to be removed.

wet snow *See* SNOW.

wet spell In Britain, a SPELL OF WEATHER lasting for at least 15 days during which at least 0.04 in (1 mm) of rain has fallen every day.

wettability The property of a surface that has an affinity for water, because its molecules attract water molecules. The surface becomes wet as water molecules adhere to it, and as the moisture penetrates the surface the substance may swell. Seaweed is the best known wettable material. At low tide, seaweed that is exposed to the air on a beach often becomes so dry it is brittle, but it recovers at once when the incoming tide wets it once more. Small particles with this property act as CLOUD CONDENSATION NUCLEI because of their capacity for capturing molecules of WATER VAPOR.

wettable aerosol *See* CONDENSATION.

wetting *See* RAIN GAUGE.

whaleback cloud   The name sailors have given to a type of LENTICULAR CLOUD that is sometimes seen in high latitudes. It forms in strong winds over islands and steep coastal cliffs and the name refers to the smooth, humped shape of the cloud top.

whippoorwill storm   *See* FROG STORM.

whirling psychrometer (sling psychrometer)   A PSY-CHROMETER in which the flow of air across the bulbs of the dry-bulb and wet-bulb thermometers is produced by manually whirling the instrument through the air. In one version, a chain is attached at the top of the board to which both thermometers are securely fixed and there is a handle on the other end of the chain. In another version, the board forms the horizontal arm of a device resembling a rattle. The top of the board is fixed to a rod, one end of which is attached to a handle in which it is free to rotate.

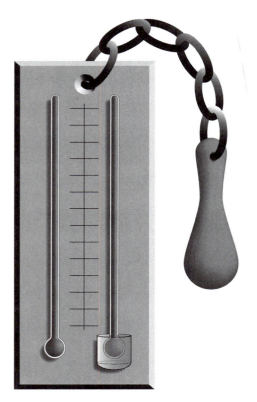

Whirling psychrometer. **An instrument that measures the wet-bulb and dry-bulb temperatures. It is swung through the air to ensure an evenly distributed flow of air over both bulbs. This process increases its accuracy.**

whirlwind   Although many people use the words *whirlwind* and TORNADO as synonym, in fact there are many differences between them; they are not at all the same.

Whirlwinds appear in the desert. They rise from the sand and dust suddenly, without warning. There is no cloud above them, no dark, menacing sky to warn of their approach, and they are seldom alone. Where there is one there are several, sometimes a small army of them. Each individual lasts for only a few minutes, but as one dies another arises. In biblical times they were feared, as much for their mysterious ways as for the damage they could cause. They are much milder than true tornadoes and can do little harm to a solidly constructed building, but they can demolish flimsy buildings and the tents in which desert dwellers often live.

Like tornadoes, whirlwinds are caused by CONVECTION, but unlike tornadoes they are not sustained by the LATENT HEAT of CONDENSATION. They occur in dry air and there is no condensation of WATER VAPOR. They develop on calm days, when there is little or no wind.

Wind mixes the air, so that if the ground surface is hotter in one place than it is in another the air in contact with it is mixed with cooler air. This mixing does not happen on still days and because the desert surface is uneven, some areas exposed to full sunshine and others shaded, and because it is made of a variety of materials, it heats unevenly. By early afternoon patches of exposed rock can be 30° F (17° C) hotter than nearby sand, because of differences in their ALBEDOS and HEAT CAPACITIES.

Air over the hot spots is heated by contact with the surface. It expands and rises by convection, creating a small region at the base where the air pressure is very slightly lower than it is farther away. Air from all sides is drawn into the low-pressure area and as the air converges its own VORTICITY makes it start to rotate. As it approaches the center the air accelerates in order to conserve its ANGULAR MOMENTUM. Then it is warmed and rises. Air is then spiraling inward and upward. This mechanism is very like a tornado, but it is driven from below rather than from a storm cloud overhead.

It is made visible by the dust and sand that is carried into its vortex and then high into the sky. Although the pressure at its center is low, the RELATIVE HUMIDITY is too low for water vapor to condense. The lack of water vapor also limits the lifespan of a whirlwind. It has no source of additional heat above ground level to maintain the upward flow of air. At the same time, relatively cool air flowing into the base of the vortex cools the hot ground, so before long it is at the same temperature as the surrounding surface. Air ceases to be drawn into the area and the whirlwind dies. The strongest whirlwinds can last for several hours, but many last for only a few seconds.

Whirlwinds vary in size. Most rise to about 100 ft (30 m). Some reach 300 ft (100 m) and a few grow to 6,000 feet (1,800 m) tall.

(You can read more about whirlwinds in Michael Allaby, *Dangerous Weather: Tornadoes* [New York: Facts On File, 1997].)

whirly A very small but violent storm, up to about 300 ft (100 m) across, that often occurs in Antarctica, especially around the time of the EQUINOXES.

white dew DEW that freezes after it has formed. It is more opaque, and therefore whiter, than HOAR FROST.

white horizontal arc An arc that is seen when sunshine falls onto a cloud that consists of ICE CRYSTALS with both vertical and horizontal faces. The faces act as mirrors. The arc passes through the shadow of the observer, which is cast onto the cloud. The phenomenon is very rare.

white noise Random data that obscure the signal from a measuring device.

whiteout The condition in which the ground, air, and sky are all a uniform white and no landscape features are visible. Persons exposed to a whiteout lose their perception of depth and quickly become disoriented, so a whiteout is extremely dangerous. There are two ways in which it can occur. In calm weather, a uniformly white snow surface may lie beneath low cloud. The cloud diffuses light passing through it, so light falls on objects evenly from all sides and there are no shadows. Consequently everything appears white. If dark objects are visible, they appear to float and it is impossible to determine their distance. Whiteout can also occur in a BLIZZARD. Again the light is diffused by clouds, but in this case there are also SNOWFLAKES between the cloud base and the ground. The snowflakes are tumbling and turning in all directions, reflecting

light in all directions as they do so. A flashlight is useless in the second type of whiteout, because the light is scattered by the falling snowflakes and so much may be reflected back to the person with the flashlight that it becomes dazzling.

white squall A SQUALL in tropical or subtropical seas that occurs suddenly and without the prior appearance of a SQUALL CLOUD. The only indication of the approaching squall is a line of white water caused by the wind.

whole gale In the BEAUFORT WIND SCALE, force 10, which is a wind that blows at 55–63 mph (88–101 km h^{-1}). In the original scale, devised for use at sea, a force 10 wind was defined as "or that with which she could scarcely bear close-reefed main-topsail and reefed foresail." On land, a whole gale breaks or uproots trees.

Wien's law A physical law that describes the relationship between the temperature of a BLACKBODY and the wavelength of its maximum emission of radiation. The wavelength varies inversely with the temperature (the higher the temperature the shorter the wavelength) and the law can be stated as $\lambda_{max} = C/T$, where λ_{max} is the wavelength of maximum emission, C is Wien's constant, and T is the temperature in kelvins. Wien's constant is $2,897 \times 10^{-6}$m ($2,897$ μm), so the law becomes $\lambda_{max} = (2,897/T) \times 10^{-6}$m. It is valid only for radiation at short wavelengths. The law was discovered in 1896 by the German physicist Wilhelm Wien (1864–1928), and for it he was awarded the 1911 Nobel Prize for Physics.

wild snow Freshly fallen snow that is very light and unstable.

Willie A TYPHOON that struck the island of Hainan, China, in September 1996. It caused floods that affected 70 percent of the streets of the capital, Haikou, and inundated 95,000 acres (38,000 ha) of farmland. At least 38 people were killed.

williwaw A violent SQUALL that occurs in the Strait of Magellan, the strip of ocean off the southernmost tip of South America that links the South Pacific and South Atlantic Oceans. Williwaws can occur at any time of year but are most frequent in winter. The same name is also given to the strong squall wind that blows as cold air down deep valleys in Alaska.

willy-nilly *See* WILLY-WILLY.

willy-willy (willy-nilly) A TROPICAL CYCLONE that strikes northwestern Australia. Willy-willies are TYPHOONS that develop in the Timor Sea, move southwestward, and then swing to an easterly direction, in a track that carries them toward the Australian coast. They are often very severe but die away as they move inland.

wilt *See* WILTING.

wilting The limpness that occurs when the cells of a plant contain insufficient water to keep them rigid. Wilting may occur when the rate of TRANSPIRATION exceeds the rate at which water is able to enter the root system of the plant from a soil containing abundant water. In this case the wilting is temporary and the plant recovers when the transpiration rate decreases. Wilting may also be due to a deficiency of water in the soil, in which case the plant does not recover unless it is given water. The percentage of water that remains in the soil after a test plant has wilted is known as the *permanent wilting percentage* (also called the *permanent wilting point, wilting coefficient,* and *wilting point*). Wilting also occurs when the vessels of a plant are blocked, often by a fungal infection, or when water is being taken from the plant vessels by a parasite. Wilting due to infection is called *wilt.*

wilting coefficient *See* WILTING.

wilting point *See* WILTING.

wind The movement of air that is due to an uneven distribution of AIR PRESSURE. Air tends to move from regions of high pressure to regions of low pressure at a speed that is proportional to the PRESSURE GRADIENT. It does not flow directly from the center of high pressure to the center of low pressure, however, as a result of the CORIOLIS EFFECT caused by the rotation of the Earth (*see* GEOSTROPHIC WIND). Local effects also produce winds such as FÖHN WINDS, ANABATIC WINDS, and KATABATIC WINDS that occur near mountains (*see also* AVALANCHE WIND); LAKE BREEZES; and LAND AND SEA BREEZES. Winds are generally thought of as blowing horizontally, but they can also have a vertical component.

windbreak A wall, fence, or other structure that is erected for the purpose of slowing or deflecting the wind.

windchill The extra feeling of cold that we experience when we are exposed to the wind is called *wind chill.* The effect can be measured and its magnitude is usually reported in degrees Fahrenheit (or Celsius). This gives the misleading impression that when a wind is blowing the air TEMPERATURE actually falls, and that is obviously wrong. Wind is simply moving air and its movement does not alter its temperature.

The confusion arises because of the use of temperature units. These are familiar and therefore easy to understand, but what happens when a person is exposed to the wind is that heat energy is removed from the body surfaces. Our internal body temperature changes very little and a drop of just a few degrees can be fatal. Heat and temperature are not the same and different units are used to measure them. Scientists measure heat in JOULES, CALORIES, or less commonly BRITISH THERMAL UNITS. These units are more difficult to understand than degrees and so temperature units are used.

Willy-willies

New Hebrides

Fiji Is.

New Caledonia

Willy-willy. **Tropical cyclones that strike northeastern Australia form over the Timor Sea. The map shows the direction in which they travel.**

We keep warm in winter by wearing clothes. These trap a layer of air. The heat of our bodies warms this air and, once it is warm, the air provides insulation. If we go outdoors into air that is much colder than our bodies, the layer of warm air inside our clothes protects us and the clothes themselves keep the warm air in place.

If we go out into a wind, however, the moving air may penetrate our clothing and blow away some of the protective layer of warm air. As warm air disappears, the body must work harder to replace it. Blood vessels near the body surface contract. This reduces the rate at which the blood is cooled by passing through cold tissues. We may stamp our feet, rub our hands, beat our arms, and eventually we start shivering. This involves rapid muscular movements that generate heat. If the wind is strong enough, however, it may remove warm air faster than our bodies can either replace it by generating more heat or compensate for its loss. That is when we start to feel cold. How cold we feel depends on the temperature of the air and the speed of the wind.

The rate at which a human body loses heat through wind chill increases rapidly as the temperature drops and the wind speed rises, until the wind speed reaches about 40 mph (64 km/h). At wind speeds faster than this there is only a small increase in wind chill.

It is heat that the body is losing, measured in joules or calories, but the effect of the wind is to cool the body at the rate it would chill if the temperature were lower. If you go outdoors on a still day when the temperature is 10° F (–12° C), that is how cold it feels. If there is a wind blowing at 10 mph (16 km h⁻¹), however, your body loses heat at the same rate as it would on a day with no wind, but with an air temperature of –9° F (–23° C). The effect of wind chill does not alter the actual temperature of the air, but it does lower the effective temperature, because it removes some of the insulating layer of warm air and so increases the rate at which the body loses heat.

You need more protection against the cold on a windy day than you do on a still day. When the temperature is below about –21° F (–29° C) on a day with no wind you need to be well wrapped up, with no bare

Wind chill temperature (°F)

Wind speed (mph)	35	30	25	20	15	10	5	0	-5	-10	-15	-20	-25	-30	-35	-40
0	35	30	25	20	15	10	5	0	-5	-10	-15	-20	-25	-30	-35	-40
5	32	27	22	16	11	6	0	-5	-10	-15	-21	-26	-31	-36	-42	-47
10	22	16	10	3	-3	-9	-15	-22	-27	-34	-40	-46	-52	-58	-64	-71
15	16	9	2	-5	-12	-18	-25	-31	-38	-45	-51	-58	-65	-72	-78	-85
20	12	4	-3	-10	-17	-24	-31	-39	-46	-53	-60	-67	-74	-81	-88	-95
25	8	1	-7	-15	-22	-29	-36	-44	-51	-59	-66	-74	-81	-88	-96	-103
30	6	-2	-10	-18	-25	-33	-41	-49	-56	-64	-71	-79	-86	-93	-101	-109
35	4	-4	-12	-20	-27	-35	-43	-52	-58	-67	-74	-82	-89	-97	-105	-113
40	3	-5	-13	-21	-29	-37	-45	-53	-60	-69	-76	-84	-92	-100	-107	-115

Windchill. To calculate the windchill, find the actual air temperature in the top row of figures. Then find the wind speed in the vertical column on the left. Follow the figures along the row and column, and the figure where they intersect is the effective temperature due to windchill. The lightly shaded figures indicate temperatures that are dangerously low. The dark shading indicates temperatures that are extremely dangerous.

skin exposed. If there is a wind, that effective temperature is reached at a much higher actual temperature. With a gentle wind of 5 mph (8 km h⁻¹) the effective temperature is –21° F (–29° C) when the actual air temperature is –15° F (–26° C). If the wind speed is 25 mph (40 km h⁻¹) that effective temperature is reached when the air temperature is +15° F (–9° C).

Weather reports and forecasts always include wind speeds and in winter they often include the "wind-chill factor," but you can work this out for yourself by using the table. First, though, you should correct the reported wind speed. Meteorologists measure wind speed well clear of the ground. Close to the ground the wind is slowed by FRICTION, especially in towns, and you should allow for this. Unless the report clearly states that the figure quoted refers to the wind speed at ground level, assume the wind speed at ground level is about two-thirds of the reported speed.

When the effective temperature falls below about –71° F (–57° C) conditions are extremely dangerous. If your body starts to lose heat in this actual or effective temperature you are likely to lose consciousness and die within a short time.

(You can read more about windchill and how to survive in the cold in Michael Allaby, *Dangerous Weather: Blizzards* [New York: Facts On File, 1997].)

wind classification *See* BEAUFORT WIND SCALE, FUJITA TORNADO INTENSITY SCALE, SAFFIR/SIMPSON HURRICANE SCALE.

wind direction The direction from which the wind blows (and *not* the direction in which it is blowing). Wind direction is measured by a WIND VANE or AEROVANE.

Windermere interstade An INTERSTADE that occurred in Britain from about 13,000 years ago until 11,000 years ago, during the latter part of the DEVENSIAN GLACIAL. The Windermere interstade coincided with the BØLLING, OLDER DRYAS, and ALLERØD periods in Scandinavia.

wind farm A number of wind turbines that are sited at the same location. Each turbine generates electrical power, and the combined output from all the turbines at the farm is fed into the public power supply. Wind farms may comprise as few as 10 turbines or as many as several hundred and the rated capacity of the individual turbines varies from about 450 kW to 5 MW. The largest wind farm in Australia is at Esperance, Western Australia. It has nine turbines and a rated capacity of 2 MW. The German Land (state) of Schleswig-Holstein is planning to build the biggest wind farm in Europe. It will be located in North Sea coastal waters near Helgoland. It will occupy 77 square miles (200 km²) and have a rated capacity of 1,200 MW. Its first phase will involve the construction of 100 turbines, rated at 4–5 MW each. They should be installed in 2004 and the entire farm is scheduled for completion by 2006. Smaller offshore wind farms are also planned for the North and Baltic Seas.

(There is more information about Australian wind farms at www.publish.csiro.au/ecos/Ecos90/Ecos90B.htm and about the German scheme at www.jxj.com/magsandj/rew/2000_01/going_to_sea_offshore_plans.html. Information about American wind farms can be obtained from the American Wind Energy Association at www.awea.org)

wind field A pattern of winds that is associated with a particular distribution of pressure.

WINDII *See* WIND IMAGING INTERFEROMETER.

wind imaging interferometer (WINDII) An INTERFEROMETER that is carried on the UPPER ATMOSPHERE RESEARCH SATELLITE. It measures the DOPPLER EFFECT on the spectral lines emitted by AIRGLOW emissions and AURORAE, from which it calculates temperatures and winds in the THERMOSPHERE.

wind power The harnessing of the wind to perform useful work. Sails were being used to propel boats along the River Nile possibly as early as 4000 B.C.E.

The earliest windmills were built in Persia (Iran) in the 7th century C.E. They were based on designs that were already used for watermills. By the 13th century windmills were appearing in many parts of Europe, their development having been strongly encouraged by the Mongol leader Genghis Khan. Until recently, mills were used principally for grinding cereal grains to make flour and meal, or to pump water. The first windmill to generate electrical power was built in Denmark in 1890. The generation of electrical power is now the principal use for windmills.

A modern "windmill" comprises rotor blades or sails. In the horizontal-axis design, which is the one most widely used, the blades are shaped as aerofoils with adjustable pitch—each blade can be turned on its own axis, thus altering the angle at which it meets the air. The horizontal axis to which the rotors are attached is free to turn on a vertical axis. This allows the blades to face into the wind at all times. Vertical-axis mills have "sails" held on radial arms.

The power that can be derived from the wind depends on the wind speed, the density of the air, and the area of the circle described by the rotor blades. It can be calculated from the equation

$$P = 0.5 \, \rho A V^3$$

where P is the power, ρ is the air density (about 0.08 pound per cubic foot [1.225 kg m^{-3}] at sea level, but less at higher elevations), A is the area swept by the rotor in square meters, and V is the wind speed in meters per second.

The proportion of this power that a wind turbine can extract is given by

$$P = 0.5 \, \rho A C_p V^3 N_g N_b$$

where C_p is the coefficient of performance of the rotor (a theoretical maximum value of 0.59, but actual value of 0.35 or less), N_g is the efficiency of the generator (from 0.5 to 0.8), and N_b is the efficiency of the gearing and bearings (up to 0.95). P is in watts (746 W = 1 horsepower; 1,000 W = 1 kilowatt; 1,000 kW = 1 megawatt).

(You can obtain more information about wind power from the American Wind Energy Association at www.awea.org. There is a general explanation of wind energy in Michael Allaby, *Elements: Air, the Nature of Atmosphere and the Climate* [New York: Facts On File, 1993].)

wind profile A diagram that shows the change in wind characteristics with height and horizontal distance. A wind speed profile, based on measurements, can be used to show the effect on the wind of an obstruction, such as a tree. The profile can then help in designing buildings or SHELTERBELTS. The wind speed at any height in the profile is given by

$$u = (u^*/k) \ln (z/z_o)$$

where u is the wind speed, u^* is the FRICTION VELOCITY, k is the VON KÁRMÁN CONSTANT, z is the height above the surface, z_o is the roughness length (*see* AERO-

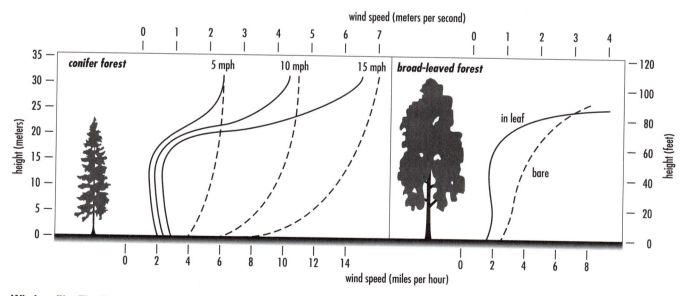

Wind profile. **The diagram compares the change of wind speed with height in a conifer forest (left) and a broad-leaved forest (right). In both cases wind speed increases sharply above the forest canopy. The broken lines in the diagram of the conifer forest show the wind speed over open ground. The two lines in the diagram of the broad-leaved forest compare the effect of the trees when bare and in full leaf.**

DYNAMIC ROUGHNESS), and ln means the natural logarithm.

wind profiler A ground-based instrument that measures the wind speed at different heights and at frequent intervals. Profilers measure the wind through the whole of the TROPOSPHERE and the lower STRATOSPHERE, with a vertical resolution ranging from about 200 ft (61 m) to about 3,300 ft (1 km), taking readings as frequently as every six minutes. The instrument works by DOPPLER RADAR. It transmits one radar beam vertically and, depending on the instrument type, two to four others at an angle of about 75° with respect to the horizon. The pulses, at a frequency of 900–1,500 MHz, are reflected by turbulent EDDIES. These move with the wind and consequently the reflected beams are Doppler-shifted. The Doppler shift can be interpreted as the speed of the wind crossing from one beam to another, and the time that elapses between the transmission of the beam and reception of its reflection indicates the height above the instrument.

wind ripple (snow ripple) A wavelike pattern on the surface of snow that is produced by the wind and forms at right angles to the WIND DIRECTION.

wind rose A diagram that shows the frequency with which the wind at a particular place blows from each direction. This reveals the direction of the PREVAILING WIND. The wind rose is constructed from measurements of the WIND DIRECTION that are made from the same point at the same time every day.

At the end of the recording period, which is often one calendar month, the daily wind directions are grouped into eight general compass directions: N, NE, E, SE, S, SW, W, and NW, according to which is nearest. They are then drawn as lines originating from a point, with north at the top of the drawing. The direction of each line is that of the wind and it is drawn to a convenient unit of length, such as 0.25 in (6 mm). If the same direction occurs on another day the line is made one unit longer. When the rose is complete it indicates the comparative frequency of each wind direction and therefore the prevailing wind for that period. The diagram also contains a circle, the radius of which is drawn to the same scale as the wind lines and represents the frequency of days on which the air

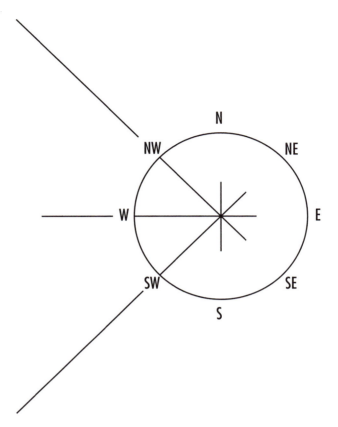

Wind rose. **Each of the thick lines represents a direction from which the wind blows. The length of each line indicates the frequency with which it blows from that direction. In this case the prevailing winds are from the southwest or northwest.**

was calm. The data can also be used to construct wind roses for each season and for the year. Obviously, the longer the record the more reliable the wind roses are likely to be.

Wind roses are very useful aids to planning. Foresters need to take account of prevailing winds when designing plantation forests in order to minimize damage from BLOWDOWN. The design of an urban development also needs to allow for the wind, to prevent the URBAN CANYON effect and exposure of open spaces to winds funneled into them (*see* FUNNELING) along streets. Wind roses are also used in planning airports so that their main runways can be aligned with the prevailing wind. This reduces the length of take-off and landing runs. On take-off it is the speed of the airflow over the wings of an aircraft that determines the amount of lift that is generated. If the aircraft takes off into the wind, that speed is equal to the sum of the forward speed of the aircraft and the speed of the wind, thus reducing the time the aircraft takes to attain the speed needed to become airborne. On landing a headwind has a strong braking effect.

windrow Loose material that has accumulated naturally to form a line. If it occurs inland or on the surface of the open sea or a lake, the material has been arranged by the wind and the orientation of the line indicates the WIND DIRECTION. If the windrow occurs on a beach it has been formed by the action of the tides.

wind scoop A shallow depression in the surface of snow that is seen at the base of walls and close to trees and other obstructions. Similar depressions form in loose sand. They are caused by eddies (*see* EDDY) produced when the wind carrying the snow or sand is deflected by the wall or obstruction.

wind shadow thermal A strong THERMAL upcurrent that develops in warm, sunny weather wherever TURBULENT FLOW greatly reduces the wind speed close to the surface. This occurs, for example, on the LEE side of hills and buildings. Because the wind speed is reduced, a shallow layer of air close to the ground is heated by contact with the surface and rises by CONVECTION. A wind shadow thermal can also rise from a field containing a crop of wheat or barley. The crop shelters the area between the ground and the top of the plants, and so the air in this layer is warmer than the air above the crop. If a gust of wind bends the plants down a body of warm air is released and rises convectively.

wind shear A change of wind VELOCITY with vertical or horizontal distance. FRICTION occurs between adjacent bodies of air that are moving in different directions or at different speeds. Where the difference is in speeds rather than direction this tends to accelerate the slower air and retard the faster air until both are moving at the same speed. Over land this would cause the wind to cease, because of friction between the lowest layer and the ground, and over the sea it would reduce the wind speed to the speed of the surface waves. This rarely happens because the wind is usually driven by a large-scale weather system or by PREVAILING WINDS acting at some distance above the surface and random movements of air molecules cause air to intermingle at the boundary between the two bodies of air. The molecules also transfer their MOMENTUM and this distributes the force driving the wind.

Above the level at which friction with the surface exerts an effect the amount of wind shear varies according to the way the air temperature changes and it gives rise to the THERMAL WIND. Where a strong upper-level wind blows across a slower lower-level wind the effect can be to start a column of rising air rotating about a vertical axis. This is the mechanism by which a MESOCYCLONE is formed, in extreme cases leading to a TORNADO.

wind-shift line A boundary that marks a large and abrupt change in the WIND DIRECTION.

wind sock A device for indicating the direction of the wind that was once a feature of every airfield and that is still seen at the smaller airfields used by light aircraft. It consists of a tapering cylinder of fabric that is open at both ends and supported on a circular frame at the larger end. The sock is attached by cords to the top of a tall pole. It fills in the slightest wind and, being free to turn in any direction, indicates the WIND DIRECTION. Unlike a WIND VANE or AEROVANE, it also indicates an absence of wind: when the air is calm the sock hangs limply. Wind socks are usually colored bright yellow or orange to make them more visible, and they

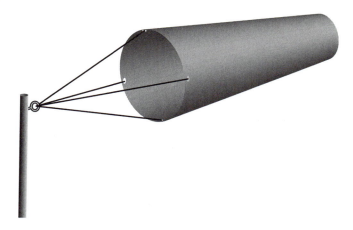

Wind sock. **The sock, made of fabric, is free to turn in any direction at the top of its pole. It can be seen from a distance.**

can be seen clearly from the air. They were especially useful in the days before airfields had runways and aircraft landed and took off from a grass surface, because they made it possible for the pilot to turn the aircraft into the wind.

wind sonde   *See* RADIOSONDE.

wind speed   The rate at which the air is moving. This is measured by using an ANEMOMETER or AEROVANE, and the measurement is of speed only, and takes no account of the WIND DIRECTION (*compare* WIND VELOCITY). There are several ways in which wind speed may be reported, depending on the purpose for which the report is issued. Ships are usually informed of the wind force on the BEAUFORT WIND SCALE. HURRICANES and other types of TROPICAL CYCLONE are allotted categories on the SAFFIR/SIMPSON HURRICANE SCALE that indicates the sustained wind speeds associated with them. TORNADO speeds are reported by the FUJITA TORNADO INTENSITY SCALE. Some public weather forecasts give the wind as a force on the Beaufort scale, but most use miles per hour (mph) or kilometers per hour (km h⁻¹). Reports to aircraft use KNOTS. Meteorologists usually use meters per second (m s⁻¹). To convert these units:

$$1 \text{ mph} = 1.619 \text{ km h}^{-1} = 0.87 \text{ knot} = 0.45 \text{ m s}^{-1}$$

$$1 \text{ knot} = 1.15 \text{ mph} - 1.85 \text{ km h}^{-1} = 0.51 \text{ m s}^{-1}$$

$$1 \text{ km h}^{-1} = 0.6214 \text{ mph} = 0.54 \text{ knot} = 0.28 \text{ m s}^{-1}$$

$$1 \text{ m s}^{-1} = 2.24 \text{ mph} = 1.95 \text{ knot} = 3.6 \text{ km h}^{-1}$$

windstorm   A STORM in which the most significant characteristic is a very strong wind. In the Tropics the areas of low pressure that produce them are not usually of frontal origin. Tropical windstorms can grow into TROPICAL CYCLONES. Windstorms in middle latitudes are often associated with deep frontal DEPRESSIONS and they can be severe. A series of windstorms crossed France and Belgium for three days between Christmas 1999 and New Year. Winds gusted to 105 mph (169 km/h), causing a major BLOWDOWN in which about 60,000 trees were damaged or destroyed in two forests on the outskirts of Paris, about 2,000 trees lining Paris streets were uprooted, and in France as a whole 160 sq mi (259 km²) of forest was destroyed. More than 120 people lost their lives.

wind stress   The force per unit area that is exerted on the land or water surface by the movement of air.

windthrow   *See* BLOWDOWN.

wind vane   The most commonly used device for measuring the direction of the wind. The practice of fastening wind vanes to the tops of church steeples and other tall buildings began in Athens in the 1st century B.C.E., with the TOWER OF THE WINDS. The wind vane consists of four fixed arms that point to the cardinal points of the compass (north, south, east, and west) and that are labeled N, S, E, and W. Above them is the vane itself, mounted so that it can move freely around a vertical axis. On one side of its axis the vane has a flat surface that is aligned by the wind. On the other side of the axis there is a pointer. The wind pushes the vane to the far side of its axis. Consequently, the pointer indicates the direction from which the wind is blowing. That is

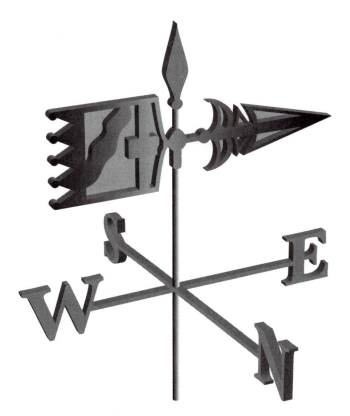

Wind vane. **The traditional wind vane, mounted on top of a church steeple, points into the direction from which the wind is blowing. This one is shaped as an arrow, and many are animal figures.**

why we name winds by the direction from which they blow and not by the direction in which they are blowing. For example, a west (or westerly) wind blows from west to east.

Often the vane is shaped as an arrow, with the flight acting as the vane and the head as the pointer. Other, more fanciful designs are also popular. These are usually figures of animals. Roosters are the favorite, and wind vanes of this design are known as *weathercocks,* but fish and other birds are also used. They all have a large tail or body to act as the vane and their heads act as the pointers.

Directions that lie between the cardinal points have to be judged by eye. For this reason, and also because a wind vane is mounted so high above the ground, the precise wind direction can be difficult to see clearly.

wind velocity   The speed and direction of the wind when these are reported together. VELOCITY is a VECTOR QUANTITY and consequently comprises two values. Wind velocity is reported in the form, direction/speed. For example, a report to an aircraft might describe the wind as 240/30, meaning that the wind is blowing from 240° at 30 KNOTS. Scientists usually report wind speed in meters per second, so in this example the wind would be 240/15.

windward   The side that faces the direction from which the wind is blowing.

wine harvest   The date when grapes are harvested, the size of the crop, and the quality of the wine that is made from them can be used to infer the weather conditions during the previous year. Wine has always been a product of great commercial importance, and during the Middle Ages many European monasteries relied on it as a source of income. Consequently, harvest records were kept meticulously and many of them have survived. Crop yields and harvest dates vary according to the variety of grape being grown, but when a grower replaced one variety with another, that was also recorded. Wine harvest records have been used to trace the climatic changes that have taken place since about the year A.D. 1000. They show, for example, that prior to 1300, 30–70 percent of wine harvests in southern Germany were described as good, but between 1400 and 1700 the proportion never rose above about 53 percent and sometimes fell to 20 percent. This deterioration

coincided with the coldest part of the LITTLE ICE AGE. Similar changes occurred in the French wine-growing regions and these have also been used in reconstructing the history of climate.

wing-tip vortices   EDDIES that develop around the wing tips of aircraft and birds and that are sometimes visible as thin streamers of cloud. They result from the way a wing generates lift.

Seen in cross section, the upper surface of a wing is more curved than the lower surface. Air flowing over the wing moves faster over the upper surface than it does over the lower surface, because it must cover a greater distance. This reduces the air pressure over the upper surface by the BERNOULLI EFFECT, so the pressure is higher on the lower surface than on the upper surface. The difference in pressure exerts an upward force on the wing. This is lift, but there is a secondary effect.

Air tends to flow from a region of high pressure to a region of low pressure. The curvature of the upper surface decreases toward the trailing edge (rear) of the wing, so near the trailing edge there is little difference in the curvature of the two surfaces and therefore little difference in the pressure on them. The center of lift, where the maximum lift is exerted, is about one-third of the distance back from the leading edge (front) of the wing. Air does not spill around the trailing edge from the high pressure below to the low pressure above, because at the trailing edge there is little difference in the two pressures.

This is not the case at the wing tips, however, because at the tips there is no boundary between the

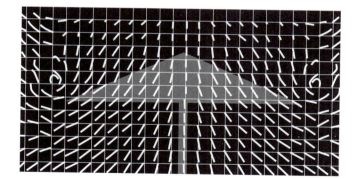

Wing-tip vortices. **The vortices that form around the tips of a delta-shaped wing, seen here in an experimental wind tunnel and indicated by small lengths of wool tied to the intersections of a wire grid.**

two pressures for the whole cross-sectional distance of the wing. Consequently, air spills over the edge of the wing tip. It then enters the LAMINAR FLOW of air over the main wing surface and is swept to the rear. This produces an eddy that describes a spiraling path behind the wing tip.

This is the wing-tip vortex. Air pressure is a little lower inside the vortex than it is outside. If the air is moist, the lower pressure may be sufficient to cause some WATER VAPOR to condense, producing a streamer of cloud behind the wing tip. This is most often seen behind high-performance aircraft that are turning steeply or pulling out of a dive. It is then that the vortices are strongest and the pressure within them is lowest.

Winnie A TYPHOON that struck Taiwan, eastern China, and the Philippines on August 18 and 19, 1997. It generated winds of up to 92 mph (148 km h⁻¹) and caused widespread flooding, especially in Taiwan and the Philippines. At least 37 people were killed in Taiwan, 140 in the Chinese provinces of Zhejiang and Jiangsu, and 16 in the Philippines. Tens of thousands of homes were destroyed in China and in the Philippines 60,000 people were forced to leave their homes.

winter ice A layer of ice that is less than one year old and forms a complete covering on a large area of the surface of the sea with a thickness of more than 8 in (20 cm).

winter storm warning A warning that is issued by the NATIONAL WEATHER SERVICE and broadcast on radio, television, and WEATHER RADIO to alert people in a particular area to the imminent arrival of severe winter weather. The warning means conditions have already begun to deteriorate or that they will do so within the next few hours.

winter storm watch A warning of approaching bad weather that is issued by the NATIONAL WEATHER SERVICE and broadcast on radio, television, and WEATHER RADIO. It provides one or two days' warning of the arrival of severe weather in a particular area, giving people time to prepare.

winter weather advisory A warning that the weather is expected to be bad enough to cause inconvenience and poor, possibly dangerous, driving conditions. It is issued by the NATIONAL WEATHER SERVICE and broadcast on radio, television, and WEATHER RADIO.

Wisconsinian glacial The most recent GLACIAL PERIOD in North America, following the SANGAMONIAN INTERGLACIAL. It began about 75,000 years ago and ended about 10,000 years ago. The last GLACIAL MAXIMUM occurred about 20,000 years ago. The Wisconsinian is named for rock deposits found in Wisconsin. Ice sheets advanced and retreated several times during the Wisconsinian, but average temperatures throughout the glacial were about 11° F (6° C) lower than those of today. The Wisconsinian was preceded by the SANGAMONIAN INTERGLACIAL and followed by the present FLANDRIAN INTERGLACIAL. The Wisconsinian is approximately equivalent in date to the glacial known as the *DEVENSIAN* in Britain, *WEICHSELIAN* in northern Europe, and *WÜRM* in the Alps.

WMO *See* WORLD METEOROLOGICAL ORGANIZATION.

Wolstonian glacial A GLACIAL PERIOD in Britain that began about 200,000 years ago and ended about 130,000 years ago. It is equivalent to the RISS GLACIAL of the European Alps and the Saalian glacial of northern Europe and partly coincides with the ILLINOIAN GLACIAL of North America. It was preceded by the HOXNIAN INTERGLACIAL and followed by the IPSWICHIAN INTERGLACIAL.

World Climate Applications and Services Program *See* WORLD CLIMATE PROGRAM.

World Climate Data and Monitoring Program *See* WORLD CLIMATE PROGRAM.

World Climate Impact Assessment and Response Strategies Program *See* WORLD CLIMATE PROGRAM.

World Climate Program (WCP) The program that was established in 1979 by the WORLD METEOROLOGICAL ORGANIZATION to collect and store climate data. The aim of the program is to provide the data necessary for economic and social planning and to improve the understanding of climate processes. Data from the WCP are also used to determine the predictability of climate and to detect climate changes. The WCP issues warnings to governments of impending climatic changes that may significantly affect their populations. The WCP has four components: the World Climate Data and Monitoring Program, the World Climate Applications and Services Program, the World Climate Impact Assessment and Response Strategies Program, and the World Climate Research Program. The WCP also supports the GLOBAL CLIMATE OBSERVING SYSTEM (GCOS), which covers all aspects of the global climate.

World Climate Research Program   *See* WORLD CLIMATE PROGRAM.

World Meteorological Organization (WMO)   The specialized United Nations (UN) agency that exists to promote the establishment of a worldwide system for gathering and reporting meteorological data, the standardization of the methods used, the development of national meteorological services in less industrialized countries, and the application of meteorological information and understanding to other fields. The WMO was founded in 1947 at the 12th meeting of its predecessor, the International Meteorological Organization (*see* INTERNATIONAL METEOROLOGICAL CONGRESS). At their 1947 meeting the directors of the International Meteorological Organization adopted the World Meteorological Convention, which authorized the creation of the WMO. The convention came into force in 1950 and in 1951 the WMO commenced its operation. Later in 1951, after discussions between the UN and WMO, the WMO became a UN agency.

The headquarters of the WMO are in Geneva, Switzerland. It has 185 members, of which 179 are member states and 6 are member territories that maintain their own weather services. The members are arranged as six regional associations, for Africa, Asia, South America, North and Central America, the Southwest Pacific, and Europe. Each association meets once every four years.

WMO policy is set by the World Meteorological Congress, which meets at least once every four years, and is implemented by an executive council with 36 members, including the WMO president and two vice presidents. The council meets at least once every year. The WMO also has eight technical commissions that deal with aeronautical meteorology, agricultural meteorology, atmospheric sciences, basic systems, climatology, hydrology, instruments and methods of observation, and marine meteorology. Each commission meets every four years.

The principal activity of the WMO centers on the WORLD WEATHER WATCH. The WMO also maintains the WORLD CLIMATE PROGRAM and the Atmospheric Research and Environment Program, which coordinates and fosters research into the composition and structure of the atmosphere, the physical and chemical properties of clouds, weather modification, tropical meteorology, and weather forecasting. This program also includes the Global Ozone Observing System that was established in the 1950s and now receives data from 140 ground-based stations. The international agreement to take steps to halt the depletion of the OZONE LAYER that resulted in the MONTREAL PROTOCOL ON SUBSTANCES THAT DEPLETE THE OZONE LAYER was based largely on data from the Global Ozone Observing System.

(You can learn more about the WMO at www.wmo.ch/web-en/wmofact.html.)

World Weather Watch (WWW)   The international program, run by the WORLD METEOROLOGICAL ORGANIZATION, that coordinates the national weather systems of the states that belong to the WMO. The WWW maintains links with four satellites in POLAR ORBIT and five in GEOSTATIONARY ORBIT, as well as about 10,000 weather stations on land, 7,000 weather ships, and 300 moored and drifting buoys that carry automated instruments. Data from these sources are sent by high-speed links to three global, 35 regional, and 183 national meteorological centers. These centers cooperate to prepare weather forecasts and analyses. The resulting information is supplied to the media and other bodies and individuals who request it. The activities of the WWW are carried out by its specialist branches. These include the TROPICAL CYCLONE PROGRAM.

Wukong   A TYPHOON that struck southern China on the weekend of September 9–10, 2000. Five people were killed, more than 1,000 houses collapsed, and 49,420 acres (30,710 ha) of rubber trees, bananas, rice, and pepper plants were destroyed.

Würm glacial   *See* DEVENSIAN GLACIAL.

WWW   *See* WORLD WEATHER WATCH.

Xangsane A TYPHOON that struck Taiwan on November 1 and 2, 2000. It had winds of 90 mph (145 km h⁻¹) and caused severe flooding. A Panamanian cargo ship, the *Spirit of Manila,* sank in the storm after breaking into three pieces. The storm killed 58 people and 31 were missing, including 23 of the crew of the *Spirit of Manila.* Xangsane (pronounced "Chang*sharn*") caused more than $2 billion of damage to crops and farms. The name means "elephant" in Thai.

xenon A rare gas that accounts for about 0.0000086 percent of the atmosphere, or 1 part in 10 million by volume. It is heavier than the other atmospheric gases. Atomic number is 54; atomic weight, 131.30; density (at sea-level pressure and 32° F [0° C]), 0.059 ounce per cubic foot (5.887 grams per liter). It melts at −169.6° F (−111.9° C) and boils at −160.6° F (−107.1° C).

xerophilous Adapted to living in places that have an arid climate.

xerophyte A plant that is adapted to an arid climate.

xerothermic An adjective that is applied to places, climates, or conditions that are hot and dry.

X rays Electromagnetic radiation that has a wavelength of 10^{-5} μm to 10^{-3} μm, between the wavelengths of ULTRAVIOLET RADIATION and GAMMA RADIATION. Less than 1 percent of the radiation emitted by the Sun is at X-ray wavelengths and all of the solar X rays reaching the Earth are absorbed in the upper atmosphere. None reaches the surface.

xylem *See* TRANSPIRATION.

yalca A severe SQUALL and snowstorm that occurs in the mountain passes of Peru.

yamase A cool, onshore easterly wind that blows in the Senriku district of Japan.

Yancy Two TYPHOONS, the first of which struck the Philippines and China in August 1990. It killed 12 people in the Philippines and 216 people in Fujian and Zhejiang Provinces, China. The second Typhoon Yancy struck Kyushu, Japan, in September 1993. Rated category 4 on the SAFFIR/SIMPSON SCALE, it generated winds of up to 130 mph (209 km h⁻¹) and killed 41 people.

Yanni A TROPICAL STORM that struck South Korea in late September and early October 1998. It flooded about one-quarter of the country's crop land and caused the deaths of at least 27 people.

yardang A ridge with a sharp crest that is shaped by strong winds that blow predominantly from one direction. A yardang forms in any rock that is only loosely structured. Yardangs are found only in very dry deserts with little plant cover and where soil has barely started to form.

Yarmouthian interglacial An INTERGLACIAL period in North America that followed the KANSAN GLACIAL and preceded the ILLINOIAN GLACIAL. It began about 230,000 years ago and ended about 170,000 years ago

and is approximately equivalent to the later part of the GÜNZ–MINDEL INTERGLACIAL in the European Alps. At various times during the Yarmouthian average temperatures were both lower and higher than those of today.

York A typhoon that struck Hong Kong on September 16, 1999, crossing the territory and reaching Guangzhou, in mainland China, later the same day. The maximum wind speed was 93 mph (150 km h⁻¹). York produced heavy rain in the province of Fujian. At least one person was killed in Hong Kong and six in Fujian.

Younger Dryas (Loch Lomond stadial) A cold, dry period, affecting the whole of northern Europe but not North America, that began about 11,000 years ago and lasted for about 1,000 years. It is recognized by soils of that date containing abundant pollen from mountain avens (*Dryas octopetala*), which is an arctic-alpine plant. When the Younger Dryas began, ICE SHEETS had probably disappeared from the whole of Scotland as the DEVENSIAN GLACIAL drew to a close, but by about 10,800 years ago the ice was hundreds of meters thick over the western Highlands and the climate of Europe was similar to that during the GLACIAL. The melting of the LAURENTIDE ICE SHEET is believed to have triggered the rapid onset of the Younger Dryas by releasing large amounts of freshwater that flowed into the North Atlantic. This shut down the ATLANTIC CONVEYOR and produced a SNOWBLITZ effect.

young ice   A layer of ice that is less than one year old and forms a complete covering on a large area of the surface of the sea with a thickness of more than about 2 in (5 cm).

Young's modulus   *See* SPEED OF SOUND.

Z *See* GREENWICH MEAN TIME.

Zack A TROPICAL STORM that struck the Philippines in October 1995. It capsized a ship sailing between islands, killing 59 people. It caused severe flooding on land. A total of at least 100 people died and 60,000 were forced to leave their homes.

Zane A TYPHOON that crossed Taiwan on September 28, 1996. It triggered mudslides and killed two people, then moved away to Okinawa.

Zeb A TYPHOON that struck the Philippines, Taiwan, and Japan in October 1998. It killed at least 74 people in the Philippines, 25 in Taiwan, and 12 in Japan.

Zelzate interstade *See* DENEKAMP INTERSTADE.

zenithal rain Rain that falls in the Tropics or subtropics every year, or every other year, during the season when the Sun is most nearly overhead—at its zenith.

zenith angle The height of the Sun above the horizon, which is measured as the angle between a line linking the observer to the Sun and the vertical (the zenith). The zenith angle (Z) is calculated from

$$\cos Z = \sin \theta \sin \delta + \cos \theta \cos \delta \cos h$$

where θ is the latitude, δ is the solar DECLINATION, and h is the hour angle (this is 0 at noon and increases by 15° ($\pi/12$) for each hour each side of noon).

zenith transmissivity A measure of the fraction of solar radiation reaching the top of the atmosphere when the Sun is directly overhead (at zenith) that penetrates to the surface. The term is used to describe the transparency of the atmosphere and is calculated in relation to the zenith even for places outside the Tropics, where the Sun is never directly overhead.

zephyr Any very gentle breeze, but especially one that blows from the west. Zephuros was the Greek god of the west wind and *Zephiros*, from which "zephyr" is immediately derived, was the Latin version of the name.

zeroth law *See* THERMODYNAMICS, LAWS OF.

zobaa A WHIRLWIND of sand, resembling a pillar, that is sometimes seen in the Egyptian desert. It is very tall and moves very rapidly.

zodiacal light A band of light that appears in the sky at night, in the east shortly before dawn, or in the west shortly after sunset. It is believed to be caused by the DIFFRACTION and REFLECTION of light by dust particles in the upper atmosphere and in space beyond the PLANE OF THE ECLIPTIC. It is most often seen in the

Tropics, where the ECLIPTIC is almost vertically overhead.

zonal average   *See* ZONAL FLOW.

zonal circulation   The movement of air in the TROPOSPHERE in a west-to-east direction that is measured over a specified period or with respect to a particular longitude.

zonal flow   A movement of air that flows in a generally west-to-east direction, approximately parallel to the lines of latitude. When winds are measured over a large area they may be separated into their zonal and meridional components (*see* MERIDIONAL FLOW). The average zonal winds, known as the *zonal average,* are strongest in the upper TROPOSPHERE over the subtropics during the three winter months (December, January, and February in the Northern Hemisphere and June, July, and August in the Southern Hemisphere). They are then blowing at an average speed of 144 km h^{-1} (89.5 mph).

zonal index   The strength of the westerly winds between latitudes 33° and 55° in both hemispheres, expressed either as the horizontal PRESSURE GRADIENT or as the corresponding GEOSTROPHIC WIND. A high zonal index indicates strong westerly winds and a continuous and almost straight JET STREAM. A low index indicates weak westerlies and the formation of a cellular pattern of air flow (*see* INDEX CYCLE).

zonal kinetic energy   The KINETIC ENERGY of the mean zonal wind (*see* ZONAL FLOW) that is calculated by averaging the zonal component of the wind along a specified latitude.

zonal wind-speed profile   A diagram in which the speed of the ZONAL FLOW is plotted against latitude.

zonda   A warm, dry wind, of the CHINOOK and FÖHN type, that occurs in winter in the lee of the Andes in Argentina. In dry weather it carries a great deal of dust and it can attain speeds of 75 mph (120 km h^{-1}).

zone of aeration   *See* GROUNDWATER.

zone of maximum precipitation   The elevation at which the PRECIPITATION that falls on a mountainside is greatest.

zone of saturation   *See* GROUNDWATER.

BIBLIOGRAPHY AND FURTHER READING

BOOKS AND ARTICLES

The following titles include all those that are mentioned at the ends of entries. I have added to this list a number of other titles that I have found helpful and that together provide an introduction to the more formal study of climatology and meteorology.

Abrahamson, John, and James Dinniss. "Ball Lightning Caused by Oxidation of Nanoparticle Networks from Normal Lightning Strikes on Soil," *Nature* 403 (2000): 519–521.

Allaby, Michael. *Ecosystem: Deserts.* New York: Facts On File, 2001.

————. *Guide to Weather.* London: Dorling Kindersley, 2000.

————. *Ecosystem: Temperate Forests.* New York: Facts On File, 1999.

————. *Dangerous Weather: Droughts.* New York: Facts On File, 1998.

————. *Dangerous Weather: Floods.* New York: Facts On File, 1998.

————. *Dangerous Weather: A Chronology of Weather.* New York: Facts On File, 1998.

————. *Dangerous Weather: Tornadoes.* New York: Facts On File, 1997.

————. *Dangerous Weather: Hurricanes.* New York: Facts On File, 1997.

————. *Dangerous Weather: Blizzards.* New York: Facts On File, 1997.

————. *How the Weather Works.* Pleasantville, N.Y.: Reader's Digest Association, 1995.

————. *Elements: Fire.* New York: Facts On File, 1993.

————. *Elements: Earth.* New York: Facts On File, 1993.

————. *Elements: Air.* New York: Facts On File, 1992.

————. *A Guide to Gaia.* New York: E. P. Dutton, 1989.

Barry, Roger G., and Richard J. Chorley. *Atmosphere, Weather and Climate,* 4th ed. London: Methuen, 1982.

Bluestein, Howard B. *Tornado Alley: Monster Storms of the Great Plains.* New York: Oxford University Press, 1999.

Bryant, Edward. *Climate Process and Change.* Cambridge: Cambridge University Press, 1997.

Calder, Nigel. "The Carbon Dioxide Thermometer and the Cause of Global Warming," *Energy and Environment* 10, no. 1 (1999): 1–18.

Critchfield, Howard J. *General Climatology.* Englewood Cliffs, N.J.: Prentice-Hall, 1960.

Eddy, John A. "The Case of the Missing Sunspots," *Scientific American,* May 1997, pp. 80–89.

Fischer, Hubertus, Martin Wahlen, Jesse Smith, Derek Mastroianni, and Bruce Deck. "Ice Core Records of Atmospheric CO_2 Around the Last Three Glacial Terminations," *Science* 283 (1999): 1,712–1,714.

Gleick, James. *Chaos: Making a New Science.* London: Heinemann, 1987.

Gurney, R. J., J. L. Foster, and C. L. Parkinson, eds. *Atlas of Satellite Observations Related to Global Change.* Cambridge: Cambridge University Press, 1993.

Henderson-Sellers, Ann, and Peter J. Robinson. *Contemporary Climatology.* London: Longman, 1986.

Herzog, Howard, Baldur Eliasson, and Olav Kaarstad. "Capturing Greenhouse Gases," *Scientific American,* February 2000, pp. 54–61.

Hidore, John J., and John E. Oliver. *Climatology, an Atmospheric Science.* New York: Macmillan, 1993.

Hoffman, Paul F., and Daniel P. Schrag. "Snowball Earth," *Scientific American* January 2000, pp. 50–57.

Hubler, Graham K. "Fluff Balls of Fire," *Nature* 403 (2000): 487–488.

Jardine, Lisa. *Ingenious Pursuits.* London: Little, Brown, 1999.

Joseph, Lawrence E. *Gaia: The Growth of an Idea.* New York: St. Martin's Press, 1990.

Kendrew, W. G. *The Climates of the Continents,* 5th ed. Oxford: Oxford University Press, 1961.

Kent, Michael. *Advanced Biology.* Oxford: Oxford University Press, 2000.

Kupchella, Charles E., and Margaret C. Hyland. *Environmental Science: Living Within the System of Nature,* 2nd ed. Needham Heights, Mass.: Allyn and Bacon, 1986.

Lamb, Hubert. *Through All the Changing Scenes of Life.* East Harling, Eng.: Taverner Publications, 1997.

———. *Climate, History and the Modern World,* 2nd ed. New York: Routledge, 1995.

Ladurie, Emmanuel Le Roy. *Times of Feast, Times of Famine: A History of Climate Since the Year 1000.* New York: Doubleday and Company, 1971.

Lockwood, John G. *World Climatology, an Environmental Approach.* London: Edward Arnold, 1974.

Lovelock, James E. *Homage to Gaia.* Oxford: Oxford University Press, 2000.

———. *The Ages of Gaia.* Oxford: Oxford University Press, 1989.

———. *Gaia: A New Look at Life on Earth.* Oxford: Oxford University Press, 1979.

Lutgens, Frederick, K., and Edward J. Tarbuck. *The Atmosphere,* 7th ed. Upper Saddle River, N.J.: Prentice-Hall, 1979.

McIlveen, Robin. *Fundamentals of Weather and Climate.* London: Chapman & Hall, 1992.

MacKenzie, James J. and Mohamed T. El-Ashry, eds. *Air Pollution's Toll on Forests and Crops.* New Haven, Conn.: Yale University Press, 1989.

Michaels, Patrick J., and Robert C. Balling. *The Satanic Gases: Clearing the Air about Global Warming.* Washington, D.C.: Cato Institute, 2000.

Morgan, J. J., and H. M. Liljestrand. *Final Report, Measurement and Interpretation of Acid Rainfall in the Los Angeles Basin.* Sacramento: California Air Resources Board, 1980.

Oke, T. R *Boundary Layer Climates,* 2nd ed. New York: Routledge, 1987.

Parkinson, Claire L. *Earth from Above.* Sausalito, Calif.: University Science Books, 1997.

Pendick, Daniel. "Cloud Dancers," in *Weather: What We Can and Can't Do About It.* New York: Scientific American, 2000, pp. 64–69.

Petit, J. R., et al. "Climate and Atmospheric History of the Past 420,000 Years from the Vostok Ice Core, Antarctica," *Nature* 399 (1999): 429–436.

Schneider, Stephen, ed. *Encyclopedia of Climate and Weather.* 2 vols. New York: Oxford University Press, 1996.

Sellers, William D. *Physical Climatology.* Chicago: University of Chicago Press, 1965.

Singer, S. Fred. *Hot Talk Cold Science: Global Warming's Unfinished Debate.* Oakland, Calif.: The Independent Institute, 1997.

Volk, Tyler. *Gaia's Body: Towards a Physiology of Earth.* New York: Copernicus/Springer-Verlag, 1998.

Wettlaufer, John S. and J. Greg Dash. "Melting Below Zero," *Scientific American,* February 2000, pp. 34–37.

APPENDIX I

CHRONOLOGY OF DISASTERS

1246–1305 DROUGHT in what is now the southwestern United States.

1281 A TYPHOON destroyed a fleet of Korean ships carrying Mongol troops on their way to invade Japan. This was the KAMIKAZE wind.

1703 On November 26 and 27 hurricane-force winds in the English Channel destroyed 14,000 homes and killed 8,000 people in southern England.

1762 In February a BLIZZARD in England lasted for 18 days and killed nearly 50 people.

1831 A HURRICANE struck Barbados, killing 1,477 people.

1865 In June a TORNADO moved through Viroqua, Wisconsin, destroying 80 buildings and killing more than 20 people.

1875 On November 15 the River Thames rose, possibly by more than 28 ft (8.5 m), causing extensive flooding in London.

1876 A CYCLONE coinciding with high, MONSOON river levels flooded islands in the Ganges Delta and on the mainland, in which 100,000 people drowned in half an hour.

1879 On December 28 the TAY BRIDGE DISASTER caused 70–90 deaths in Scotland.

1887 In September and October the Yellow River, China, flooded about 10,000 square miles (26,000 km²). Between 900,000 and 2.5 million people died.

1888 On March 11–13 BLIZZARDS with winds up to 70 mph (113 km/h) struck the eastern United States. More than 400 people died, including 200 in New York City.

1925 In March a series of possibly seven tornadoes developed over Missouri and crossed Illinois and Indiana, killing 689 people.

1931 The Yangtze River, China, rose 97 ft (29.6 m) after heavy rain. About 3.7 million people died, some in the floods but most as a result of the famine that followed.

1954 On October 12 Hurricane HAZEL killed 1,175 people (1,000 of them in Haiti).

1956 On June 27 Hurricane AUDREY killed nearly 400 people.

1957 In August Hurricane DIANE killed more than 190 people.

1959 In September Typhoon VERA killed nearly 4,500 people and left 1.5 million homeless.

1966 On November 3 the River Arno flooded Florence, Italy, causing extensive damage to historic buildings and works of art, killing 35 people, and leaving 5,000 homeless.

1969   On August 17–18 Hurricane CAMILLE killed about 275 people.

1970   In November a CYCLONE killed about 500,000 people in Bangladesh.

1973   On January 10 a tornado killed 60 people and injured more than 300 in San Justo, Argentina.

1974   On September 20 Hurricane FIFI killed about 5,000 people in Honduras. Cyclone TRACY struck Darwin, Australia, on December 25.

1976   On September 8–13 Typhoon FRAN killed 104 people and made 325,000 homeless in Japan.

1977   On November 19 a cyclone and STORM SURGE washed away 21 villages and damaged 44 more in Andhra Pradesh, India, killing an estimated 20,000 people and making more than 2 million homeless.

1978   On April 16 a tornado killed nearly 500 people and injured more than 1,000 in Orissa, India. On October 26 Typhoon RITA killed nearly 200 people and destroyed 10,000 homes in the Philippines. On November 23 a cyclone killed at least 1,500 people and destroyed more than 500,000 buildings in Sri Lanka and southern India.

1979   On April 10 a tornado killed 59 people and injured 800 at Wichita Falls, Texas. On May 12–13 a cyclone killed more than 350 people in India. On August 11 heavy rain caused a dam to break, flooding the town of Morvi, India, and killing up to 5,000 people. In August Hurricane DAVID killed more than 1,000 people in the Caribbean and eastern United States.

1980   In August Hurricane ALLEN killed more than 270 people. A heatwave killed 1,265 people in the United States; in Texas temperatures exceeded 38° C (100° F) almost every day.

1981   On July 12–14 monsoon rains caused the Yangtze River, China, to flood, killing about 1,300 people and leaving 1.5 million homeless. On September 1 Typhoon AGNES killed 120 people in South Korea. On November 24 Typhoon IRMA killed more than 270 people and left 250,000 homeless in the Philippines.

1982   On January 23–24 floods killed at least 600 people and left 2,000 missing in Peru. On June 3 monsoon floods in Sumatra, Indonesia, killed at least 225 people and left 3,000 homeless. In September monsoon floods in Orissa, India, killed at least 1,000 people.

1983   In June floods killed at least 935 people in Gujarat, India.

1984   Between January 31 and February 2 Cyclone DOMOINA killed at least 124 people. On September 2–3 Typhoon IKE killed more than 1,300 people in the Philippines. In November Typhoon AGNES killed at least 300 people in the Philippines and left 100,000 homeless.

1985   On May 25 a cyclone and storm surge killed an estimated 2,540 people, but possibly as many as 11,000 on islands off Bangladesh. On May 31 tornadoes killed 88 people and caused extensive damage in Pennsylvania, Ohio, New York, and Ontario.

1986   On March 17 Cyclone HONORINNIA killed 32 people and left 20,000 homeless in Madagascar. On September 4 a typhoon killed 400 people in Vietnam.

1987   In August floods killed more than 1,000 people in Bangladesh. On November 26 Typhoon NINA killed 500 people in the Philippines.

1988   In August and September monsoon floods inundated 75 percent of Bangladesh, killing more than 2,000 people and leaving at least 30 million homeless. On September 12–17 Hurricane GILBERT killed at least 260 people in the Caribbean and Gulf of Mexico and generated nearly 40 tornadoes in Texas. On October 24–25 Typhoon RUBY killed about 500 people in the Philippines. On November 7 Typhoon SKIP killed at least 129 people in the Philippines. On November 29 a cyclone killed up to 3,000 people in Bangladesh and eastern India.

1989 On April 26 a tornado in Bangladesh killed up to 1,000 people and injured 12,000. On September 17–21 Hurricane HUGO killed more than 40 people in the Caribbean and eastern United States. On November 4–5 Typhoon GAY killed 365 people in Thailand.

1990 On May 9 a cyclone killed at least 962 people in Andhra Pradesh, India. In August Typhoon YANCY killed 228 people in the Philippines and China. On August 28 a tornado killed 29 people and injured 300 at Plainfield, Illinois.

1991 On March 10 floods killed more than 500 people and left 150,000 homeless in Mulanje, Malawi. On April 26 more than 70 tornadoes killed 26 people and injured more than 200 in Kansas. On April 30 a cyclone killed at least 131,000 people on coastal islands off Bangladesh. On May 7 a tornado killed 100 people at Tungi, Bangladesh. In June flash floods killed up to 5,000 people in Jowzjan Province.

1992 In July floods killed more than 1,000 people in Fujian and Zheijiang Provinces, China. On August 23–26 Hurricane ANDREW killed 38 people and caused extensive damage in the Bahamas, Florida, and Louisiana. On September 11–16 monsoon rains caused the Indus River to flood, killing at least 500 people in India and more than 2,000 in Pakistan.

1993 On January 8 a tornado killed 32 people and injured more than 1,000 in Bangladesh. On March 12–15 a blizzard killed at least 238 people in the eastern United States, 4 in Canada, and 3 in Cuba. Between October 31 and November 2 mudslides killed 400 peo-

ple and destroyed 1,000 homes in Honduras.

1994 On February 2–4 Cyclone GERALDA killed 70 people and left 500,000 homeless in Madagascar. On August 20–21 Typhoon TED killed about 1,000 people in Zheijiang Province, China. On November 13–19 Tropical Storm GORDON killed 537 people in the Caribbean, Florida, and South Carolina.

1995 Beginning in July, floods affected 5 million people, nearly 25 percent of the population, in North Korea. On November 3 Typhoon ANGELA killed more than 700 people and left more than 200,000 homeless in the Philippines.

1996 On May 13 a tornado in Bangladesh destroyed 80 villages in less than half an hour, killing more than 440 people and injuring more than 32,000. On September 10 Typhoon SALLY killed more than 130 people and destroyed nearly 400,000 homes in Guangdong, China.

1996–7 In December and January floods in California, Idaho, Nevada, Oregon, and Washington caused at least 29 deaths and forced 125,000 people to leave their homes.

1997 In January a cold wave crossed Europe in January, killing at least 228 people. On May 2 a SANDSTORM killed 12 people and injured 50 in Egypt. On May 27 tornadoes in central Texas destroyed about 60 homes and killed 30 people. On July 2 THUNDERSTORMS and tornadoes in southern Michigan destroyed 339 homes and business premises, killed 16 people, and injured more than 100. On August 18–19 Typhoon WINNIE killed nearly 200 people in China, Taiwan, and the Philippines. On October 8–10 Hurricane PAULINE killed 217 people and left 20,000 homeless in southern Mexico. On October 12 a tornado killed at least 25 people and injured thousands who had gathered for a religious ceremony at Tongi, Bangladesh. In November Typhoon LINDA killed at least 484 people in Vietnam, Cambodia, and Thailand.

1998

On February 23 tornadoes in Florida killed at least 42 people, injured more than 260, and left hundreds homeless. On March 20 tornadoes killed at least 14 people and injured 80 in Georgia, and killed 2 and injured at least 22 in North Carolina. In March a cyclone killed at least 200 people and made 10,000 homeless in West Bengal and Orissa, India. On April 8–9 tornadoes killed 39 people in Mississippi, Alabama, and Georgia. In May and early June a heat wave killed at least 2,500 people in India. From June to August the Yangtze River, China, flooded, killing 3,656 people and affecting an estimated 230 million. On September 21–28 Hurricane GEORGES killed more than 330 people in the Caribbean and the U.S. Gulf Coast. In September and early October Tropical Storm YANNI killed 27 people in South Korea. In October Typhoon ZEB killed 111 people in the Philippines, Taiwan, and Japan. In late October Hurricane MITCH killed more than 8,600 people, left 12,000 unaccounted for, and made more than 1.5 million homeless in Central America. In late October Typhoon BABS killed at least 132 people and made about 320,000 homeless in the Philippines. On November 19–23 Typhoon DAWN caused floods in Vietnam that forced 200,000 people from their homes and killed more than 100.

1999

In August Typhoon OLGA caused extensive flooding in South Korea. In September Hurricane FLOYD killed 49 people in the Bahamas and the eastern coast of the United States.

2000

In February the worst floods in 50 years devastated Mozambique, destroying about 200,000 homes. On February 22 Cyclone ELINE struck Mozambique, with winds of up to 162 mph (260 km h^{-1}). Eline moved to Madagascar, which

was also struck by Tropical Storm Gloria on March 4–5. The two storms left at least 500,000 people homeless on the island and killed at least 137. Shortly after midnight on February 14 tornadoes sweeping through southwestern Georgia killed 18 people and injured about 100. In May, severe flooding combined with a tidal surge killed at least 140 people and left about 20,000 homeless on West Timor, Indonesia. Between late July and early October the Mekong Delta, in Vietnam, Laos, and Cambodia, experienced the worst flooding in 40 years, killing at least 315 people. In September and October flooding killed more than 900 people in India and about 150 in Bangladesh, and left some 5 million homeless in the two countries. On November 1–2 Typhoon Xangsane caused severe flooding on Taiwan.

2001

Floods early in the year killed at least 52 people in Mozambique and left more than 80,000 homeless. In January a blizzard in northern China killed 20 people and left thousands with no access to food supplies. In February tornadoes killed five people in Mississippi and one in Arkansas. In April the Mississippi River burst its banks, flooding parts of Minnesota, Iowa, Illinois, and Wisconsin. A 165-ft (50-m) dam of ice blocks caused the River Lena, in Siberia, to flood in May, washing away thousands of homes and killing at least five people. On May 28, 18 people were injured and many buildings damaged by a tornado in Ellicott, Colorado. Weekend storms on July 7–8 caused widespread flooding in West Virginia. In July, Typhoon Utor caused floods and mudslides in which 23 people died in the Philippines and one in Taiwan. Two days of rain in South Korea, also in July, caused 40 deaths. On July 29 and 30 Typhoon Toraji swept through Taiwan, causing at least 72 deaths and leaving more than 130 people unaccounted for. Monsoon floods in late July trapped nearly 50,000 people in inundated villages in Bangladesh.

APPENDIX II

CHRONOLOGY OF DISCOVERY

c. 340 B.C.E.	ARISTOTLE wrote *Meteorologica,* the oldest known work on meteorology and possibly the first. It gave us the word *meteorology.*
140–131 B.C.E.	Han Ying, in China, wrote *Moral Discourses Illustrating the Han Text of the "Book of Songs."* This contained the first known description of the hexagonal structure of SNOWFLAKES.
first century B.C.E.	The TOWER OF THE WINDS was built in Athens. It was possibly the first attempt to forecast the weather systematically on the basis of observations.
c. 55 B.C.E.	Lucretius Carus (c. 94–55 B.C.E.) proposed that thunder is the sound of great clouds crashing together. Although he was wrong he may have been the first person to notice that thunder is always associated with big, solid-looking clouds.
first century	HERO OF ALEXANDRIA demonstrated that air is a substance.
1555	OLAUS MAGNUS published a book containing the first European depictions of ICE CRYSTALS and snowflakes.
1586	SIMON STEVINUS showed that the pressure a liquid exerts on a surface depends on the height of the liquid above the surface and the area of the surface on which it presses, but it does not depend on the shape of the vessel containing the liquid.
1593	GALILEO GALILEI invented his AIR THERMOSCOPE.
1611	Johannes Kepler (1571–1630) published *A New Year's Gift, or On the Six-Cornered Snowflake,* in which he described snowflakes.
1641	FERDINAND II invented a THERMOMETER consisting of a sealed tube containing liquid.
1643	EVANGELISTA TORRICELLI invented the BAROMETER.
1646	BLAISE PASCAL demonstrated that atmospheric pressure decreases with height.
1654	FERDINAND II improved on his thermometer, producing the design that would lead to the mercury thermometer invented by DANIEL FAHRENHEIT in 1714.
1660	ROBERT BOYLE published his discovery of the relationship between the volume occupied by a gas and the pressure under which the gas is held.
1686	EDMUND HALLEY proposed the first explanation for the TRADE WINDS.
1687	GUILLAUME AMONTONS invented the HYGROMETER.

1714   DANIEL FAHRENHEIT invented the mercury thermometer and the temperature scale that bears his name.

1735   GEORGE HADLEY proposed his model of the circulation of the atmosphere to explain the direction from which the trade winds blow.

1738   DANIEL BERNOULLI demonstrated that when the velocity of a flowing fluid increases its internal pressure decreases.

1742   ANDERS CELSIUS proposed the temperature scale that bears his name.

1752   BENJAMIN FRANKLIN performed his experiment with a kite, demonstrating that storm clouds carry electric charge and that a lightning stroke is a giant spark.

1761   JOSEPH BLACK demonstrated that when ice melts it absorbs heat with no rise in its own temperature. He later showed that heat is absorbed or released when water vaporizes and condenses. He called this *LATENT HEAT*.

1783   HORACE BÉNÉDICT DE SAUSSURE invented the hair hygrometer.

1806   ADMIRAL FRANCIS BEAUFORT proposed a scale for classifying wind forces.

1820   JOHN DANIELL invented the dew point hygrometer.

1824   JOHN DANIELL showed the importance of maintaining a humid atmosphere in hothouses growing tropical plants.

1827   JEAN-BAPTISTE FOURIER wrote what may have been the first account of the GREENHOUSE EFFECT.

1835   GASPARD DE CORIOLIS explained why anything moving over the surface of the Earth, but not attached to it, is deflected by inertia acting at right angles to its direction of motion.

1840   LOUIS AGASSIZ discovered that GLACIERS move and that they once covered a much larger area than they do now.

1842   MATTHEW MAURY discovered the shape of storms from data gathered from ships at sea.

CHRISTIAN DOPPLER discovered the effect bearing his name, that the pitch of a sound rises and light becomes bluer if the source is approaching, and the pitch of a sound falls and light becomes redder if the source is receding.

1844   The world's first telegraph line opened between Baltimore and Washington, D.C.

1846   JOSEPH HENRY was elected secretary of the Smithsonian Institution and used his position to obtain weather reports from all over the United States.

1851   The first weather map was exhibited at the Great Exhibition in London, England.

1855   Urbain Jean Joseph Leverrier (1811–1877) began supervising the installation of a network to gather meteorological data from observatories across Europe.

1856   WILLIAM FERREL found that winds blowing close to the equator are deflected by VORTICITY, not the CORIOLIS EFFECT, and that once established they continue to rotate in order to conserve ANGULAR MOMENTUM. He also proposed that in the Northern Hemisphere winds blow counterclockwise around areas of low pressure.

1857   C.D.H. BUYS BALLOT proposed the law bearing his name (but discovered earlier by WILLIAM FERREL).

1861   The Meteorological Department of the Board of Trade issued the first British storm warnings for coastal areas on February 6 and for shipping on July 31.

JOHN TYNDALL showed that certain atmospheric gases absorb heat and therefore that the chemical composition of the atmosphere affects climate.

1863   The first network of meteorological stations to be linked to a central point by telegraph opened in France.

FRANCIS GALTON devised a method for mapping weather systems and coined the term ANTICYCLONE.

1869   The first daily weather bulletins began to be issued from Cincinnati Observatory on September 1.

1871   The first three-day weather forecasts were issued by the Weather Bureau.

1874   The International Meteorological Congress was founded.

1875   A weather map appeared in a newspaper, The Times of London, for the first time.

1884   S.P. LANGLEY published a paper on the climatic effect of the absorption of heat by atmospheric gases.

1891   The U.S. Weather Bureau was founded.

1893   EDWARD MAUNDER discovered the link between solar activity and the LITTLE ICE AGE.

1896   SVANTE ARRHENIUS linked climatic changes to the atmospheric concentration of carbon dioxide.

The International Meteorological Congress published the first edition of the INTERNATIONAL CLOUD ATLAS.

1902   L. P. TEISSERENC DE BORT discovered the STRATOSPHERE.

VILHELM BJERKNES published one of the first scientific studies of weather forecasting.

1905   V. W. EKMAN discovered the deflection of winds and ocean currents changes with vertical distance from the surface.

1913   CHARLES FABRY discovered the OZONE LAYER.

1918   W. P. KÖPPEN published a system for classifying climates.

1918   VILHELM BJERKNES established the existence of AIR MASSES.

1922   L. F. RICHARDSON described a method for NUMERICAL FORECASTING.

1923   Gilbert Walker described the high-level flow of air from west to east close to the equator. He also described the SOUTHERN OSCILLATION that is linked to EL NIÑO events.

1930   MILUTIN MILANKOVICH proposed a link between the onset and ending of GLACIAL PERIODS and variations in the Earth's orbit and rotation.

1931   WILSON BENTLEY published more than 2,000 photographs of snowflakes.

C. W. THORNTHWAITE published a system for classifying climates.

1940   CARL-GUSTAV ROSSBY discovered large-wavelength undulations in the westerly winds of the upper atmosphere.

1946   VINCENT SCHAEFFER discovered that pellets of dry ice (solid carbon dioxide) can trigger the formation of ice crystals.

1949   RADAR was used for the first time to obtain meteorological data.

1951   An international system for the classification of snowflakes was adopted.

1959   The U.S. Weather Bureau began publishing a temperature–humidity index as an indication of how comfortable the air will feel on a warm day.

1960	The first weather satellite, *Tiros 1*, was launched.
1964	The *Nimbus 1* weather satellite was launched.
1966	The first satellite to be placed in a GEO-STATIONARY ORBIT was launched on December 6.
1971	The FUJITA TORNADO INTENSITY SCALE was published.
1973	DOPPLER RADAR was used successfully for the first time to study a TORNADO.
1974	The first of the GEOSTATIONARY OPERATIONAL ENVIRONMENTAL SATELLITE (GOES) spacecraft was launched.

	F. SHERWOOD ROWLAND and MARIO MOLINA proposed that chlorofluorocarbons (CFCs) might deplete stratospheric OZONE.
1985	Depletion of the ozone layer over Antarctica was discovered by J. C. Farman, B. G. Gardiner, and J. D. Shanklin.
1993	Using powerful computers and advanced climate models, the National Weather Service was able to predict a major storm five days in advance.
1996	Technological advances meant five-day forecasts were as accurate as three-day forecasts had been in 1980.

Appendix III

THE GEOLOGICAL TIME SCALE

Eon	Era	Sub-era	Period	Epoch	Start (Ma*)
Phanerozoic	Cenozoic	Quaternary	Pleistogene	Holocene	0.01
				Pleistocene	1.64
		Tertiary	Neogene	Pliocene	5.2
				Miocene	23.3
			Paleogene	Oligocene	35.4
				Eocene	56.5
				Paleocene	65.0
	Mesozoic		Cretaceous		145
			Jurassic		208
			Triassic		245
	Paleozoic	Upper Paleozoic	Permian		290
			Carboniferous		362.5
			Devonian		408.5
		Lower Paleozoic	Silurian		439
			Ordovician		510
			Cambrian		570
Proterozoic					2,500
Archaean					4,000
Priscoan					4,600

(Together, the Proterozoic, Archaean, and Priscoan constitute what was formerly known as the *Precambrian*.)

*Ma, millions of years ago.

APPENDIX IV

TORNADOES OF THE PAST

Many tornadoes occur in remote areas, far from human habitations, but if they strike in populated areas they can cause great devastation.

1140: Warwickshire, England. Extensive damage.

July 1558: Nottingham, England. Extensive damage and some deaths.

October 1638: Devon, England. Between 5 and 50 deaths.

May 1840: Natchez, Mississippi. 317 killed.

June 1865: Viroqua, Wisconsin. More than 20 killed.

December 1879: Scotland. The Tay Bridge destroyed by two tornadoes that struck simultaneously. Between 75 and 90 killed.

February 1884: Mississippi, Alabama, North Carolina, South Carolina, Tennessee, Kentucky, Indiana. More than 800 killed.

May 1896: Missouri and Illinois. 300 killed.

June 1899: New Richmond, Wisconsin. 117 killed, at least 150 injured.

May 1902: Goliad, Texas. 114 killed.

March 1925: Missouri, Illinois, and Indiana. 689 killed by up to seven tornadoes. A separate tornado at Annapolis, Maryland, overturned passenger trains and lifted 50 motor cars, carried them over houses, and dropped them in fields.

March 1932: Alabama. 268 killed.

April 1936: 216 killed at Tupelo, Mississippi, and 203 at Gainesville, Georgia, by two separate tornadoes.

June 1944: Ohio, Pennsylvania, West Virginia, and Maryland. 150 killed.

April 1947: Texas, Oklahoma, and Kansas. 169 killed.

March 1952: Arkansas, Missouri, and Texas. 208 killed.

May 1953: Texas. 114 killed.

June 1953: 143 killed in Michigan and Ohio and 90 around Worcester, Massachusetts, by two separate tornadoes.

May 1955: Kansas, Missouri, Oklahoma, and Texas. 115 killed.

April 1965: Indiana, Illinois, Ohio, Michigan, and Wisconsin. 271 killed.

February 1971: Mississippi Delta. 110 killed.

January 1973: San Justo, Argentina. 60 killed.

April 1974: More than 300 killed during the SUPER-OUTBREAK.

June 1974: Oklahoma, Kansas, and Arkansas. 24 killed by several tornadoes.

January 1975: Mississippi. 12 killed, 200 injured when a tornado struck a shopping mall.

April 1976: Bangladesh. 19 killed, more than 200 injured.

April 1977: Bangladesh. More than 600 killed, 1,500 injured.

May 1977: Moundou, Chad. 13 killed, 100 injured.

March 1978: Delhi, India. 32 killed, 700 injured.

April 1978: Orissa State, India. Nearly 500 killed, more than 1,000 injured. 100 believed killed in West Bengal by another tornado.

April 1979: Texas and Oklahoma. 59 killed, 800 injured.

August 1979: Irish Sea. 18 killed when tornadoes struck yachts taking part in the Fastnet Race between England and Ireland.

May 1980: Kalamazoo, Michigan. 5 killed, at least 65 injured.

April 1981: Bangladesh. About 70 killed, 1,500 injured. Orissa State, India. More than 120 killed.

April 1982: Kansas, Oklahoma, and Texas. 7 killed.

May 1982: Marion, Illinois. 10 killed.

April 1983: Fujian Province, China. 54 killed. Bangladesh. 12 killed, 200 injured.

May 1983: Texas, Tennessee, Missouri, Georgia, Louisiana, Mississippi, and Kentucky. 24 killed by at lest 59 tornadoes. Vietnam. 76 killed.

March 1984: North and South Carolina. More than 70 killed.

April 1984: Water Valley, Mississippi. 15 killed. Oklahoma. 14 killed. Kentucky, Louisiana, Tennessee, Ohio, Maryland, and West Virginia. 14 killed.

June 1984: Russia. Hundreds killed.

October 1984: Maravilha, Brazil. 10 killed.

May 1985: Ohio, Pennsylvania, New York, and Ontario. 90 killed.

May 1987: Saragosa, Texas. 29 killed.

July 1987: Heilongjiang Province, China. 16 killed, more than 400 injured. Edmonton, Alberta, Canada. 25 killed.

April 1989: Bangladesh. Up to 1,000 killed, 12,000 injured.

May 1989: Texas, Virginia, North Carolina, Louisiana, South Carolina, and Oklahoma. 23 killed, more than 100 injured.

November 1989: Huntsville, Alabama. 18 killed.

June 1990: Indiana, Illinois, and Wisconsin. 13 killed.

May 1991: Bangladesh. 13 killed.

January 1993: Bangladesh. 32 killed, more than 1,000 injured.

April 1993: West Bengal, India. 100 killed.

March 1994: Alabama, Georgia, and North and South Carolina. 42 killed.

May 1996: Bangladesh. More than 440 killed, more than 32,000 injured.

March 1997: Arkansas. 25 killed.

May 1997: Central Texas. 30 killed, 24 in Jarrell.

July 1997: Kiangsu Province, China. 21 killed, more than 200 injured. Southern Michigan. 16 killed, more than 100 injured.

October 1997: Bangladesh. 25 killed, thousands injured.

May 1999: Kansas, Oklahoma, Texas. 46 people killed and about 900 injured by more than 76 tornadoes, at least one classified F-5.

July 2000: Alberta, Canada. 10 killed when a tornado swept through a trailer park near Edmonton.

October 2000: Bognor Regis, England. Four people injured and hundreds of homes damaged.

December 2000: Tuscaloosa, Alabama. 11 people killed.

February 2001: Mississippi. Five killed. Arkansas. One person killed.

May 2001: Ellicott, Colorado. 18 killed.

Appendix V

WEBSITES

The following Web addresses also appear at the ends of particular entries. It is impossible to list web addresses alphabetically: most of them begin *www*! Instead I have listed them in the alphabetical order of the entries to which they relate. To help identify them, I have added the headword to the entry at the end of the address. Where there is more than one web address I have added the headword at the end of the final address. In some cases, to save space, I have listed entries by their acronyms rather than their full titles. If you are unfamiliar with an acronym you should look it up in the A-to-Z section, where you will find the full name.

www.allernet.com/default.asp

www.usatoday.com/weather/health/pollen.wpusa.htm.

www.isao.bo.cnr.it/~aerobio/aia/AIANET.html (aeroallergens)

www.geocities.com/ResearchTriangle/Campus/9792/PAAA.html (aerobiology)

research.amnh.org/ichthyology/neoich/collectors/agassiz.html

www.nceas.ucsb.edu/~alroy/lefa/Lagassiz.html

www.uinta6.k12.wy.us/WWW/MS/8grade/Info%20Access/SPANTLGY/aga ssiz.htm (Agassiz)

www.legal.gsa.gov/legal14air.htm (Air Quality Acts)

www.agu.org/pubs/abs/gl/97GL02694/97GL02694.html

www.ees.lanl.gov/staff/cal/acen.html

jedac.ucsd.edu/wbwhite/ray/paper.html. (Antarctic Circumpolar Waves)

www.nobel.se/chemistry/laureates/1903/arrhenius-bio.html (Arrhenius)

www-bprc.mps.ohio-state.edu/~bdasye/balligh.html (ball lightning)

www-groups.dcs.st-and.ac.uk/~history/Mathematicians/Bernoulli_Daniel.html (Bernoulli)

www.earthbase.org/home/timeline/1984/bhopal/text.html (Bhopal)

www.astro.uu.se/history/Celsius_eng.html (Celsius)

www.santesson.com/engtemp.html (Celsius temperature scale)

www.opb.org/ofg/1001/missoula/sitemap.htm

www.sentex.net/~tcc/scabland.html

www.gi.alaska.edu/ScienceForum/ASF11/1160.html

www.spokaneoutdoors.com/scabland.htm (Channeled Scablands)

www.soest.hawaii.edu/MET/Faculty/bwang/bw/pubs/52.html (CISO)

users.pld.com/hailman/master.html (cloud seeding)

www.ncdc.noaa.gov/ogp/papers/kutzbac.html

www.ngdc.noaa.gov/paleo/paleo.html. (COHMAP)

www.mac-med.com/M%26C%20FILES/09maccs.html

camille-f.gsfc.nasa.gov/912/geerts/cwx/notes/chap11/gustave.html (Coriolis)

www.windpower.dk/tour/wres/coriolis.htm (Coriolis effect)

www.nobel.se/chemistry/laureates/1995/crutzen-auto-bio.html

www.mpch-mainz.mpg.de/~air/crutzen/vita.html (Crutzen)

www.cinemedia.net/SFCV-RMIT-Annex/rnaughton/DANIELL_BIO.html

www.bioanalytical.com/calendar/97/daniell.htm (Daniell)

aa.usno.navy.mil/AA/data/docs/RS_OneYear.html

www.sunrisesunset.com/custom_srss_calendar.asp (day)

www.ngdc.noaa.gov/dmsp/descriptions/dmsp_desc.html

www.laafb.af.mil/SMC/CI/overview/index.html (DMSP)

dizzy.library.arizona.edu/library/teams/set/earthsci/treering /html

www-personal.umich.edu/~dushanem/whatis.html

www.sonic.net/bristlecone/dendro.html. (dendrochronology)

usatoday.com/weather/wdenalt.html (density altitude)

www.unccd.de/main2.html (desertification)

www.unep.ch/earthw.html (Earthwatch)

taylor.math.ualberta.ca/~eifl/teaching/ekman/index.html (Ekmann spiral)

sputnik.infospace.ru/goms/engl.goms_1.htm (Elektro satellite)

es.rice.edu/ES/humsoc/Galileo/People/medici/html (Ferdinand)

www.who.int/ (fire)

www.intranet.ca/~jedr/fitzroy.htm

www.sciencemuseum.org.uk/collections.exhiblets/weather/fitzroy.htm (FitzRoy)

www.fsl.noaa/gov/ (Forecast Systems Laboratory)

www.unfcc.de/ (Framework Convention on Climate Change)

nsidc.org/NSIDC/CATALOG/ENTRIES/nsi-0063.html (freezing index)

www.cimm.jcu.edu.au/hist/stats/galton/index.htm (Galton)

www.unece.org/env/1rtap/env_eb1.htm. (Geneva Convention on Long-Range Transboundary Air Pollution)

www.glaciers./net/

www.geo.unizh.ch/wgms/index1.htm (glacier)

neonet.nlr.nl/ceos-idn/campaigns/GARPFGGE.html (GARP)

www.plas.bee.qut.edu.au/wwwjsc/psb320/320envmon01.htm (GEMS)

gust.sr.unh.edu/GISP2/ (GISP)

www.esf.org/life/lp/old/grip/lp_013a.htm (GRIP)

www.dwd.de (Grosswetterlage)

es.rice.edu/ES/humsoc/Galileo/Catalog/Files/halley.html

www.astro.uni-bonn.de/~pbrosche/persons/pers_halley.htm l

www-groups.dcs.st-and.ac.uk/~history/Mathematicians/Halley.html (Halley)

www.redcross.org/disaster/safety/heat.html (heat wave)

www.nws.noaa.gov/om/hydro.htm. (Integrated Hydrometeorological Services Core)

orpheus.ucsd.edu/speccoll/weather/27.htm

www.wmo.ch/web/catalogue/New%20HTML/frame/engfil/407.html (International Cloud Atlas)

www.ciesin.org/TG/HDP/igbp.html (IGBP)

www.ugs.state.ut.us/pi-39/pi39pg1.htm. (Lake Bonneville)

www.cru.uea.ac.uk/~mikeh/datasets/uk/lamb.htm. (Lamb's classification)

www.aber.ac.uk/~jpg/volcano/lecture2.html. (Lamb's dust veil index)

geo.arc.nasa.gov/sge/landsat/lpchron.html (Landsat)

www.runet.edu/~wkovarik/papers/leadinfo.html (lead)

www.cam.org/~gouletc/decl_faq.html feature. geography.wisc.edu/sco/maps/m_magnet.html (magnetic declination)

humbabe.arc.nasa.gov/mgcm/faq/faq.html

humbabe.arc.nasa.gov/mgcm/faq/marsfacts.html (Mars)

www.nssl.noaa.gov/~doswell/microburst/Handbook.html (microburst)

www.dukenews.duke.edu/Environ/RADCARB.html

earthobservatory.nasa.gov/Study/BOREAS/

www.bren.ucsb.edu/~keller/papers/climch33.html (missing carbon)

www.sciam.com/1197issue/1197profile.html (Molina)

www.nhc.noaa.gov/ (National Hurricane Center)

www.nssl.noaa.gov/ (National Severe Storms Laboratory)

nsidc.org/index.html (National Snow and Ice Data Center)

www.south-pole.com/p0000091.htm (Nordenskjöld)

podaac.jpl.nasa.gov/info/ (PO.DAAC)

vulcan.wr.usgs.gov/Glossary/PlateTectonics/ description_plate_ tectonics.html (plate tectonics)

www.atm.ch.cam.ac.uk/tour/psc.html

www.awi-potsdam.de/www-pot/atmo/psc/psc.html (polar stratospheric clouds)

www.geo.arizona.edu/palynology/plns1295.html (pollen analysis)

www.vos.noaa.gov/pmo11.html (Port Meteorological Officer)

www.envf.port.ac.uk/geog/teaching/quatgern/q8b.htm (post-glacial)

www.rlaha.ox.ac.uk/orau/ (radiocarbon dating)

maths.paisley.ac.uk.lfr.htm

www.mpae.gwdg.de/EGS/egs_info/richardson.htm (Richardson)

fsr10.ps.uci.edu/GROUP/people/drowland.html

www.nobel.se/chemistry/laureates/1995/ rowland-autobio.html (Rowland)

www.jpl.nasa.gov:2031/SENSOR_DOCS/smmr.html (Scanning Multichannel Microwave Radiometer)

winds.jpl.nasa.gov/mission/quikscat/quikindex.html (Sea Winds)

www.wsmr.army.mil/paopage/Pages/solar.htm

www.sandia.gov/Renewable_Energy/solarthermal. furnaces.html

howstuffworks.com/solar-cell.htm (solar energy)

www.jpl.nasa.gov:2031/SENSOR_DOCS/ssmi.html (Special Sensor Microwave Imager)

www.crh.noaa.gov/techpapers/service/tsp-10 /0sevwea-stbind.ht ml (stability index)

www.unibg.it/rls/essays/memport/mp-9.htm (Stevenson screen)

www.spc.noaa.gov/products/wwa. (storm warning)

www.concord.org/haze/ref.html (Sun photometer)

www.met.fsu.edu/Classes/Common/sfc.html

www.zetnet.co.uk/sigs/weather/Met_Codes/codes.html

www.usatoday.com/weather/wpcodes.htm. (synoptic code)

nsidc.org/NSIDC/CATALOG/ENTRIES/nsi-0063.html (thawing index)

edcwww.cr.usgs.gov/glis/hyper/guide/landsat_tm (Thematic Mapper)

www.ccar.colorado.edu/research/topex/html/topex.html

www.tsgc.utexas.edu/spacecraft/topex.intro.html (Topex/Poseidon)

www.nws.noaa.gov/om/tropical.htm. (Tropical Cyclone Program)

www.nhc.noaa.gov/aboutintro.html (Tropical Prediction Center)

www.tyndall.uea.ac.uk/tyndall.htm, www.irsa.ie/Resources/Heritage.tyndall.html

earthobservatory.nasa.gov/Library/Giants/Tyndall/ (Tyndall)

www.nws.mbay.net/history.html

www.crh.noaa.gov/lmk/history1.htm (US Weather Bureau)

ruby.kordic.re.kr/~vr/CyberAstronomy/Venus/HTML/ index.html

www.planetary.brown.edu/tepswg/venus_strategy. html

www.eurekalert.org/releases/uco-nvm021899.html (Venus)

www.wx-fx.com/viroqua.html. (Viroqua)

www.vos.noaa.gov/wmo.html

www.srh.noaa.gov/bro/vos.htm (Voluntary Observing Ship)

library.thinkquest.org/20901/overview_2.htm

www.cotf.edu/ete/modules/elnino/cratmosphere.html (Walker circulation)

www.pangaea.org/wegener.htm

www.pbs.org/wgbh/aso/databank/entries/bowege.html

www.ucmp.berkeley.edu/history/wegener.html

www.dkonline.com/science/private/earthquest/contents/hall.htm l

pubs.usgs.gov/publications/text/wegener.html (Wegener)

www.publish.csiro.au/ecos/Ecos90/Ecos90B.htm

www.jxj.com/magsandj/rew/2000_01/going_to_sea_offshore_plans. html

www.awea.org (wind farm)

www.awea.org (wind power)

www.wmo.ch/web-en/wmofact.html (World Meteorological Organization)

INDEX